Francesco Altomare, Mirella Cappelletti Montano, Vita Leonessa, Ioan Raşa
Markov Operators, Positive Semigroups and Approximation Processes

De Gruyter Studies in Mathematics

Edited by
Carsten Carstensen, Berlin, Germany
Nicola Fusco, Napoli, Italy
Fritz Gesztesy, Columbia, Missouri, USA
Niels Jacob, Swansea, United Kingdom
Karl-Hermann Neeb, Erlangen, Germany

Volume 61

Francesco Altomare, Mirella Cappelletti Montano, Vita Leonessa, Ioan Raşa

Markov Operators, Positive Semigroups and Approximation Processes

DE GRUYTER

Mathematics Subject Classification 2010
35A35, 35B40, 35K20, 35K65, 41-02, 41A36, 41A63, 46-02, 46E15, 47-02, 47B65, 47D07, 47F05, 60J60.

Authors
Prof. Francesco Altomare
Università degli Studi di Bari Aldo Moro
Dipartimento di Matematica
Via E. Orabona 4
70125 Bari
Italy
francesco.altomare@uniba.it

Dr. Mirella Cappelletti Montano
Università degli Studi di Bari Aldo Moro
Dipartimento di Matematica
Via E. Orabona 4
70125 Bari
Italy
mirella.cappellettimontano@uniba.it

Dr. Vita Leonessa
Università degli Studi della Basilicata
Dipartimento di Matematica, Informatica ed Economia
Campus di Macchia Romana
Viale dell'Ateneo Lucano 10
85100 Potenza
Italy
vita.leonessa@unibas.it

Prof. Ioan Raşa
Technical University of Cluj-Napoca
Department of Mathematics
Str. Memorandumului 28
400114 Cluj-Napoca
Romania
Ioan.Rasa@math.utcluj.ro

ISBN 978-3-11-037274-8
e-ISBN (PDF) 978-3-11-036697-6
e-ISBN (EPUB) 978-3-11-038641-7
Set-ISBN 978-3-11-036698-3
ISSN 0179-0986

Library of Congress Cataloging-in-Publication Data
A CIP catalog record for this book has been applied for at the Library of Congress.

Bibliographic information published by the Deutsche Nationalbibliothek
The Deutsche Nationalbibliothek lists this publication in the Deutsche Nationalbibliografie; detailed bibliographic data are available on the Internet at http://dnb.dnb.de.

© 2014 Walter de Gruyter GmbH, Berlin/Munich/Boston
Printing and binding: CPI books GmbH, Leck
♾ Printed on acid-free paper
Printed in Germany

www.degruyter.com

To our parents and families

Preface

> If you have built castles in the air, your work need not be lost; that is where they should be. Now put the foundation under them.
>
> *Henry David Thoreau*

During the last twenty years important progresses have been made, from the point of view of constructive approximation theory, in the study of initial-boundary value differential problems of parabolic type governed by positive C_0-semigroups of operators. The main aim of this approach is to construct suitable positive approximation processes whose iterates strongly converge to the semigroups which, as it is well-known, in principle furnish the solutions to the relevant initial-boundary value differential problems. By means of such kind of approximation it is then possible to investigate, among other things, preservation properties and the asymptotic behavior of the semigroups, i.e., spatial regularity properties and asymptotic behavior of the solutions to the differential problems.

This series of researches, for a survey on which we refer to [25] and the references therein, has its roots in several studies developed between 1989 and 1994 which are documented in Chapter 6 of the monograph [18]. These studies were concerned with special classes of second-order elliptic differential operators acting on spaces of smooth functions on finite dimensional compact convex subsets, which are generated by a positive projection. The projections themselves offer the tools to construct an approximation process whose iterates converge to the relevant semigroup, making possible the development of a qualitative analysis as above.

This theory disclosed several interesting applications by stressing the relationship among positive semigroups, initial-boundary value problems, Markov processes and approximation theory, and by offering, among other things, a unifying approach to the study of diverse differential problems. Nevertheless, over the subsequent years, it has naturally arisen the need to extend the theory by developing a parallel one for positive operators rather than for positive projections and by trying to include in the same project of investigations more general differential operators having a first-order term.

The aim of this research monograph is to accomplish such an attempt by considering complete second-order (degenerate) elliptic differential operators whose leading coefficients are generated by a positive operator, by means of which it is possible to construct suitable approximation processes which approximate the relevant semigroups. Some aspects of the theory are treated also in infinite dimensional settings.

The above described more general framework guarantees to notably enlarge the class of differential operators by including, in particular, those which can be obtained by usual operations with positive operators such as convex combinations, compositions, tensor products and so on. Moreover, this non-trivial generalization discloses new challenging problems as well.

However, the approximation processes which we construct in terms of the given positive operator seem to have an interest in their own also for the approximation of continuous functions. For these reasons, a special emphasis is placed upon various aspects of this theme as well.

We are happy to thank several friends and colleagues and, particularly, Michele Campiti and Rainer Nagel for their critical reading of the manuscript.

We also wish to express our appreciation to Niels Jacob for his interest in this work. We thank him and Walter de Gruyter & Co. Publishing House for accepting to publish this monograph in the Series Studies in Mathematics as well as for producing it according with their usual high standards.

September 2014

Francesco Altomare
Mirella Cappelletti Montano
Vita Leonessa
Ioan Raşa

Contents

Introduction

In recent years several investigations have been devoted to the study of large classes of (mainly degenerate) initial-boundary value evolution problems in connection with the possibility to obtain a constructive approximation to the associated positive C_0-semigroups by means of iterates of suitable positive linear operators which also constitute approximation processes in the underlying Banach function space.

Usually, as a consequence of a careful analysis of the preservation properties of the approximating operators, such as monotonicity, convexity, Hölder continuity and so on, it is possible to infer similar preservation properties for the relevant semigroups and, in turn, some spatial regularity properties of the solutions to the evolution problems.

In this research monograph we present the main lines of a theory which finds its roots in the above-mentioned research field. It is mainly concerned with a class of (abstract) degenerate differential operators, Markov semigroups and approximation processes which can be generated by an arbitrary Markov operator T acting on the space $\mathscr{C}(K)$, K being a convex compact subset of a (not necessarily finite-dimensional) locally convex space. These mathematical subjects are strongly interrelated and the study of the relationship between them constitutes one of the distinguishing features of the book.

Several particular examples of such differential operators are involved in evolution problems arising in population genetics, financial mathematics and other fields, so that they seem to be worthy of a comprehensive and thorough study. Furthermore, our general framework allows us to apply the theory to a variety of situations which seem to have interest both in theoretical and applied problems.

In Chapter 3 we begin by introducing and studying a fundamental sequence of positive linear operators associated with the Markov operator T, called Bernstein-Schnabl operators. These operators generalize the classical Bernstein operators defined on the unit interval, on multidimensional simplices and on hypercubes, and they share with them several preservation properties which we investigate in detail.

They are defined as

$$B_{n,T}(f)(x) := \int_K \cdots \int_K f\left(\frac{x_1 + \ldots + x_n}{n}\right) d\tilde{\mu}_x^T(x_1) \cdots d\tilde{\mu}_x^T(x_n)$$

($f \in \mathscr{C}(K)$, $x \in K$, $n \geq 1$), where $\tilde{\mu}_x^T$ is the unique Borel measure on K such that

$$T(f)(x) = \int_K f \, d\tilde{\mu}_x^T$$

for every $f \in \mathscr{C}(K)$ and $x \in K$.

The main interest in their study lies not only in the fact that they are an approximation process for continuous functions on K, but also because, by means

of them, it is possible to constructively approximate suitable Markov semigroups of operators.

However, other than this last aspect, we carry out a detailed study of the approximation properties of Bernstein-Schnabl operators by also pointing out some estimates of the rate of convergence. Then we study some preservation properties such as the preservation of Hölder continuous functions as well as of convex functions.

Motivated by similar properties of Bernstein operators on $\mathscr{C}([0,1])$, we also investigate the monotonicity and the majorizing properties for these operators, which lead us to consider other classes of convex functions such as T-axially convex and T-convex ones.

Subsequently, in Chapter 4 we introduce the (abstract) differential operator which is the other main object of investigation of the monograph. When K is finite dimensional, this differential operator is, indeed, a second-order elliptic differential operator which degenerates on a suitable subset of K containing its extreme points and, in some cases, on the whole boundary of K. It is defined as

$$W_T(u) := \frac{1}{2} \sum_{i,j=1}^{d} \alpha_{ij} \frac{\partial^2 u}{\partial x_i \partial x_j}$$

($u \in \mathscr{C}^2(K)$, $K \subset \mathbf{R}^d$, $d \geq 1$), where for each $i, j = 1, ..., d$, $\alpha_{ij} := T(pr_i pr_j) - pr_i pr_j$ and each pr_i stands for the $i-th$ coordinate function.

One of the main aims of this monograph is to show that, under suitable assumptions on T, the associated (abstract) differential operator is closable and its closure generates a Markov semigroup $(T(t))_{t\geq 0}$ on $\mathscr{C}(K)$ which, in turn, is the transition semigroup of a suitable right-continuous normal Markov process with state space K.

The difficulties in studying such a problem (in a finite dimensional setting) with the methods of the theory of partial differential equations, lie in the fact that the boundary of K is generally non-smooth due to the presence of possible sides and corners.

After showing that the above-mentioned differential operators are related to the Bernstein-Schnabl operators associated with T via the asymptotic formula

$$\lim_{n\to\infty} n(B_{n,T}(u) - u) = W_T(u) \qquad (u \in \mathscr{C}^2(K)),$$

we obtain the generation result by means of a Trotter-Schnabl-type theorem by also showing that the Markov semigroup can be described as a limit of suitable iterates of Bernstein-Schnabl operators.

This representation/approximation formula allows us to transfer several previously studied preservation properties of Bernstein-Schnabl operators to the Markov semigroup by stressing, as a consequence, some spatial regularity properties of the solutions to the relevant initial-boundary value evolution problem as

well as some additional properties of the transition functions of the corresponding Markov process.

In some special cases we exactly describe the asymptotic behavior of the Markov semigroup, i.e., we determine $\lim_{t\to+\infty} T(t)$, and we characterize the saturation class of the sequence $(B_{n,T})_{n\geq 1}$ and the Favard class of $(T(t))_{t\geq 0}$.

The main results rely on the crucial assumption that the Markov operator T maps polynomials into polynomials of the same degree. This preservation property seems to have an independent interest on its own and, in the final part of Section 4.3 and in Appendix A.1 we discuss several situations where it is satisfied. In particular, we present a characterization of ellipsoids and balls of $\mathbf{R}^d$ in terms of it.

In the Appendix A.2 we touch some particular aspects of the problem to estimate the rate of convergence of the iterates of Bernstein-Schnabl operators towards the relevant semigroups. More complete results concerning the general context of this monograph would be highly significant.

In the last chapter we broaden our investigation in the same above-mentioned spirit by studying both multiplicative perturbations of the differential operator W_T of the form λW_T ($\lambda \in \mathscr{C}([0,1])$, λ strictly positive) and first-order additive perturbations of it of the form

$$V_T(u) = W_T(u) + \sum_{i=1}^{d} \beta_i \frac{\partial u}{\partial x_i} + \gamma u = \frac{1}{2} \sum_{i,j=1}^{d} \alpha_{ij} \frac{\partial^2 u}{\partial x_i \partial x_j} + \sum_{i=1}^{d} \beta_i \frac{\partial u}{\partial x_i} + \gamma u$$

($u \in \mathscr{C}^2(K)$), where the functions $\beta_1, \ldots, \beta_d, \gamma \in \mathscr{C}(K)$ satisfy suitable assumptions.

This class of complete differential operators is sufficiently large to include several interesting differential operators which have been widely studied on the hypercube, on the simplex, on the balls and on the ellipsoids. A particular class of them arises in the theory of Fleming-Viot processes on the d-dimensional simplex. They are involved, for instance, in Wright-Fisher diffusion model in population genetics which are affected by mutation, migration and selection or other factors, as well as in some recent models from mathematical finance on volatility-stabilized markets.

Also for the above-mentioned differential operators we establish a generation result by means of a Trotter-Schnabl-type theorem by properly modifying the Bernstein-Schnabl operators with other approximation processes. These modifications allow us to infer some useful properties of the relevant Feller semigroups.

Several examples scattered throughout the monograph illustrate and highlight the general results in more concrete cases and in more specific contexts, by recapturing results previously obtained with different methods.

The results of this monograph generalize those documented in the book of Altomare and Campiti ([18, Chapter 6 and the references contained in the relevant notes]) and in the paper by Altomare, Cappelletti Montano and Diomede ([19]),

which are concerned with similar problems in the special case where T is a Markov projection.

The more general framework of this monograph guarantees to notably enlarge the class of differential operators by including, in particular, those which can be obtained by usual operations with Markov operators such as convex combinations, compositions, tensor products and so on.

Moreover, this non-trivial generalization discloses new challenging and more difficult problems as well. A pioneer work pre-announcing the theory developed here can be tracked in a paper by Altomare, Leonessa and Raşa ([28]) and in the papers [21] and [22].

This monograph is mainly addressed to research mathematicians, especially to those who could be interested in the interrelations in some way or another between the modern approximation theory by positive linear operators, the theory of positive C_0-semigroups of operators and evolution equations as well as the analytic theory of Markov processes. Furthermore, we strive to make the presentation and the proofs of the theorems as clear and detailed as possible so that the book is accessible to graduate students too and it could serve as textbook for a graduate level course.

As for prerequisites, the reader is expected to be familiar with the basic principles of real and functional analysis as well as of the approximation theory by positive linear operators and of the theory of C_0-semigroups of operators. For the reader's convenience and to make the exposition self-contained we collect several parts of these prerequisites in Chapters 1 and 2, often without proofs. However, these two chapters are more than a mere list of results and in different parts they contain topics which, as far as we know, appear in a book for the first time and which are provided with complete proofs.

Finally, we close every chapter with a section containing historical notes and giving credits and detailed references to supplementary results. Any inaccuracy or omission for historical details or in assigning priorities is unintentional and we apologize in advance for possible errors.

Guide to the reader and interdependence of sections

The core of the monograph concerns the interrelations between some aspects of the modern approximation theory by positive linear operators and the theory of positive C_0-semigroups of operators and evolution equations which can be associated with a Markov operator. The relevant material is contained in Chapters 3-5.

However, the reader interested in deepening only the study of positive approximation processes related to a Markov operator could refer to Chapters 1 and 3, Section 4.1, Subsection 4.5.1 and Sections 5.1-5.5.

The study of differential operators associated with a Markov operator, their multiplicative and first-order additive perturbations, as well as their generation properties, are developed in Chapter 4 (after the necessary preliminaries of Chapter 2, provided the reader does not posses a solid knowledge of the basic aspects of the theory of C_0-semigroups of linear operators), and in Sections 5.1 and 5.6-5.8.

However, a complete interdependence of sections is shown in the following diagram where

- On the line corresponding to a given section on the vertical axis, we have indicated all the sections on which the section depends.
- On the column corresponding to a given section on the horizontal axis, we have indicated all the sections which depend on it.
- The symbol $\bullet$ indicates a main dependence while the symbol $\circ$ denotes a minor dependence.

	1.1	1.2	1.3	1.4	1.5	1.6	2.1	2.2	2.3	3.1	3.2	3.3	3.4	3.5	4.1	4.2	4.3	4.4	4.5	5.1	5.2	5.3	5.4	5.5	5.6	5.7	5.8
1.1																											
1.2	•																										
1.3	•	•																									
1.4	•																										
1.5	∘																										
1.6	∘																										
2.1																											
2.2							•																				
2.3	∘						•																				
3.1	•																										
3.2		∘	∘			•				•																	
3.3						•				•																	
3.4										•																	
3.5	•									•			•														
4.1					•					•																	
4.2	•									•					•												
4.3	•						•	•	•	•					•	•											
4.4			•				•		•	•		∘	∘	∘			•										
4.5									•		•		•	•	•	•	•	•									
5.1										•		∘	∘		•		•										
5.2										•																	
5.3						•															•						
5.4						•						•	•								•	•					
5.5					•																•						
5.6	•						•	•	•												•		•	•			
5.7							•	•	•									•	•		•		•	•	•		
5.8	•						•		•			•	•								•				•		

Interdependence of sections

Notation

Here we assemble some notation and terminology which will be used throughout the book without no explicit mention. For some unexplained terminology concerning topological as well as measure-theoretical concepts we refer, e.g., to [94] and [50].

As usual, we denote by $\mathbf{N}$ the set of all natural numbers, by $\mathbf{R}$ the set of real numbers and by $\mathbf{C}$ the set of complex numbers. The letter $\mathbf{K}$ stands either for the field $\mathbf{R}$ or $\mathbf{C}$. In particular, if $z = x + iy \in \mathbf{C}$, then $\mathrm{Re}(z) := x$ and $\mathrm{Im}(z) := y$ denote, respectively, the *real part* and the *imaginary part* of z.

If $k \in \mathbf{N}$, $k \geq 1$ and $l \in \mathbf{N}$, we shall denote by $\binom{k}{l}$ the *binomial coefficient*, i.e.,

$$\binom{k}{0} = 1$$

and, for every $l \geq 1$,

$$\binom{k}{l} = \frac{k!}{l!\,(k-l)!},$$

where the symbol $p!$ $(p \geq 1)$ stands for $p! := p \cdot (p-1) \cdots 2 \cdot 1$ and $0! := 1$.

For any real numbers a and b, $a < b$, the *interval* $[a, b]$, $]a, b]$, $[a, b[$, $]a, b[$ is the subset of those $x \in \mathbf{R}$ satisfying $a \leq x \leq b$ (resp., $a < x \leq b, a \leq x < b, a < x < b$).

For every $\lambda \in \mathbf{R}$, the symbol $[\lambda]$ stands for the integer part of λ.

The symbol $\mathbf{R}^+$ denotes the subset of all positive elements of $\mathbf{R}$, whereas we set $\mathbf{R}^d := \underbrace{\mathbf{R} \times \cdots \times \mathbf{R}}_{d \text{ times}}$ $(d \geq 1)$.

We denote by $\langle \cdot, \cdot \rangle$ the *scalar product* in $\mathbf{R}^d$, i.e., for every $x, y \in \mathbf{R}^d$, $x = (x_1, \dots, x_d)$, $y = (y_1, \dots, y_d)$,

$$\langle x, y \rangle := \sum_{i=1}^{d} x_i y_i.$$

Further, we denote by $\|\cdot\|_2$ the *Euclidean norm* in $\mathbf{R}^d$, i.e., $\|x\|_2 := \sqrt{\sum_{i=1}^{d} x_i^2}$ $(x = (x_1, \dots, x_d) \in \mathbf{R}^d)$.

More generally, for every $p \geq 1$, the symbol $\|\cdot\|_p$ stands for the l_p*-norm* in $\mathbf{R}^d$, i.e., $\|x\|_p := \sqrt[p]{\sum_{i=1}^{d} |x_i|^p}$ $(x = (x_1, \dots, x_d) \in \mathbf{R}^d)$; moreover, $\|\cdot\|_\infty$ is the l_∞*-norm* in $\mathbf{R}^d$, i.e., $\|x\|_\infty := \max_{1 \leq i \leq d} |x_i|$ $(x = (x_1, \dots, x_d) \in \mathbf{R}^d)$.

If X is a compact subset of $\mathbf{R}^d$, we shall set $r(X) := \max\{\|x\|_2 \mid x \in X\}$.

The symbol $(a_{ij})_{\substack{1 \leq i \leq n \\ 1 \leq j \leq m}}$ denotes a matrix of objects with n rows and m columns; moreover, if $n = m$, then $\det(a_{ij})$ denotes the determinant of the matrix $(a_{ij})_{\substack{1 \leq i \leq n \\ 1 \leq j \leq n}}$.

If A is a matrix, the symbol A^t stands for the corresponding *transpose* matrix.

If M is a subset of a linear space X, $\text{span}(M)$ denotes the linear subspace generated by M.

If X is a topological space, then $\overline{Y}$, $\text{int}(Y)$, ∂Y denote the closure, the interior and the boundary of a subset Y of X. A *cluster* point of Y is any point $x \in X$ which belongs to $\overline{Y \setminus \{x\}}$.

The symbol $\mathscr{F}(X)$ stands for the space of all real-valued functions on X. If $f \in \mathscr{F}(X)$ and $Y \subset X$, the restriction of f to Y will be denoted by $f|_Y$.

We denote by $\mathscr{B}(X)$ the linear subspace of all functions $f : X \longrightarrow \mathbf{R}$ which are bounded, endowed with the norm $\|\cdot\|_\infty$ of the uniform convergence (briefly, the sup-norm) defined by

$$\|f\|_\infty := \sup_{x \in X} |f(x)| \qquad (f \in \mathscr{B}(X)),$$

with respect to which it is a Banach space.

If $f \in \mathscr{F}(X)$, then we denote by $\text{Supp}(f)$ the set

$$\text{Supp}(f) := \overline{\{x \in X \mid f(x) \neq 0\}}.$$

Moreover, the symbol $\mathscr{C}(X)$ denotes the linear space of all continuous real-valued functions on X. If X is compact, then $\mathscr{C}(X) \subset \mathscr{B}(X)$ and $\mathscr{C}(X)$, endowed with sup-norm, is a Banach space. If, in addition, X is metrizable, then $\mathscr{C}(X)$ is separable.

Denoting again by X an arbitrary set, the symbol $\mathbf{1}$ stands for the function of constant value 1 on X and, if M is a subset of X, $\mathbf{1}_M$ denotes the characteristic function of M, defined by setting, for every $x \in X$,

$$\mathbf{1}_M(x) := \begin{cases} 1 & \text{if } x \in M; \\ 0 & \text{if } x \notin M. \end{cases}$$

For every $f, g \in \mathscr{F}(X)$, we denote by the symbol $\sup(f, g)$ (resp., $\inf(f, g)$) the function on X defined by

$$\sup(f, g)(x) := \sup(f(x), g(x)) \qquad (x \in X)$$

(resp.,

$$\inf(f, g)(x) := \inf(f(x), g(x)) \qquad (x \in X)).$$

More generally, if $f_1, \dots, f_n \in \mathscr{F}(X)$ $(n \geq 3)$, we define the function $\sup\limits_{1 \leq i \leq n} f_i$ (resp., $\inf\limits_{1 \leq i \leq n} f_i$) as

$$\left(\sup_{1 \leq i \leq n} f_i\right)(x) := \sup_{1 \leq i \leq n} f_i(x) \qquad (x \in X)$$

(resp.,

$$\left(\inf_{1 \leq i \leq n} f_i\right)(x) := \inf_{1 \leq i \leq n} f_i(x) \qquad (x \in X)).$$

In particular, if $f \in \mathscr{F}(X)$, then by $|f|$ we mean the (positive) real-valued function on X defined as

$$|f| := \sup(f, -f).$$

A linear subspace E of $\mathscr{F}(X)$ is said to be a *lattice subspace* if

$$|f| \in E \qquad \text{for every } f \in E.$$

For instance, $\mathscr{B}(X)$ and $\mathscr{C}(X)$ are lattice subspaces.

Note that, if E is a lattice subspace, then $\sup(f,g)$, $\inf(f,g) \in E$ for every $f, g \in E$.

This follows at once by the elementary identities

$$\sup(f,g) = \frac{f+g+|f-g|}{2} \qquad \text{and} \qquad \inf(f,g) = \frac{f+g-|f-g|}{2}.$$

More generally, if $f_1, \dots, f_n \in E$, $n \geq 3$, then $\sup\limits_{1\leq i\leq n} f_i, \inf\limits_{1\leq i\leq n} f_i \in E$.

We say that a linear subspace E of $\mathscr{F}(X)$ is a *subalgebra* if

$$f \cdot g \in E \qquad \text{for every } f, g \in E,$$

or, equivalently, if $f^2 \in E$ for every $f \in E$. In this case, if $f \in E$ and $n \geq 1$, then $f^n \in E$ and hence for every real polynomial $Q(x) = \alpha_1 x + \alpha_2 x^2 + \dots + \alpha_n x^n$ ($x \in \mathbf{R}$) vanishing at 0, the function

$$Q(f) := \alpha_1 f + \alpha_2 f^2 + \dots + \alpha_n f^n$$

belongs to E as well. If E contains the constant functions, then $P(f) \in E$ for every real polynomial P.

If $X \subset \mathbf{R}^d$ has non-empty interior, we denote by $\mathscr{C}^2(X)$ the space of all real-valued (continuous) functions $f : X \longrightarrow \mathbf{R}$ which are twice-continuously differentiable on $\operatorname{int}(X)$ and whose partial derivatives of order ≤ 2 can be continuously extended to X. For $u \in \mathscr{C}^2(X)$ and $i, j = 1, \dots, d$, we shall continue to denote by $\frac{\partial u}{\partial x_i}$ and $\frac{\partial^2 u}{\partial x_i \partial x_j}$ the continuous extensions to X of the partial derivatives $\frac{\partial u}{\partial x_i}$ and $\frac{\partial^2 u}{\partial x_i \partial x_j}$.

Given an arbitrary compact space X, we denote by B_X the σ-algebra of all Borel subsets of X, i.e., the σ-algebra generated by the system of all open subsets of X. Of course, open subsets and closed subsets of X are Borel subsets.

A *Borel measure* on X is a finite measure on B_X. Those Borel measures $\tilde{\mu}$ on X such that $\tilde{\mu}(X) = 1$ are also called *probability Borel measures* on X.

A measure $\tilde{\mu} : B_X \longrightarrow [0, +\infty]$ is said to be *inner regular* if

$$\tilde{\mu}(B) = \sup\{\tilde{\mu}(K) \mid K \subset B, K \text{ compact}\} \qquad \text{for every } B \in B_X,$$

and *outer regular* if

$$\tilde{\mu}(B) = \inf\{\tilde{\mu}(U) \mid B \subset U, U \text{ open}\} \qquad \text{for every } B \in B_X.$$

A measure $\tilde{\mu}$ is said to be *regular* if it is both inner and outer regular. If X is metrizable, then every Borel measure on X is regular ([50, Theorem 29.12]).

A property P of points of X is said to hold *$\tilde{\mu}$-almost everywhere* (shortly, $\tilde{\mu}$-a.e.) if the subsets of those points $x \in X$ for which $P(x)$ is false is contained in a set N such that $\tilde{\mu}(N) = 0$.

If $\tilde{\mu}$ is a regular Borel measure, we shall denote by $\mathrm{Supp}(\tilde{\mu})$ the support of $\tilde{\mu}$, i.e., the complement of the largest open subset of X on which $\tilde{\mu}$ is zero.

It can be proved that, if $f \in \mathscr{C}(X)$ and $f = 0$ on $\mathrm{Supp}(\tilde{\mu})$, then $\int_X f\,d\tilde{\mu} = 0$. Conversely, if $f \in \mathscr{C}(X)$, $f \geq 0$, and if $\int_X f\,d\tilde{\mu} = 0$, then $f = 0$ on $\mathrm{Supp}(\tilde{\mu})$.

Finally, for every $x \in X$ we denote by ε_x the unit mass concentrated at x, i.e., for every $B \in B_X$,

$$\varepsilon_x(B) := \begin{cases} 1 & \text{if } x \in B; \\ 0 & \text{if } x \notin B. \end{cases}$$

Another noteworthy and useful result concerning regular Borel measures is shown below. Consider a regular Borel measure $\tilde{\mu}$ and $p \in [1, +\infty[$. As usual, we denote by $\mathscr{L}^p(X, \tilde{\mu})$ the linear subspace of all B_X-measurable functions $f : X \longrightarrow \mathbf{R}$ such that $|f|^p$ is $\tilde{\mu}$-integrable.

If $f \in \mathscr{L}^p(X, \tilde{\mu})$, it is customary to set

$$\mathscr{N}_p(f) := \left(\int_X |f|^p \, d\tilde{\mu} \right)^{1/p}.$$

The functional $\mathscr{N}_p : \mathscr{L}^p(X, \tilde{\mu}) \longrightarrow \mathbf{R}$ is a seminorm and the convergence with respect to it is the usual *convergence in p-th-mean* . Setting

$$\begin{aligned} \mathscr{N} &:= \{f \in \mathscr{L}^p(X, \tilde{\mu}) \mid \mathscr{N}_p(f) = 0\} \\ &= \{f \in \mathscr{F}(X) \mid f \text{ is } B_X - \text{measurable and } f = 0\ \tilde{\mu} - \text{a.e.}\}, \end{aligned}$$

the quotient linear space

$$L^p(X, \tilde{\mu}) := \mathscr{L}^p(X, \tilde{\mu}) / \mathscr{N}$$

endowed with the norm $\| \cdot \|_p$ defined by

$$\|\tilde{f}\|_p := \mathscr{N}_p(f) \qquad (\tilde{f} \in L^p(X, \tilde{\mu}))$$

is a Banach space (here $\tilde{f} := \{g \in \mathscr{L}^p(X, \tilde{\mu}) \mid f = g\ \tilde{\mu} - \text{a.e.}\}$). Note that, if $\tilde{\mu}$ is a Borel measure, then $\mathscr{C}(X) \subset \mathscr{L}^p(X, \tilde{\mu})$ for every $p \in [1, +\infty[$.

Moreover, if $\tilde{\mu}$ is a regular Borel measure, then for every $p \in [1, +\infty[$, $\mathscr{C}(X)$ is dense in $\mathscr{L}^p(X, \tilde{\mu})$ with respect to the convergence in the p-th-mean (and, hence, in $L^p(X, \tilde{\mu})$ with respect to $\| \cdot \|_p$) (for a proof, see [50, Theorem 29.14]).

We now proceed to recall the definition of nets. For more details we refer, e.g., to [140, Section 2.1].

Consider a set I and a partial ordering $\leq$ on I; then the pair $(I, \leq)$ is said to be a *directed set* if for every $i, j \in I$ there exists $\lambda \in I$ such that $i \leq \lambda$ and $j \leq \lambda$.

A *net* on a set X is a family $(x_i)_{i\in I}^{\leq}$ of elements of X such that on the set I there is a partial ordering $\leq$ with respect to which $(I, \leq)$ is a directed set.

If X is a topological space and $(x_i)_{i\in I}^{\leq}$ is a net on X, we say that $(x_i)_{i\in I}^{\leq}$ converges to a point $x_0 \in X$ if for every neighborhood V of x_0 there exists $i_0 \in I$ such that $x_i \in V$ for every $i \geq i_0$. The point x_0 is called a *limit* of the net $(x_i)_{i\in I}^{\leq}$.

If X is a *Hausdorff space*, i.e., for every pair of distinct points $x_1, x_2 \in X$ there exist neighborhoods U_1 and U_2 of x_1 and x_2, respectively, such that $U_1 \cap U_2 = \emptyset$, then x_0 is unique and we write

$$\lim_{i\in I}{}^{\leq} x_i = x_0.$$

Finally, let $(E, \|\cdot\|)$ be a normed space and let $(F, \|\cdot\|')$ be a Banach space over $\mathbf{K}$; we denote by $\mathscr{L}(E, F)$ the space of all bounded linear operators from E into F. If $E = F$, we shall denote by $\mathscr{L}(E)$ the space $\mathscr{L}(E, E)$.

$\mathscr{L}(E, F)$ turns out to be a Banach space, once endowed with the *operator norm* $\|\cdot\|$ defined by

$$\|B\| := \sup_{f\in E\setminus\{0\}} \frac{\|B(f)\|'}{\|f\|} = \sup_{\substack{f\in E\\ \|f\|\leq 1}} \|B(f)\|' \qquad (B \in \mathscr{L}(E, F)).$$

1 Positive linear operators and approximation problems

The role of functional analysis has been decisive exactly in connection with classical problems. Almost all problems are on the applications, where functional analysis enables one to focus on a specific set of concrete analytical tasks and organize material in a clear and transparent form so that you know what the difficulties are.

Felix E. Browder

This is the first of two preparatory chapters which are devoted to review those concepts and results which are essential to develop all the material of the subsequent chapters.

After recalling the elementary properties of positive linear operators and functionals (Radon measures) on spaces of continuous functions, we review some properties of Choquet boundaries, Choquet simplices and Bauer simplices as well as several results concerning the convergence of nets of positive linear operators, including some Korovkin-type theorems.

In the final sections we present general asymptotic formulae for positive linear operators and estimates of the rate of convergence of positive approximation processes in terms of suitable moduli of smoothness.

Several of the most important results are given with proofs. Alternatively, we give explicit references.

1.1 Positive linear functionals and operators

In this introductory section we shall present some basic notation and properties of positive linear operators and positive linear functionals on spaces of continuous functions.

To make the exposition self-contained we also review some prerequisites on positive Radon measures as well as on Choquet boundaries and Choquet and Bauer simplices, which are presented without pretence of completeness and without proofs.

Throughout this section, unless otherwise stated, the symbol X designates a topological compact Hausdorff space.

A linear functional $\mu : \mathscr{C}(X) \longrightarrow \mathbf{R}$ is said to be *positive* if

$$\mu(f) \geq 0 \qquad \text{for every } f \in \mathscr{C}(X), f \geq 0. \tag{1.1.1}$$

If Y is another compact Hausdorff space, we say that a linear operator $T : \mathscr{C}(X) \longrightarrow \mathscr{C}(Y)$ is *positive* if

$$T(f) \geq 0 \qquad \text{for every } f \in \mathscr{C}(X), f \geq 0. \tag{1.1.2}$$

Every positive linear operator $T : \mathscr{C}(X) \longrightarrow \mathscr{C}(Y)$ gives rise to a family $(\mu_y^T)_{y \in Y}$ of positive linear functionals on $\mathscr{C}(X)$ defined by

$$\mu_y^T(f) := T(f)(y) \quad \text{for every } f \in \mathscr{C}(X) \text{ and } y \in Y. \tag{1.1.3}$$

Below we state some elementary properties of both positive linear functionals and positive linear operators.

In what follows, the symbol $\mathbf{F}$ stands either for the field $\mathbf{R}$ or for the space $\mathscr{C}(Y)$, Y being an arbitrary compact Hausdorff space.

Consider a positive linear operator $T : \mathscr{C}(X) \longrightarrow \mathbf{F}$. Then

(i) For every $f, g \in \mathscr{C}(X)$, $f \leq g$,

$$T(f) \leq T(g). \tag{1.1.4}$$

(ii) For every $f \in \mathscr{C}(X)$,

$$|T(f)| \leq T(|f|). \tag{1.1.5}$$

(iii) (Cauchy-Schwarz inequality) For every $f, g \in \mathscr{C}(X)$,

$$T(|fg|) \leq \sqrt{T(f^2)T(g^2)}. \tag{1.1.6}$$

In particular, for every $f \in \mathscr{C}(X)$,

$$T(|f|)^2 \leq T(\mathbf{1})T(f^2). \tag{1.1.7}$$

(iv) (Hölder inequality) For every $f, g \in \mathscr{C}(X)$ and for every $p, q > 1$, $1/p + 1/q = 1$,

$$T(|fg|) \leq T(|f|^p)^{1/p} T(|g|^q)^{1/q}. \tag{1.1.8}$$

(v) T is continuous and

$$\|T\| = \|T(\mathbf{1})\|_\infty, \tag{1.1.9}$$

where the symbol $\|T\|$ stands for the operator norm of T, which, as usual, is defined by

$$\|T\| := \sup_{\substack{f \in \mathscr{C}(X) \\ \|f\|_\infty \leq 1}} \|T(f)\|_\infty.$$

Thus, if $\mu : \mathscr{C}(X) \longrightarrow \mathbf{R}$ is a positive linear functional, then μ is continuous and $\|\mu\| = \mu(\mathbf{1})$. In this case, the norm $\|\mu\|$ of μ is defined by

$$\|\mu\| := \sup_{\substack{f \in \mathscr{C}(X) \\ \|f\|_\infty \leq 1}} |\mu(f)|.$$

If Y is another compact Hausdorff space, a positive linear operator $T : \mathscr{C}(X) \longrightarrow \mathscr{C}(Y)$ is said to be a *Markov operator* if $T(\mathbf{1}) = \mathbf{1}$. Thus, in this case, $\|T\| = 1$.

1.1.1 Positive Radon measures

In the sequel, the positive linear functionals on $\mathscr{C}(X)$ will be also called *positive Radon measures* on X. The set of them will be denoted by $M^+(X)$ and the subset of all $\mu \in M^+(X)$ such that $\mu(\mathbf{1}) = 1$ will be denoted by $M_1^+(X)$ and it will be referred to as the set of all *probability Radon measures* on X.

A simple example of a positive Radon measure is furnished by the *Dirac measure* at a point $a \in X$, that is defined by

$$\delta_a(f) := f(a) \qquad (f \in \mathscr{C}(X)). \tag{1.1.10}$$

A positive combination of Dirac measures is called a (positive) *discrete* measure. In other words, a Radon measure $\mu \in M^+(X)$ is discrete if there exist finitely many points $a_1, \dots, a_n \in X$, $n \geq 1$, and finitely many (positive) real numbers $\lambda_1, \dots, \lambda_n$ such that

$$\mu = \sum_{i=1}^{n} \lambda_i \delta_{a_i}, \tag{1.1.11}$$

i.e.,

$$\mu(f) = \sum_{i=1}^{n} \lambda_i f(a_i) \qquad (f \in \mathscr{C}(X)). \tag{1.1.12}$$

In this case, $\|\mu\| = \sum\limits_{i=1}^{n} \lambda_i$ and μ is also said to be *supported on* the set $\{a_1, \dots, a_n\}$.

There is a strong relationship between positive Radon measures and regular Borel measures on X.

If $\tilde{\mu}$ is a Borel measure on X, then every $f \in \mathscr{C}(X)$ is $\tilde{\mu}$-integrable; therefore, we can consider the positive Radon measure $I_{\tilde{\mu}}$ on X defined by

$$I_{\tilde{\mu}}(f) := \int_X f \, d\tilde{\mu} \qquad (f \in \mathscr{C}(X)). \tag{1.1.13}$$

Moreover, $\|I_{\tilde{\mu}}\| = \tilde{\mu}(X)$.

As a matter of fact, the previous formula describes all the positive Radon measures on X, as the following result shows (see [50, Section 29]).

Theorem 1.1.1. (Riesz representation theorem.) *If μ is a positive Radon measure on X, then there exists a unique regular Borel measure $\tilde{\mu}$ such that*

$$\mu(f) := \int_X f \, d\tilde{\mu} \qquad \textit{for every } f \in \mathscr{C}(X).$$

Moreover, $\|\mu\| = \tilde{\mu}(X)$.

As an immediate consequence, on account of (1.1.3), we get that, if $T : \mathscr{C}(X) \longrightarrow \mathscr{C}(Y)$ is a positive linear operator, then there exists a unique family $(\tilde{\mu}_y^T)_{y \in Y}$ of

Borel measures on X such that

$$T(f)(y) = \int_X f \, d\tilde{\mu}_y^T \qquad (f \in \mathscr{C}(X), y \in Y). \tag{1.1.14}$$

The family $(\tilde{\mu}_y^T)_{y \in Y}$ will be referred to as the *continuous selection of Borel measures associated with* T.

If T is a Markov operator, then each $\tilde{\mu}_y^T$ $(y \in Y)$ is a probability Borel measure on X. In this case, from Hölder's inequality, it also follows that

$$|T(f)|^p \leq T(|f|)^p \leq T(|f|^p) \tag{1.1.15}$$

for every $f \in \mathscr{C}(X)$ and $p \in [1, +\infty[$. In particular, if $f = T(f)$, then

$$|f|^p \leq T(|f|^p). \tag{1.1.16}$$

Note also that, if $(f_n)_{n \geq 1}$ is an equibounded sequence in $\mathscr{C}(X)$ that converges pointwise on X to some $f \in \mathscr{C}(X)$, then from (1.1.14) and from the Lebesgue convergence theorem it follows that

$$\lim_{n \to \infty} T(f_n) = T(f) \quad \text{pointwise on } Y. \tag{1.1.17}$$

Next, we discuss a characterization of discrete positive Radon measures. Usually, this characterization is proved by using the notion of support of Radon measures (see [59, Chapter III, Section 2] and [77, Vol. I, Section 11]). Below we follow a simple and direct path (see [11, Theorems 11.5-11.7]). We start with the following result which is important in its own right.

Theorem 1.1.2. *Let $\mu \in M^+(X)$ and consider a closed subset Y of X such that*

$$\mu(\varphi) = 0 \text{ for every } \varphi \in \mathscr{C}(X), \operatorname{Supp}(\varphi) \subset X \setminus Y. \tag{1.1.18}$$

Then $\mu(f) = \mu(g)$ for every $f, g \in \mathscr{C}(X)$ such that $f = g$ on Y.

We point out that there always exists a closed subset Y of X satisfying (1.1.18). The smallest of them is called the *support* of the Radon measure μ and it will be denoted by $\operatorname{Supp}(\mu)$ (see [59, Chapter II, Section 2] and [77, Vol. I, Section 11]). Thus, if $f \in \mathscr{C}(X)$ and $f = 0$ on $\operatorname{Supp}(\mu)$, then $\mu(f) = 0$. Conversely, if $f \in \mathscr{C}(X)$, $f \geq 0$, and if $\mu(f) = 0$, then $f = 0$ on $\operatorname{Supp}(\mu)$.

Furthermore, if $\tilde{\mu}$ denotes the Borel measure corresponding to μ via Theorem 1.1.1, then $\operatorname{Supp}(\tilde{\mu}) = \operatorname{Supp}(\mu)$.

An important example of a subset Y satisfying (1.1.18) is shown below.

Corollary 1.1.3. *Let $\mu \in M^+(X)$ and consider an arbitrary family $(f_i)_{i \in I}$ of positive functions in $\mathscr{C}(X)$ such that $\mu(f_i) = 0$ for every $i \in I$. Then the subset*

$$Y := \{x \in X \mid f_i(x) = 0 \text{ for every } i \in I\} \tag{1.1.19}$$

satisfies (1.1.18).

Therefore, if $f, g \in \mathscr{C}(X)$ and $f(x) = g(x)$ for every $x \in Y$, then $\mu(f) = \mu(g)$.

By means of the previous results, it is easy to reach the announced characterization of discrete Radon measures.

Theorem 1.1.4. *Given $\mu \in M^+(X)$ and distinct points $a_1, \dots, a_n \in X$, $n \geq 1$, the following statements are equivalent:*

(i) There exist $\lambda_1, \dots, \lambda_n \in [0, +\infty[$ such that $\mu = \sum_{i=1}^{n} \lambda_i \delta_{a_i}$.

(ii) If $\varphi \in \mathscr{C}(X)$ and $\mathrm{Supp}(\varphi) \cap \{a_1, \dots, a_n\} = \emptyset$, then $\mu(\varphi) = 0$.

(iii) For every $x \in X \setminus \{a_1, \dots, a_n\}$ there exists $f \in \mathscr{C}(X)$, $f \geq 0$, such that $f(x) > 0$, $f(a_i) = 0$ for every $i = 1, \dots, n$, and $\mu(f) = 0$.

(iv) $\mathrm{Supp}(\mu) = \{a_1, \dots, a_n\}$.

On $M^+(X)$ we shall consider the *vague topology*, which is, by definition, the coarsest topology on $M^+(X)$ for which all the mappings φ_f $(f \in \mathscr{C}(X))$ are continuous, where

$$\varphi_f(\mu) := \mu(f) \qquad \text{for every } \mu \in M^+(X). \tag{1.1.20}$$

Thus, the vague topology is a locally convex topology and it is, in fact, the restriction to $M^+(X)$ of the weak∗-topology of the Banach space $\mathscr{C}(X)$.

If $(\mu_i)_{i \in I}^{\leq}$ is a net in $M^+(X)$, then $(\mu_i)_{i \in I}^{\leq}$ converges to a certain $\mu \in M^+(X)$ with respect to the vague topology if $\lim_{i \in I^{\leq}} \mu_i(f) = \mu(f)$ for every $f \in \mathscr{C}(X)$. In this case, we also say that $(\mu_i)_{i \in I}^{\leq}$ converges vaguely to μ.

Note that, if $T : \mathscr{C}(X) \longrightarrow \mathscr{C}(Y)$ is a positive linear operator, then, by referring to the family $(\mu_y^T)_{y \in Y}$ in $M^+(X)$ defined by (1.1.3), the mapping $y \in Y \mapsto \mu_y^T \in M^+(X)$ is continuous.

If X is metrizable, then $M^+(X)$ is metrizable and separable.

Moreover, the following result holds.

Theorem 1.1.5. *Given a subset $\mathscr{U}$ of $M^+(X)$, the following statements are equivalent:*

(i) $\mathscr{U}$ is relatively compact in $M^+(X)$ with respect to the vague topology.

(ii) For every $f \in \mathscr{C}(X)$, $\sup_{\mu \in \mathscr{U}} |\mu(f)| < +\infty$.

(iii) $\sup_{\mu \in \mathscr{U}} \|\mu\| < +\infty$.

In particular, if X is metrizable, then every sequence in $M^+(X)$ which is pointwise bounded has a subsequence which converges vaguely to some $\mu \in M^+(X)$.

For a proof of Theorem 1.1.5 and for more details on the vague topology, we refer, e.g., to [50, § 30] or to [77, Vol. I, Section 12].

We now proceed to discuss a useful method to construct positive Radon measures and positive linear operators on spaces of continuous functions defined on product spaces. For the sake of simplicity we shall limit ourselves to consider compact spaces only.

Let $(X_i)_{1\le i\le d}$ be a finite family of compact Hausdorff spaces and consider the product space $\prod_{i=1}^{d} X_i$ endowed with the product topology, with respect to which it is compact and Hausdorff.

For each $j = 1, \dots, d$ we shall denote by $pr_j : \prod_{i=1}^{d} X_i \longrightarrow X_j$ the *j-th coordinate function*, which is defined by

$$pr_j(x) := x_j \qquad \text{for every } x = (x_i)_{1\le i\le d} \in \prod_{i=1}^{d} X_i. \tag{1.1.21}$$

By a common notation, if $X \subset \prod_{i=1}^{d} X_i$, the restriction of each pr_j to X will be denoted by pr_j as well.

If finitely many functions $f_i : X_i \longrightarrow \mathbf{R}$, $1 \le i \le d$, are given, we shall denote by $\bigotimes_{i=1}^{d} f_i : \prod_{i=1}^{d} X_i \longrightarrow \mathbf{R}$ the new function defined by

$$\left(\bigotimes_{i=1}^{d} f_i\right)(x) := \prod_{i=1}^{d} f_i(x_i) \qquad \text{for every } x = (x_i)_{1\le i\le d} \in \prod_{i=1}^{d} X_i. \tag{1.1.22}$$

Thus

$$\bigotimes_{i=1}^{d} f_i = \prod_{i=1}^{d} f_i \circ pr_i. \tag{1.1.23}$$

Furthermore, if $j = 1, \dots, d$ and $f_j : X_j \longrightarrow \mathbf{R}$, then

$$f_j \circ pr_j = \bigotimes_{i=1}^{d} f_{i,j}, \tag{1.1.24}$$

where $f_{i,j} := \mathbf{1}$ if $i \ne j$ and $f_{i,j} := f_j$ if $i = j$.

Clearly, if $f_i \in \mathscr{C}(X_i)$ for every $i = 1, \dots, d$, then $\bigotimes_{i=1}^{d} f_i \in \mathscr{C}\left(\prod_{i=1}^{d} X_i\right)$.

We shall denote by $\bigotimes_{i=1}^{d} \mathscr{C}(X_i)$ the linear subspace generated by

$$\left\{\bigotimes_{i=1}^{d} f_i \;\middle|\; f_i \in \mathscr{C}(X_i),\ i = 1, \dots, d\right\}.$$

In fact, by the Stone-Weierstrass theorem, $\bigotimes_{i=1}^{d} \mathscr{C}(X_i)$ is dense in $\mathscr{C}\left(\prod_{i=1}^{d} X_i\right)$ with respect to the sup-norm.

Now, for every $i = 1, \dots, d$, fix $\mu_i \in M^+(X_i)$. Then there exists a uniquely determined positive Radon measure ν on $\prod_{i=1}^{d} X_i$ such that for every $(f_i)_{1\le i\le d} \in \prod_{i=1}^{d} \mathscr{C}(X_i)$

$$\nu\left(\bigotimes_{i=1}^{d} f_i\right) = \prod_{i=1}^{d} \mu_i(f_i). \tag{1.1.25}$$

Such a positive linear functional is called the *tensor product* of the family $(\mu_i)_{1\le i\le d}$ and it is denoted by $\bigotimes_{i=1}^{d}\mu_i$. Thus, if $f_i \in \mathscr{C}(X_i)$, $1 \le i \le d$, then

$$\left(\bigotimes_{i=1}^{d}\mu_i\right)\left(\bigotimes_{i=1}^{d}f_i\right) = \prod_{i=1}^{d}\mu_i(f_i). \tag{1.1.26}$$

For every $i = 1, \dots, d$, denote by $\tilde{\mu}_i$ the regular Borel measure on X_i corresponding to μ_i via the Riesz representation theorem. Analogously, consider the regular Borel measure $\tilde{\mu}$ on $\prod_{i=1}^{d} X_i$ corresponding to $\bigotimes_{i=1}^{d}\mu_i$. By (1.1.26) and by Fubini theorem we infer that $\tilde{\mu}$ is the *product Borel measure of the family* $(\tilde{\mu}_i)_{1\le i\le d}$ (see [50, Definition 23.4]), i.e.,

$$\tilde{\mu} = \bigotimes_{i=1}^{d}\tilde{\mu}_i \tag{1.1.27}$$

and hence, again by Fubini theorem, for every $f \in \mathscr{C}\left(\prod_{i=1}^{d} X_i\right)$,

$$\left(\bigotimes_{i=1}^{d}\mu_i\right)(f) = \int_{X_1}\cdots\int_{X_d} f(x_1, \dots, x_d)\, d\tilde{\mu}_1(x_1)\cdots d\tilde{\mu}_d(x_d). \tag{1.1.28}$$

From (1.1.26) it also follows that

$$\left\|\bigotimes_{i=1}^{d}\mu_i\right\| = \prod_{i=1}^{d}\|\mu_i\| \tag{1.1.29}$$

and hence, in particular, $\bigotimes_{i=1}^{d}\mu_i \in M_1^+\left(\prod_{i=1}^{d} X_i\right)$ whenever $\mu_i \in M_1^+(X_i)$ for every $i = 1, \dots, d$.

Some useful properties of tensor products of measures are listed below (for a proof see [77, Vol. I, Section 13]).

Proposition 1.1.6. *Let $(X_i)_{1\le i\le d}$ be a finite family of compact Hausdorff spaces and, for every $i = 1, \dots, d$, fix $\mu_i \in M^+(X_i)$. The following statements hold true:*

(1) (Commutativity property.) If $\sigma : \{1, \dots, d\} \longrightarrow \{1, \dots, d\}$ is a permutation, then $\bigotimes_{i=1}^{d}\mu_{\sigma(i)} = \bigotimes_{i=1}^{d}\mu_i$.

(2) (Associativity property.) If $(I_k)_{1\le k\le n}$ is a partition of $\{1, \dots, d\}$, then $\bigotimes_{k=1}^{n}\left(\bigotimes_{i\in I_k}\mu_i\right) = \bigotimes_{i=1}^{d}\mu_i$.

(3) The mapping $(\mu_i)_{1\le i\le d} \in \prod_{i=1}^{d} M^+(X_i) \mapsto \bigotimes_{i=1}^{d}\mu_i \in M^+\left(\prod_{i=1}^{d} X_i\right)$ is continuous with respect to the product topology (of the vague topologies) and the vague topology on $M^+\left(\prod_{i=1}^{d} X_i\right)$.

(4) $\mathrm{Supp}\left(\bigotimes_{i=1}^{d}\mu_i\right) = \prod_{i=1}^{d}\mathrm{Supp}(\mu_i)$.

By using tensor products of measures we can also construct positive linear operators on spaces of continuous functions on the product space.

Let $(X_i)_{1\le i\le d}$ and $(Y_i)_{1\le i\le d}$ be two finite families of compact Hausdorff spaces. For every $i = 1, \dots, d$, let us consider a positive linear operator $T_i : \mathscr{C}(X_i) \longrightarrow \mathscr{C}(Y_i)$. Then, we define a linear operator $T : \mathscr{C}\left(\prod_{i=1}^{d} X_i\right) \longrightarrow \mathscr{C}\left(\prod_{i=1}^{d} Y_i\right)$ as follows: for every $f \in \mathscr{C}\left(\prod_{i=1}^{d} X_i\right)$ and $y = (y_1, \dots, y_d) \in \prod_{i=1}^{d} Y_i$,

$$T(f)(y) := \left(\bigotimes_{i=1}^{d}\mu_{y_i}^{T_i}\right)(f), \tag{1.1.30}$$

where the Radon measures $\mu_{y_i}^{T_i} \in M^+(Y_i)$ are defined as in (1.1.3). For every $f \in \mathscr{C}\left(\prod_{i=1}^{d} X_i\right)$, $T(f)$ is continuous by virtue of Proposition 1.1.6.

The operator T is positive and it is denoted by $\bigotimes_{i=1}^{d} T_i$. It is also called the *tensor product of the family* $(T_i)_{1\le i\le d}$.

Thus, $\bigotimes_{i=1}^{d} T_i : \mathscr{C}\left(\prod_{i=1}^{d} X_i\right) \longrightarrow \mathscr{C}\left(\prod_{i=1}^{d} Y_i\right)$ and for every $f \in \mathscr{C}\left(\prod_{i=1}^{d} X_i\right)$ and $y = (y_1, \dots, y_d) \in \prod_{i=1}^{d} Y_i$,

$$\begin{aligned}\left(\bigotimes_{i=1}^{d} T_i\right)(f)(y) &= \left(\bigotimes_{i=1}^{d}\mu_{y_i}^{T_i}\right)(f)\\ &= \int_{X_1}\cdots\int_{X_d} f(x_1, \dots, x_d)\, d\tilde{\mu}_{y_1}^{T_1}(x_1)\cdots d\tilde{\mu}_{y_d}^{T_d}(x_d).\end{aligned} \tag{1.1.31}$$

In particular, taking (1.1.3) and (1.1.26) into account, for every $(f_i)_{1\le i\le d} \in \prod_{i=1}^{d}\mathscr{C}(X_i)$,

$$\left(\bigotimes_{i=1}^{d} T_i\right)\left(\bigotimes_{i=1}^{d} f_i\right) = \bigotimes_{i=1}^{d} T_i(f_i). \tag{1.1.32}$$

Therefore, $\bigotimes_{i=1}^{d} T_i$ is a Markov operator whenever T_i is a Markov one for every $i = 1, \dots, d$.

Remark 1.1.7. We finally remark that, if for every $i = 1, \dots, d$ a linear subspace D_i of $\mathscr{C}(X_i)$ is given along with a linear operator $A_i : D_i \longrightarrow \mathscr{C}(X_i)$, then there exists a unique linear operator $A : \bigotimes_{i=1}^{d} D_i \longrightarrow \mathscr{C}\left(\prod_{i=1}^{d} X_i\right)$ such that, for every

$(f_i)_{1\le i\le d} \in \prod_{i=1}^{d} D_i$,

$$A\left(\bigotimes_{i=1}^{d} f_i\right) = \bigotimes_{i=1}^{d} A_i(f_i), \tag{1.1.33}$$

where $\bigotimes_{i=1}^{d} D_i$ denotes the linear subspace generated by $\left\{\bigotimes_{i=1}^{d} f_i \mid f_i \in D_i,\ 1 \le i \le d\right\}$ (see also Subsection 1.1.2 below).

The operator A will be denoted by $\bigotimes_{i=1}^{d} A_i$ and it will be again referred to as the *tensor product of the family* $(A_i)_{1\le i\le d}$.

1.1.2 Choquet boundaries

Another useful tool which we shall use in the sequel is the notion of Choquet boundary.

Given a linear subspace H of $\mathscr{C}(X)$, the *Choquet boundary* of H is the subset of all points $x \in X$ such that, if $\mu \in M^+(X)$ and $\mu(h) = h(x)$ for every $h \in H$, then $\mu(f) = f(x)$ for every $f \in \mathscr{C}(X)$, i.e., $\mu = \delta_x$ (see (1.1.10)).

It will be denoted by $\partial_H X$. Clearly, $\partial_H X = \partial_{\overline{H}} X$. If H contains the constants and separates the points of X, then $\partial_H X$ is non-empty and every $h \in H$ attains its minimum and maximum on $\partial_H X$ (see, e.g, [18, Corollary 2.6.5]).

Note also that if $\partial_H X \neq \emptyset$, then the unique $\mu \in M^+(X)$ which vanishes on H is the null measure.

A characterization of the points in $\partial_H X$ is given below (for a proof, see [18, Theorem 2.5.2]).

Theorem 1.1.8. *Given a subspace H of $\mathscr{C}(X)$ and $x \in X$, the following statements are equivalent:*
(i) $x \in \partial_H X$.
(ii) For every $f \in \mathscr{C}(X)$ and $\varepsilon > 0$ there exist $h, k \in H$ such that $h \le f \le k$ and $k(x) - h(x) \le \varepsilon$.
(iii)(1) For every $\varepsilon > 0$ and for every closed subset Y of X such that $x \notin Y$ there exists $k \in H$ such that $0 \le k$, $1 \le k$ on Y and $k(x) \le \varepsilon$.
(2) There exists $h \in H$ such that $h(x) \neq 0$.

From the above theorem the following useful result can be derived as well (for more details see [18, Theorem 2.5.4]).

Theorem 1.1.9. *Fix $x \in X$ and consider a linear subspace H of $\mathscr{C}(X)$ such that*
(i) There exists $h \in H$ with $h(x) \neq 0$.
(ii) For every $y \in X$, $y \neq x$, there exists $k \in H$, $k \ge 0$, such that $k(x) = 0$ and $k(y) > 0$.
Then $x \in \partial_H X$.

An important example of Choquet boundary is the set $\partial_e K$ of the *extreme points* of a convex compact subset K (of some locally convex Hausdorff space). They are defined as those points $x_0 \in K$ such that $K \setminus \{x_0\}$ is convex, i.e., if $x_1, x_2 \in K$ and $\lambda \in \mathbf{R}$, $0 < \lambda < 1$, and if $x_0 = \lambda x_1 + (1-\lambda)x_2$, then $x_0 = x_1 = x_2$.

Denote by $A(K)$ the linear subspace of all continuous *affine* functions $u : K \longrightarrow \mathbf{R}$, i.e., the subspace of all functions $u \in \mathscr{C}(K)$ such that

$$u(\lambda x + (1-\lambda)y) = \lambda u(x) + (1-\lambda)u(y) \tag{1.1.34}$$

$(x, y \in K,\ 0 \le \lambda \le 1)$.

More generally, if $u \in A(K)$, then

$$u\left(\sum_{i=1}^{n} \lambda_i x_i\right) = \sum_{i=1}^{n} \lambda_i u(x_i) \tag{1.1.35}$$

for every $x_1, \dots, x_n \in K$, $n \ge 2$, and $\lambda_1, \dots, \lambda_n \in \mathbf{R}^+$, $\sum\limits_{i=1}^{n} \lambda_i = 1$.

Clearly, $A(K)$ contains the constants and, by the Hahn-Banach theorem, it separates the points of K. As a matter of fact, it turns out that

$$\partial_{A(K)} K = \partial_e K \tag{1.1.36}$$

(for a proof see, e.g., [18, Proposition 2.6.3]).

We conclude the section by studying some Choquet boundaries for the product of a finite family of compact spaces.

Let $(X_i)_{1\le i\le d}$ be a finite family of compact Hausdorff spaces and denote by $X := \prod\limits_{i=1}^{d} X_i$ the (compact) product space of the family $(X_i)_{1\le i\le d}$, endowed with the product topology. As usual, we denote by $pr_i : X \longrightarrow X_i$ the i-th coordinate function from X into X_i (see (1.1.21)).

We recall that, if $f_i \in \mathscr{C}(X_i)$ for every $i = 1, \dots, d$, we denote by $\bigotimes\limits_{i=1}^{d} f_i : X \longrightarrow \mathbf{R}$ the continuous function defined by (1.1.22) (or by (1.1.23)).

Suppose that, for every $i = 1, \dots, d$, a subspace H_i of $\mathscr{C}(X_i)$ is assigned and that every H_i contains the constants and separates the points of X_i. We set

$$\sum_{i=1}^{d} H_i := \operatorname{span}(\{h_i \circ pr_i \mid h_i \in H_i,\ i = 1, \dots, d\}) \tag{1.1.37}$$

and

$$\bigotimes_{i=1}^{d} H_i := \operatorname{span}\left(\left\{\bigotimes_{i=1}^{d} h_i \mid h_i \in H_i,\ i = 1, \dots, d\right\}\right). \tag{1.1.38}$$

Here we recall that the symbol span(M) applied to a subset of functions M denotes the linear subspace generated by M.

We also consider the space

$$\bigodot_{i=1}^{d} H_i := \operatorname{span}(\{f \in \mathscr{C}(X) \mid f_{j,\overline{x}} \in H_j \text{ for every } j = 1, \dots, d \text{ and } \overline{x} = (\overline{x}_i)_{1 \le i \le d} \in \prod_{i=1}^{d} X_i\}); \tag{1.1.39}$$

here $f_{j,\overline{x}}$ denotes the continuous function on X_j defined by

$$f_{j,\overline{x}}(x_j) := f((y_i)_{1 \le i \le d}) \qquad (x_j \in X_j), \tag{1.1.40}$$

where $y_i = \overline{x}_i$ if $i \neq j$ and $y_i = x_j$ if $i = j$.

Clearly,

$$\sum_{i=1}^{d} H_i \subset \bigotimes_{i=1}^{d} H_i \subset \bigodot_{i=1}^{d} H_i. \tag{1.1.41}$$

According to the following proposition, we see that, in fact, these three subspaces of $\mathscr{C}(X)$ have the same Choquet boundary (for a proof, see [18, Proposition 2.6.8]).

Proposition 1.1.10. *Under the above assumptions,*

$$\partial_{\sum_{i=1}^{d} H_i} X = \partial_{\bigotimes_{i=1}^{d} H_i} X = \partial_{\bigodot_{i=1}^{d} H_i} X = \prod_{i=1}^{d} \partial_{H_i} X_i. \tag{1.1.42}$$

Corollary 1.1.11. *If $(X_i)_{1 \le i \le d}$ is a finite family of convex compact sets and if we consider the subspaces $A(X_i) \subset \mathscr{C}(X_i)$ $(i = 1, \dots, d)$, then $\bigodot_{i=1}^{d} A(X_i)$ coincides with the subspace of all continuous multiaffine functions on $X = \prod_{i=1}^{d} X_i$ (i.e., those continuous functions which are affine with respect to each variable). Furthermore, according to Proposition 1.1.10 and* (1.1.36),

$$\partial_{\bigodot_{i=1}^{d} A(X_i)} X = \prod_{i=1}^{d} \partial_e X_i. \tag{1.1.43}$$

1.1.3 Bauer simplices

Choquet simplices and Bauer simplices play an important role in the theory of integral representations for convex compact sets. For more details we refer to the excellent monograph of J. Lukeš, J. Malý, I. Netuka and J. Spurný ([135]) or to the short survey in Altomare-Campiti ([18, Section 1.5 and the references therein]).

Before presenting the definition of a Choquet simplex, we recall that the *convex hull* of a subset B of a (real or complex) vector space E is, by definition, the

smallest convex subset of E containing B and it is denoted by $\mathrm{co}(B)$. It is easy to show that

$$\mathrm{co}(B) = \left\{ \sum_{i=1}^{n} \lambda_i x_i \mid n \geq 1,\, x_i \in B,\, \lambda_i \geq 0,\, i = 1, \dots, n,\, \sum_{i=1}^{n} \lambda_i = 1 \right\}. \quad (1.1.44)$$

Given a convex compact subset K of a locally convex Hausdorff space X, set $G(K) := \{\lambda K + a \mid \lambda \geq 0,\ a \in X\}$. Then K is said to be a *Choquet simplex* if the intersection of two arbitrary elements of $G(K)$ belongs to $G(K)$ provided that it is non-empty.

In $\mathbf{R}^d$, $d \geq 1$, the Choquet simplices are convex hulls of any $d+1$ affinely independent points, (we recall that p points $x_1, \dots, x_p \in \mathbf{R}^d$ are said to be *affinely independent* if for every $\lambda_1, \dots, \lambda_p \in \mathbf{R}$ satisfying $\sum_{i=1}^{p} \lambda_i x_i = 0$ and $\sum_{i=1}^{p} \lambda_i = 0$, we get $\lambda_1 = \dots = \lambda_p = 0$).

Therefore, the subset

$$K_d := \left\{ (x_1, \dots, x_d) \in \mathbf{R}^d \mid x_i \geq 0 \text{ for every } i = 1, \dots, d \text{ and } \sum_{i=1}^{d} x_i \leq 1 \right\}, \quad (1.1.45)$$

being the convex hull of $\{v_0, \dots, v_d\}$, where

$$v_0 := (0, \dots, 0),\ v_1 := (1, 0, \dots, 0), \dots, v_d := (0, \dots, 0, 1), \quad (1.1.46)$$

is a Choquet simplex in $\mathbf{R}^d$ and it is called the *canonical simplex of* $\mathbf{R}^d$.

Note that, if this is the case, $\partial_e K_d = \{v_0, \dots, v_d\}$.

A *Bauer simplex* is a Choquet simplex such that the subset of its extreme points is closed.

Thus, K_d is a Bauer simplex in $\mathbf{R}^d$. The subset of all probability Radon measures on a compact space X is a Bauer simplex in the dual space of $\mathscr{C}(X)$.

According to the next theorem, when K is a Bauer simplex, then on $\mathscr{C}(K)$ there exists a natural positive *projection* $T : \mathscr{C}(K) \longrightarrow \mathscr{C}(K)$, i.e., a positive linear operator such that $T \circ T = T$.

Theorem 1.1.12. *Given a convex compact subset K, the following statements are equivalent:*

(a) K is a Bauer simplex.

(b) For every $x \in K$ there exists a unique probability Borel measure $\tilde{\mu}_x$ on K such that $\tilde{\mu}_x(K \setminus \overline{\partial_e K}) = 0$ and

$$\int_K h\, d\tilde{\mu}_x = h(x) \qquad \text{for every } h \in A(K).$$

(c) Every continuous function $f : \partial_e K \longrightarrow \mathbf{R}$ can be continuously extended to a (unique) function $\widehat{f} \in A(K)$.

(d) There exists a (unique) positive projection $T : \mathscr{C}(K) \longrightarrow \mathscr{C}(K)$ such that $T(\mathscr{C}(K)) = A(K)$.

Moreover, if one of these statements holds true, then, for every $f \in \mathscr{C}(K)$ and $x \in K$,

$$T(f)(x) = \int_K f \, d\tilde{\mu}_x = \widehat{f|_{\partial_e K}}(x). \tag{1.1.47}$$

Given a Bauer simplex K, the positive projection $T : \mathscr{C}(K) \longrightarrow \mathscr{C}(K)$ given by (1.1.47) is referred to as the *canonical positive projection associated with* K. Thus, for every $f \in \mathscr{C}(K)$, $T(f)$ is the unique continuous affine function on K which coincides with f on $\partial_e K$.

In the case $K = K_d$, $d \geq 1$, the canonical projection is given by

$$T_d(f)(x) := \left(1 - \sum_{i=1}^{d} x_i\right) f(v_0) + \sum_{i=1}^{d} x_i f(v_i) \tag{1.1.48}$$

($f \in \mathscr{C}(K_d), x = (x_1, \dots, x_d) \in K_d, v_0, \dots, v_d$ as in (1.1.46)).

In particular, for $d = 1$,

$$T_1 := (1 - x) f(0) + x f(1) \tag{1.1.49}$$

($f \in \mathscr{C}([0, 1]), 0 \leq x \leq 1$).

1.2 Korovkin-type approximation theorems

In this section we shall discuss some useful criteria for ascertaining whether a given sequence of positive linear operators on $\mathscr{C}(X)$ converges strongly to a given positive linear operator on $\mathscr{C}(X)$.

Usually, these criteria are concerned with special subsets of $\mathscr{C}(X)$ which guarantee the required convergence provided that it holds on them. These kinds of results are called *Korovkin-type theorems* and they refer to the Russian mathematician P. P. Korovkin who in 1953 discovered such a property for the functions $\mathbf{1}, x, x^2$ in the space $\mathscr{C}([0,1])$ and for the identity operator ([125; 126]; see also [18], [11]).

We begin by recalling the following fundamental definition.

Definition 1.2.1. *Let X and Y be compact Hausdorff spaces and consider a positive linear operator $T : \mathscr{C}(X) \longrightarrow \mathscr{C}(Y)$. A subset M of $\mathscr{C}(X)$ is said to be a* Korovkin subset *of $\mathscr{C}(X)$ for T if for every net $(L_i)_{i\in I}^{\leq}$ of positive linear operators from $\mathscr{C}(X)$ into $\mathscr{C}(Y)$ satisfying*

(i) $\sup_{i\in I} \|L_i\| < +\infty$

and

(ii) $\lim_{i\in I}^{\leq} L_i(g) = T(g)$ *for every $g \in M$, uniformly on Y,*

it turns out that

$$\lim_{i\in I}^{\leq} L_i(f) = T(f)$$

uniformly on Y, for every $f \in \mathscr{C}(X)$.

The subset M is said to be a sequential Korovkin subset *of $\mathscr{C}(X)$ for T if it satisfies a similar property with respect only to sequences of positive linear operators rather than nets.*

Note that a subset M is a Korovkin subset if and only if the linear subspace $\mathrm{span}(M)$ generated by M is a Korovkin subset. In the sequel, a linear subspace which is a Korovkin subset will be referred to as a *Korovkin subspace* of $\mathscr{C}(X)$ for T.

The next result furnishes a characterization of Korovkin subsets for positive linear operators. It was obtained by C. A. Micchelli ([142]) and, independently, by M. D. Rusk ([178]) (for a generalization to locally compact Hausdorff spaces see [11, Theorem 5.5] or [18, Theorem 3.1.3]).

Theorem 1.2.2. *Let X and Y be compact Hausdorff spaces. Given a positive linear operator $T : \mathscr{C}(X) \longrightarrow \mathscr{C}(Y)$ and a subset M of $\mathscr{C}(X)$, the following statements are equivalent:*

(i) M is a Korovkin subset of $\mathscr{C}(X)$ for T.

(ii) If $\mu \in M^+(X)$ and $y \in Y$ satisfy $\mu(g) = T(g)(y)$ for every $g \in M$, then $\mu(f) = T(f)(y)$ for every $f \in \mathscr{C}(X)$.

Moreover, if X and Y are metric spaces, then conditions (i) and (ii) are also equivalent to

(iii) M is a sequential Korovkin subset of $\mathscr{C}(X)$ for T.

As an application, consider distinct points $a_1, \dots, a_n \in X$, $n \geq 1$, and $\alpha_1, \dots, \alpha_n \in \mathbf{R}^+$. Fix $h_0 \in \mathscr{C}(X)$ such that

$$h_0(a_1) = \dots = h_0(a_n) = 0 \quad \text{and} \quad h_0(x) > 0 \quad \text{for} \quad x \in X \setminus \{a_1, \dots, a_n\}. \tag{1.2.1}$$

Further, consider $h_1, \dots, h_n \in \mathscr{C}(X)$ such that

$$\det\left(h_i(a_j)\right) \neq 0. \tag{1.2.2}$$

Corollary 1.2.3. *The subset $\{h_0, \dots, h_n\}$ is a Korovkin subset for the operator $T : \mathscr{C}(X) \longrightarrow \mathscr{C}(X)$ defined by*

$$T(f) = \sum_{i=1}^{n} \alpha_i f(a_i)\mathbf{1} \qquad (f \in \mathscr{C}(X)). \tag{1.2.3}$$

Proof. Considering $\mu \in M^+(X)$ and $x \in X$ satisfying $\mu(h_j) = T(h_j)(x)$ for every $j = 0, \dots, n$, we obtain in particular that $\mu(h_0) = 0$. Therefore, on account of (1.2.1) and Theorem 1.1.4, there exist $\lambda_1, \dots, \lambda_n \in \mathbf{R}^+$ such that $\mu = \sum_{i=1}^{n} \lambda_i \delta_{a_i}$. On the other hand, for $j = 1, \dots, n$,

$$\sum_{i=1}^{n} \alpha_i h_j(a_i) = T(h_j)(x) = \mu(h_j) = \sum_{i=1}^{n} \lambda_i h_j(a_i)$$

and hence, by (1.2.2), $\alpha_i = \lambda_i$ for every $i = 1, \dots, n$. Thus, $\mu(f) = T(f)(x)$ for every $f \in \mathscr{C}(X)$ and so the result follows from Theorem 1.2.2. □

Remark 1.2.4. From the preceding corollary, in the particular case $n = 1$, it follows that, if $h_0 \in \mathscr{C}(X)$, $h_0 \geq 0$ and h_0 vanishes only at a point $a \in X$, then, for every $\alpha \geq 0$, $\{\mathbf{1}, h_0\}$ is a Korovkin subset for the positive linear operator

$$T(f) = \alpha f(a)\mathbf{1} \qquad (f \in \mathscr{C}(X)). \tag{1.2.4}$$

Korovkin subsets of $\mathscr{C}(X)$ for the identity operator deserve to be treated separately because of their usefulness in the classical approximation theory. These subsets will be briefly called *Korovkin subsets* of $\mathscr{C}(X)$. Thus, a subset M of $\mathscr{C}(X)$ is said to be a Korovkin subset of $\mathscr{C}(X)$ if for every net $(L_i)_{i\in I}^{\leq}$ of positive linear operators from $\mathscr{C}(X)$ into $\mathscr{C}(X)$ satisfying

(i) $\quad \sup_{i\in I} \|L_i\| < +\infty$

and

(ii) $\lim_{i \in I}^{\leq} L_i(g) = g$ for every $g \in M$, uniformly on X,

it turns out that

$$\lim_{i \in I}^{\leq} L_i(f) = f$$

uniformly on X, for every $f \in \mathscr{C}(X)$.

Similarly, replacing nets of positive linear operators by sequences, it can be stated the definition of a *sequential Korovkin subset* of $\mathscr{C}(X)$.

When T is the identity operator, then Theorem 1.2.2 turns into the following result.

Theorem 1.2.5. *Given a compact Hausdorff space X and a subset M of $\mathscr{C}(X)$, the following statements are equivalent:*
(i) M is a Korovkin subset of $\mathscr{C}(X)$.
(ii) If $\mu \in M^+(X)$ and $x \in X$ satisfy $\mu(g) = g(x)$ for every $g \in M$, then $\mu(f) = f(x)$ for every $f \in \mathscr{C}(X)$, i.e., $\mu = \delta_x$.
(iii) $\partial_{\text{span}(M)} X = X$.
Moreover, if X is metrizable, then conditions (i)-(iii) are also equivalent to
(iv) M is a sequential Korovkin subset of $\mathscr{C}(X)$.

Combining Theorems 1.2.2 and 1.2.5 we immediately obtain the following useful property of Korovkin subsets.

Corollary 1.2.6. *Let X and Y be compact Hausdorff spaces. Every Korovkin subset M of $\mathscr{C}(X)$ is a Korovkin subset for any positive linear operator $T : \mathscr{C}(X) \longrightarrow \mathscr{C}(Y)$ of the form*

$$T(f) := \lambda(f \circ \varphi),$$

where $\lambda \in \mathscr{C}(Y)$, $\lambda \geq 0$, and $\varphi : Y \longrightarrow X$ is continuous.

Proof. Let $\mu \in M^+(X)$ and $y \in Y$ satisfying $\mu(g) = \lambda(y)g(\varphi(y))$ for every $g \in M$. If $\lambda(y) = 0$, then $\mu = 0$ on M and hence on $\mathscr{C}(X)$, i.e., $\mu(f) = \lambda(y)f(\varphi(y))$ for every $f \in \mathscr{C}(X)$.

If $\lambda(y) > 0$, it is sufficient to apply property (ii) of Theorem 1.2.5 to $\mu/\lambda(y)$ and $\varphi(y)$ and we obtain at once that $\mu(f) = \lambda(y)f(\varphi(y))$ for every $f \in \mathscr{C}(X)$. Therefore, the result follows from Theorem 1.2.2. □

We now proceed to discuss some useful criteria to explicitly determine Korovkin subsets of $\mathscr{C}(X)$.

The following result is a direct consequence of Theorem 1.1.9 and Theorem 1.2.5, (iii).

Proposition 1.2.7. *Let M be a subset of $\mathscr{C}(X)$ and assume that for every $x, y \in X$, $x \neq y$, there exists $h \in \text{span}(M)$, $h \geq 0$, such that $h(x) = 0$ and $h(y) > 0$. Then $\partial_{\text{span}(M)} X = X$ and hence M is a Korovkin subset of $\mathscr{C}(X)$.*

In the sequel, if M is a subset of $\mathscr{C}(X)$, we shall set

$$M^2 := \{f^2 \mid f \in M\}. \tag{1.2.5}$$

Note that, if in addition M separates the points of X, then for every $x, y \in X$, $x \neq y$, choosing $f \in M$ such that $f(x) \neq f(y)$, the function $h := (f - f(x)\mathbf{1})^2 \in \operatorname{span}(\{\mathbf{1}\} \cup M \cup M^2)$ is positive, vanishes at x and $h(y) > 0$.

Moreover, if $M := \{h_n \mid n \geq 1\}$ is finite or countable and if $\Phi := \sum_{n=1}^{\infty} h_n^2$ is uniformly convergent on X, then the series $h := \sum_{n=1}^{\infty} (h_n - h_n(x)\mathbf{1})^2 \in \overline{\operatorname{span}(\{\mathbf{1}\} \cup M \cup \{\Phi\})}$ satisfies the same property. Therefore, from Proposition 1.2.7 we obtain the following result, due to Schempp (see [184]).

Theorem 1.2.8. *If M is a subset of $\mathscr{C}(X)$ separating the points of X, then $\{\mathbf{1}\} \cup M \cup M^2$ is a Korovkin subset of $\mathscr{C}(X)$.*

Moreover, if $M = \{h_n \mid n \geq 1\}$ is finite or countable and if $\Phi := \sum_{n=1}^{\infty} h_n^2$ is uniformly convergent on X, then $\{\mathbf{1}\} \cup M \cup \{\Phi\}$ is a Korovkin subset of $\mathscr{C}(X)$.

Remark 1.2.9. If X is metrizable and H is a linear subspace of $\mathscr{C}(X)$ separating the points of X, then there always exists a sequence $(h_n)_{n\geq 1}$ in H such that the series $\sum_{n=1}^{\infty} h_n^2$ converges uniformly on X.

Indeed, H is separable (because $\mathscr{C}(X)$ is separable) and hence there exists a countable dense family $(\varphi_n)_{n\geq 1}$ of H. Then, it is sufficient to set $h_n := \frac{\varphi_n}{2^n(\|\varphi_n\|_\infty + 1)}$ for every $n \geq 1$.

If X is a compact subset of $\mathbf{R}^d$, $d \geq 1$, then the subset $M := \{pr_1, \dots, pr_d\}$ of the coordinate functions on X (see (1.1.21)) separates the points of X. Therefore, from Theorem 1.2.8, we obtain the following result which is due to Volkov ([206]) and which, for $d = 1$, reduces to the classical Korovkin theorem.

Theorem 1.2.10. *If X is a compact subset of $\mathbf{R}^d$, $d \geq 1$, then*

$$\left\{\mathbf{1}, pr_1, \dots, pr_d, \sum_{i=1}^{d} pr_i^2\right\}$$

is a Korovkin subset of $\mathscr{C}(X)$.

In particular, $\{\mathbf{1}, e_1, e_2\}$ is a Korovkin subset in $\mathscr{C}([a, b])$ $(a, b \in \mathbf{R}, a < b)$, where

$$e_i(t) := t^i \qquad (i \geq 1, t \in [a, b]). \tag{1.2.6}$$

1.3 Further convergence criteria for nets of positive linear operators

In this section we state some criteria concerning the convergence of nets of positive linear operators on $\mathscr{C}(X)$ towards a positive linear operator. The main assumptions refer to the subset of interpolation points of the limit operator and its possible representation by means of suitable functions which always exist if X is metrizable. As an application, we discuss some results about the asymptotic behavior of the iterates of Markov operators (for some additional results in this respect see the next Section 1.4 as well).

Let X be a compact Hausdorff space and consider a Markov operator $T : \mathscr{C}(X) \longrightarrow \mathscr{C}(X)$, i.e., a positive linear operator such that $T(\mathbf{1}) = \mathbf{1}$.

Denote by $(\tilde{\mu}_x^T)_{x\in X}$ the continuous selection of probability Borel measures on X associated with T, i.e.,

$$T(f)(x) = \int_X f\, d\tilde{\mu}_x^T \qquad (f \in \mathscr{C}(X), x \in X). \tag{1.3.1}$$

We shall set

$$M_T := \{h \in \mathscr{C}(X) \mid T(h) = h\}. \tag{1.3.2}$$

Clearly, M_T is contained in the range of T which will be also denoted by

$$H := T(\mathscr{C}(X)) = \{T(f) \mid f \in \mathscr{C}(X)\}. \tag{1.3.3}$$

The subspace M_T contains the constants and hence, if it separates the points of X, then its Choquet boundary $\partial_{M_T} X$ is not empty (see Section 1.2). We also recall that each $g \in M_T$ attains its minimum and its maximum on $\partial_{M_T} X$, so that $g = 0$ if $g = 0$ on $\partial_{M_T} X$.

In the sequel, the following subset

$$\partial_T X := \{x \in X \mid T(f)(x) = f(x) \text{ for every } f \in \mathscr{C}(X)\} \tag{1.3.4}$$

will play an important role. It will be referred to as the *set of interpolation points for T*.

Theorem 1.3.1. *Consider a Markov operator $T : \mathscr{C}(X) \longrightarrow \mathscr{C}(X)$ such that the subspace M_T defined by (1.3.2) separates the points of X. Then*

$$\emptyset \neq \partial_{M_T} X \subset \partial_T X \subset \partial_H X. \tag{1.3.5}$$

Moreover, if V is an arbitrary subset of M_T separating the points of X, then

$$\partial_T X = \{x \in X \mid T(h^2)(x) = h^2(x) \text{ for every } h \in V\}. \tag{1.3.6}$$

Finally, if $(h_n)_{n\geq 1}$ is a finite or countable family in M_T separating the points of X and such that the series $\Phi := \sum\limits_{n=1}^{\infty} h_n^2$ is uniformly convergent, then

$$\partial_T X = \{x \in X \mid T(\Phi)(x) = \Phi(x)\}. \tag{1.3.7}$$

Proof. Fixing $x \in \partial_{M_T} X$ and considering the measure $\tilde{\mu}_x^T$ defined as in (1.3.1), then $\int_X h\,d\tilde{\mu}_x^T = h(x)$ for every $h \in M_T$ and hence $\tilde{\mu}_x^T = \varepsilon_x$, i.e., $x \in \partial_T X$.

On the other hand, given $x \in \partial_T X$, then for every $y \in X$, $y \neq x$, we may choose $h \in M_T$ such that $h(x) \neq h(y)$. Set $f := T((h - h(x))^2) \in H$. Then $f \geq 0$ and, by (1.1.16), $f(y) \geq (h(y) - h(x))^2 > 0$. Therefore, $x \in \partial_H X$ by Theorem 1.1.9.

The equality (1.3.6) easily follows by observing that, if $x \in X$ and $T(h^2)(x) = h^2(x)$ for every $h \in V$, then $\int_X g\,d\tilde{\mu}_x^T = g(x)$ for every $g \in \{\mathbf{1}\} \cup V \cup V^2$ and hence $\tilde{\mu}_x^T = \varepsilon_x$ by Theorems 1.2.5 and 1.2.8.

Similarly, (1.3.7) follows from the second part of Theorem 1.2.8. □

Remark 1.3.2. According to Remark 1.2.9, if X is metrizable, then there always exists a sequence $(h_n)_{n \geq 1}$ in the linear space generated by V such that the series $\sum_{n=1}^{\infty} h_n^2$ converges uniformly on X. Therefore, the function $\Phi := \sum_{n=1}^{\infty} h_n^2$ satisfies (1.3.7) and, in addition, $\Phi \leq T(\Phi)$ because of (1.1.16).

The following example shows that both the inclusions in (1.3.5) may be strict.

Example 1.3.3. *Let $\lambda \in \mathscr{C}([0,1])$ satisfying $\lambda(1/2) = 0$ and $0 < \lambda(x) \leq 1/2$ for every $x \in [0,1]$, $x \neq 1/2$. Consider the Markov operator $T : \mathscr{C}([0,1]) \longrightarrow \mathscr{C}([0,1])$ defined by*

$$T(f)(x) = \lambda(x)((1-x)f(0) + xf(1)) + (1 - \lambda(x))f(x) \tag{1.3.8}$$

($f \in \mathscr{C}([0,1]), x \in [0,1]$).

Then, $\partial_{M_T}[0,1] = \{0,1\}$, $\partial_T[0,1] = \{0,1/2,1\}$ and $\partial_H[0,1] = [0,1]$ (see (1.3.3)).

Consider, indeed, $h \in \mathscr{C}([0,1])$ such that $T(h) = h$; then, for every $x \in [0,1]$, $x \neq 1/2$, $h(x) = (1-x)h(0) + xh(1)$. By continuity, we then obtain that h is affine in $[0,1]$. Conversely, each affine function h on $[0,1]$ satisfies $T(h) = h$ and, hence, it belongs to M_T.

Thus, $M_T = A([0,1])$ and, therefore, $\partial_{M_T}[0,1] = \{0,1\}$ (see (1.1.34) and (1.1.36)).

As regards the second equality, obviously $\{0,1/2,1\} \subset \partial_T[0,1]$. Conversely, let $x \in \partial_T[0,1]$ and assume that $x \notin \{0,1/2,1\}$. Then choosing $f \in \mathscr{C}([0,1])$ such that $f(0) = f(1) = 0$ and $f(x) \neq 0$, we obtain $T(f)(x) = (1 - \lambda(x))f(x) \neq f(x)$, which leads to a contradiction.

Finally, note that $H = T(\mathscr{C}([0,1])) = \mathscr{C}([0,1])$ because, if $g \in \mathscr{C}([0,1])$, then $g = T(f)$, where $f = \frac{2-\lambda}{1-\lambda} g - \frac{1}{1-\lambda} T(g)$. Accordingly, $\partial_H[0,1] = [0,1]$.

Below we discuss some cases where in (1.3.5) we may have equalities.

Proposition 1.3.4. *Under the same assumptions of Theorem 1.3.1, the following statements are equivalent:*

(a) There exists a subset V of M_T separating the points of X such that $T^2(h^2) = T(h^2)$ for every $h \in V$, i.e., $T(V^2) \subset M_T$.
(b) T is a projection, i.e., $T^2(f) = T(f)$ for every $f \in \mathscr{C}(X)$.
Moreover, if X is metrizable, then statement (b) is equivalent to
(c) There exists a finite or countable family $(h_n)_{n\geq 1}$ in M_T separating the points of X such that the series $\Phi := \sum_{i=1}^{n} h_n^2$ is uniformly convergent and $T^2(\Phi) = T(\Phi)$.
Finally, if (a), (b) or (c) holds true, then $M_T = H$ and hence $\partial_{M_T} X = \partial_T X = \partial_H X$. Moreover, for every $x \in X$,

$$\operatorname{Supp}(\tilde{\mu}_x^T) \subset \partial_T X = \partial_H X \tag{1.3.9}$$

and, for every $f, g \in \mathscr{C}(X)$,

$$T(f) = T(g) \qquad \text{provided } f = g \text{ on } \partial_H X. \tag{1.3.10}$$

Proof. We have only to prove the implication (a)⇒(b). From (1.3.6) it follows that

$$\partial_T X = \{x \in X \mid T(h^2)(x) - h^2(x) = 0 \text{ for every } h \in V\}$$

and, by (1.1.16), $T(h^2) - h^2 \geq 0$ for each $h \in V$.

Now, if $x \in X$, then from the assumption we get

$$\int_X (T(h^2) - h^2)\, d\tilde{\mu}_x^T = 0 \qquad (h \in V).$$

Therefore, because of Corollary 1.1.3, for every $f \in \mathscr{C}(X)$,

$$\int_X (T(f) - f)\, d\tilde{\mu}_x^T = 0$$

since $f = T(f)$ on $\partial_T X$ and this shows that $T^2(f) = T(f)$.

A similar reasoning can be used to prove the implication (c)⇒(b).

Finally, (1.3.9) follows from (1.3.7), taking into account that

$$\int_X (T(\Phi) - \Phi)\, d\tilde{\mu}_x^T = 0$$

for every $x \in X$.

As regards (1.3.10), it suffices to point out that $T(f - g)$ belongs to H and it vanishes on $\partial_H X$, so that it must be equal to zero. □

Theorem 1.3.5. *Let $T : \mathscr{C}(X) \longrightarrow \mathscr{C}(X)$ be a Markov operator such that the subset $\partial_T X$ defined by (1.3.4) is non-empty and assume that there exists $\Psi \in \mathscr{C}(X)$, $\Psi \geq 0$, such that $\partial_T X = \{x \in X \mid \Psi(x) = 0\}$ (for example $\Psi = T(\Phi) - \Phi$ as in (1.3.7)).*

Consider a net $(L_i)_{i\in I}^{\leq}$ of positive linear operators from $\mathscr{C}(X)$ into itself such that

(i) $\lim\limits_{i\in I}^{\leq} L_i(\mathbf{1}) = \mathbf{1}$ *pointwise (resp., uniformly) on* X*;*
(ii) $\lim\limits_{i\in I}^{\leq} L_i(\Psi) = 0$ *pointwise (resp., uniformly) on* X.
Then $\lim\limits_{i\in I}^{\leq} L_i(T(f)-f) = 0$ *pointwise (resp., uniformly) on* X *for every* $f \in \mathscr{C}(X)$.

Accordingly, if $(L_i(T(f)))_{i\in I}^{\leq}$ *converges pointwise (resp., uniformly) on* X *to some function* $S(f)$ *for every* $f \in \mathscr{C}(X)$*, then*

$$\lim_{i\in I}^{\leq} L_i(f) = S(f)$$

pointwise (resp., uniformly) on X *for every* $f \in \mathscr{C}(X)$*. In particular,* $S \circ T = S$.

Proof. Given $f \in \mathscr{C}(X)$ and $\varepsilon > 0$, the open set $U := \{x \in X \mid |T(f)(x) - f(x)| < \varepsilon\}$ contains $\partial_T X$ and its complement $X \setminus U$ is compact. Letting $\alpha := \min\limits_{x\in X\setminus U} \Psi(x)$, we get

$$|T(f) - f| \leq \varepsilon\mathbf{1} + \frac{2\|f\|_\infty}{\alpha}\Psi.$$

Accordingly, for any $i \in I$,

$$|L_i(T(f) - f)| \leq \varepsilon L_i(\mathbf{1}) + \frac{2\|f\|_\infty}{\alpha} L_i(\Psi)$$

and, hence, the result follows. □

As a special case of Theorem 1.3.5 we get the following result.

Corollary 1.3.6. *Assume that* X *is metrizable and let* $T : \mathscr{C}(X) \longrightarrow \mathscr{C}(X)$ *be a Markov operator such that the subspace* M_T *defined by* (1.3.2) *separates the points of* X*. Furthermore, set* $H := T(\mathscr{C}(X))$ *and consider* $\Phi \in \mathscr{C}(X)$ *such that* $\Phi \leq T(\Phi)$ *and*

$$\partial_H X = \{x \in X \mid T(\Phi)(x) = \Phi(x)\}.$$

Given a Markov operator $S : \mathscr{C}(X) \longrightarrow \mathscr{C}(X)$ *such that* $S \circ T = S$*, if* $(L_i)_{i\in I}^{\leq}$ *is a net of positive linear operators from* $\mathscr{C}(X)$ *into itself and if*
(i) $\lim\limits_{i\in I}^{\leq} L_i(h) = S(h)$ *pointwise (resp., uniformly) on* X *for every* $h \in H$*;*
(ii) $\lim\limits_{i\in I}^{\leq} L_i(\Phi) = S(\Phi)$ *pointwise (resp., uniformly) on* X*,*
then $\lim\limits_{i\in I}^{\leq} L_i(f) = S(f)$ *pointwise (resp., uniformly) on* X *for every* $f \in \mathscr{C}(X)$*. In other words,* $H \cup \{\Phi\}$ *is a Korovkin subset of* $\mathscr{C}(X)$ *for* S*. In particular, if* T *is a projection, then* $H \cup \{\Phi\}$ *is a Korovkin subset of* $\mathscr{C}(X)$ *for* T.

Remark 1.3.7. In [11, Theorem 10.3] it was proved that, when considering the uniform convergence, condition (ii) can be replaced by requiring that
(ii)' $\lim\limits_{i\in I}^{\leq} L_i(h^2) = T(h^2)$ uniformly on X for every $h \in V$, where V is an arbitrary subset of H separating the points of X.

As an application of Corollary 1.3.6 we show the following result which will be useful to study the asymptotic behaviour of the Markov semigroup associated with Bernstein-Schnabl operators (see Chapter 4).

To this end, we recall that, if L is a linear operator on $\mathscr{C}(X)$, for every $m \geq 0$, we denote by L^m the *iterate of* L *of order* m, i.e.,

$$L^m := \begin{cases} I & \text{if } m = 0; \\ L \circ L^{m-1} & \text{if } m \geq 1, \end{cases} \tag{1.3.11}$$

where I is the *identity operator* on $\mathscr{C}(X)$.

Theorem 1.3.8. *Let X be a compact metric space and consider a Markov projection $T : \mathscr{C}(X) \longrightarrow \mathscr{C}(X)$ such that its range $H := T(\mathscr{C}(X))$ separates the points of X. Further, consider $\Phi \in \mathscr{C}(X)$ of the form $\Phi = \sum\limits_{n=1}^{\infty} h_n^2$ as in Theorem 1.3.1, so that $\Phi \leq T(\Phi)$ and $\partial_H X = \{x \in X \mid T(\Phi)(x) = \Phi(x)\}$.*

Let $(L_n)_{n\geq 1}$ be a sequence of positive linear operators on $\mathscr{C}(X)$ such that $L_n(h) = h$ for every $h \in H$ and $n \geq 1$, and set

$$a_{n,p} := \max_{x \in X}\{T(\Phi)(x) - \Phi(x) - pn(L_n(\Phi)(x) - \Phi(x))\} \tag{1.3.12}$$

$(n \geq 1, p \geq 1)$.

Finally, let $(T(t))_{t\geq 0}$ be a family of positive linear operators on $\mathscr{C}(X)$ such that for every $t \geq 0$ there exists a sequence $(k(n))_{n\geq 1}$ of positive integers satisfying $k(n)/n \to t$ and

$$T(t)(f) = \lim_{n\to\infty} L_n^{k(n)}(f) \tag{1.3.13}$$

uniformly on X for every $f \in \mathscr{C}(X)$, where each $L_n^{k(n)}$ denotes the iterate of L_n of order $k(n)$.

If $\lim\limits_{p\to\infty} a_{n,p} = 0$ uniformly with respect to $n \geq 1$, then

$$\lim_{t\to+\infty} T(t)(f) = T(f) \tag{1.3.14}$$

uniformly on X for every $f \in \mathscr{C}(X)$.

Proof. From (1.3.13) it follows that $T(t)(h) = h$ for every $h \in H$ and $t \geq 0$. Therefore, on account of Corollary 1.3.6, in order to get the result it is sufficient to show that $\lim\limits_{t\to+\infty} T(t)(\Phi) = T(\Phi)$ uniformly on X. To this end, we preliminary observe that, by (1.1.16), $\Phi \leq L_n(\Phi) \leq L_n(T(\Phi)) = T(\Phi)$ for every $n \geq 1$. Furthermore, since H is invariant under each L_n, then

$$\partial_H X \subset \{x \in X \mid L_n(f)(x) = f(x) \text{ for every } f \in \mathscr{C}(X)\}$$

and hence, for every $n, p \geq 1$,

$$T(\Phi) - \Phi \geq a_{n,p} \geq \max_{x\in\partial_H X}\{T(\Phi)(x) - \Phi(x) - pn(L_n(\Phi)(x) - \Phi(x))\} = 0.$$

From the inequality $\frac{1}{pn}(T(\Phi) - \Phi - a_{n,p}) \leq L_n(\Phi) - \Phi$ we infer that, for every $n, p \geq 1$,

$$\left(1 - \frac{1}{pn}\right)\Phi + \frac{1}{pn}(T(\Phi) - a_{n,p}) \leq L_n(\Phi) \leq T(\Phi)$$

and therefore iterating, for each $k \geq 1$,

$$\left(1-\frac{1}{pn}\right)^k \Phi + \left(1-\left(1-\frac{1}{pn}\right)^k\right)(T(\Phi)-a_{n,p}) \leq L_n^k(\Phi) \leq T(\Phi).$$

For a given $\varepsilon > 0$, fix $p \geq 1$ such that $a_{n,p} \leq \varepsilon/2$ for every $n \geq 1$ and choose $t^* > 0$ such that

$$e^{-t/p}\Phi + \left(1-e^{-t/p}\right)(T(\Phi)-\varepsilon/2) \geq T(\Phi)-\varepsilon$$

for any $t \geq t^*$. Accordingly, if $t \geq t^*$ and if $(k(n))_{n\geq 1}$ is a sequence of positive integers satisfying $k(n)/n \to t$ and (1.3.13), then

$$\left(1-\frac{1}{pn}\right)^{k(n)} \Phi + \left(1-\left(1-\frac{1}{pn}\right)^{k(n)}\right)(T(\Phi)-\varepsilon/2) \leq$$
$$\left(1-\frac{1}{pn}\right)^{k(n)} \Phi + \left(1-\left(1-\frac{1}{pn}\right)^{k(n)}\right)(T(\Phi)-a_{n,p}) \leq L_n^{k(n)}(\Phi) \leq T(\Phi).$$

Letting n to tend to ∞, we obtain

$$e^{-t/p}\Phi + \left(1-e^{-t/p}\right)(T(\Phi)-\varepsilon/2) \leq T(t)(\Phi) \leq T(\Phi)$$

and hence

$$T(\Phi)-\varepsilon \leq T(t)(\Phi) \leq T(\Phi).$$

Thus, $\lim_{t\to+\infty} T(t)(\Phi) = T(\Phi)$ uniformly on X. □

Below we show some further applications of Theorem 1.3.5 which concern the asymptotic behaviour of iterates of Markov operators.

Consider two Markov operators $S : \mathscr{C}(X) \longrightarrow \mathscr{C}(X)$ and $T : \mathscr{C}(X) \longrightarrow \mathscr{C}(X)$ such that $S \circ T = T$ and the subspace $M_T := \{h \in \mathscr{C}(X) \mid T(h) = h\}$ separates the points of X. Then

$$M_T \subset H := T(\mathscr{C}(X)) \subset \{h \in \mathscr{C}(X) \mid S(h) = h\} \tag{1.3.15}$$

and hence, by Theorem 1.3.1,

$$\emptyset \neq \partial_{M_T} X \subset \partial_T X \subset \partial_H X \subset \partial_S X. \tag{1.3.16}$$

Theorem 1.3.9. *Assume that X is metrizable. Let $S : \mathscr{C}(X) \longrightarrow \mathscr{C}(X)$ and $T : \mathscr{C}(X) \longrightarrow \mathscr{C}(X)$ be Markov operators and assume that the subspace $M_T := \{h \in \mathscr{C}(X) \mid T(h) = h\}$ separates the points of X. Then the following statements are equivalent:*

(a) $\lim_{n\to\infty} S^n(f) = T(f)$ *uniformly on X for every $f \in \mathscr{C}(X)$.*

(b) $\lim_{n\to\infty} S^n(f) = T(f)$ *pointwise on X for every $f \in \mathscr{C}(X)$.*

(c) $S \circ T = T$ *and $\partial_S X \subset \partial_T X$, i.e., for every $x \in X \setminus \partial_T X$ there exists $f \in \mathscr{C}(X)$ such that $S(f)(x) \neq f(x)$.*

(d) $S \circ T = T$ *and for every sequence* $(h_n)_{n\geq 1}$ *in* M_T *separating the points of* X *such that the series* $\Phi := \sum_{n=1}^{\infty} h_n^2$ *is uniformly convergent on* X, *one gets* $\Phi \leq S(\Phi)$ *and* $\{x \in X \mid S(\Phi)(x) = \Phi(x)\} \subset \partial_T X$.

(e) $S \circ T = T$ *and there exists* $\Phi \in \mathscr{C}(X)$ *such that* $\Phi \leq S(\Phi)$ *and* $\{x \in X \mid S(\Phi)(x) = \Phi(x)\} \subset \partial_T X$.

Moreover, if one of the statements above holds true, then T *necessarily is a Markov projection,* $T \circ S = T$ *and, finally*

$$\partial_T X = \partial_S X = \partial_H X. \tag{1.3.17}$$

Proof. On account of Remark 1.2.9, we only need to prove the implications (b)⇒(c)⇒(d) and (e)⇒(a). Moreover, the final part of the statement is a direct consequence of statement (a) and formula (1.3.16).

(b)⇒(c). Given $f \in \mathscr{C}(X)$, since the sequence $(S^n(f))_{n\geq 1}$ is equibounded, from (1.1.17) it follows that $S(T(f)) = \lim_{n\to\infty} S^{n+1}(f)$ pointwise on X, i.e., $S(T(f)) = T(f)$. Now, considering $x \in \partial_S X$, clearly

$$T(f)(x) = \lim_{n\to\infty} S^n(f)(x) = f(x)$$

for every $f \in \mathscr{C}(X)$, i.e., $x \in \partial_T X$.

(c)⇒(d). Consider a sequence $(h_n)_{n\geq 1}$ in M_T as in statement (c). On account of Theorem 1.3.1 (see, also, (1.3.15)), we get $\Phi \leq S(\Phi)$ and $\{x \in X \mid S(\Phi)(x) = \Phi(x)\} = \partial_S X$ and hence the result follows.

(e)⇒(a). First, observe that the sequence $(S^n(\Phi))_{n\geq 1}$, being increasing and bounded, is pointwise convergent on X. Setting $\Psi := S(\Phi) - \Phi$, we get that

$$S^n(\Psi) = S^{n+1}(\Phi) - S^n(\Phi) \to 0$$

pointwise on X. Therefore, by Theorem 1.3.5, $\lim_{n\to\infty} S^n(f) = T(f)$ pointwise on X for every $f \in \mathscr{C}(X)$. In particular, $\lim_{n\to\infty} S^n(\Phi) = T(\Phi)$ pointwise on X. Since $T(\Phi)$ is continuous and $(S^n(\Phi))_{n\geq 1}$ is increasing, by Dini's Theorem, $S^n(\Phi) \to T(\Phi)$ uniformly on X as well. Hence, by applying at once Theorem 1.3.1, we conclude that $S^n(f) \to T(f)$ uniformly on X for every $f \in \mathscr{C}(X)$. □

Remarks 1.3.10.

1. We point out that the implication (e)⇒(a) of the preceding theorem holds true without necessarily assuming that X is metrizable and that the subspace $M_T := \{h \in \mathscr{C}(X) \mid T(h) = h\}$ separates the points of X. It is enough to assume that $\partial_T X$ is non-empty.
2. If $T : \mathscr{C}(X) \longrightarrow \mathscr{C}(X)$ is a Markov projection whose range separates the points of X, considering an arbitrary function $\Phi \in \mathscr{C}(X)$ such that $\Phi \leq T(\Phi)$ and $\partial_H X = \{x \in X \mid T(\Phi) = \Phi\}$, then an example of a Markov operator $S : \mathscr{C}(X) \longrightarrow \mathscr{C}(X)$ satisfying statement (c) of Theorem 1.3.9 is

$$S := \lambda T + (\mathbf{1} - \lambda)I,$$

where I denotes the identity operator on $\mathscr{C}(X)$ and $\lambda \in \mathscr{C}(X)$ satisfies $0 < \lambda(x) \leq 1$ for every $x \in X$.

1.4 Asymptotic behaviour of Lipschitz contracting Markov semigroups

In this section we discuss another useful method to investigate the convergence of nets of Markov operators which, however, applies to one-parameter semigroups of operators as well as to the iterates of a single Markov operator. The main result involves the behaviour of the operators on the Lipschitz continuous functions and on the relevant Lipschitz seminorms.

Let (X, ρ) be a compact metric space and consider a Markov operator $T : \mathscr{C}(X) \longrightarrow \mathscr{C}(X)$. A (regular) probability measure $\tilde{\mu}$ on X is said to be *T-invariant* or, simply, *invariant* (if no confusion can arise) if

$$\int_X T(f)\, d\tilde{\mu} = \int_X f\, d\tilde{\mu} \qquad \text{for every } f \in \mathscr{C}(X). \tag{1.4.1}$$

There always exists an invariant probability Borel measure (see, e.g., [127, Section 5.1, p. 178]).

Furthermore, if $\tilde{\mu}$ is such a measure, then for every $f \in \mathscr{C}(X)$ and $p \in [1, +\infty[$, on account of (1.1.15) it follows that

$$\int_X |T(f)|^p\, d\tilde{\mu} \le \int_X T(|f|^p)\, d\tilde{\mu} = \int_X |f|^p\, d\tilde{\mu}$$

and hence T extends to a unique bounded linear operator $T_p : L^p(X, \tilde{\mu}) \longrightarrow L^p(X, \tilde{\mu})$ such that $\|T_p\| \le 1$. Moreover, T_p is positive as $\mathscr{C}(X)$ is a dense sublattice of $L^p(X, \tilde{\mu})$ and, if $1 \le p < q < +\infty$, then $T_p = T_q$ on $L^q(X, \tilde{\mu})$.

From now on, for a given $p \in [1, +\infty[$, if no confusion can arise, we shall denote by $\widetilde{T}$ the operator T_p.

In the sequel, given a probability Borel measure $\tilde{\mu}$, we shall denote by $\Lambda(\tilde{\mu})$ the subset of all Markov operators T on $\mathscr{C}(X)$ for which $\tilde{\mu}$ is an invariant measure.

Below we list some simple properties of this subset, that can be easily verified.

Proposition 1.4.1. *For a probability Borel measure $\tilde{\mu}$ on X the following properties hold:*

(1) If $S, T \in \Lambda(\tilde{\mu})$, then $S \circ T \in \Lambda(\tilde{\mu})$ and $\widetilde{S \circ T} = \widetilde{S} \circ \widetilde{T}$ on $L^p(X, \tilde{\mu})$ for every $p \in [1, +\infty[$.

(2) If $(T_i)_{i \in I}^{\le}$ is a net in $\Lambda(\tilde{\mu})$ and if there exists $T(f) = \lim_{i \in I}^{\le} T_i(f)$ uniformly on X for every $f \in \mathscr{C}(X)$, then $T \in \Lambda(\tilde{\mu})$ and $\widetilde{T}(f) = \lim_{i \in I}^{\le} \widetilde{T}_i(f)$ in $L^p(X, \tilde{\mu})$, whenever $f \in L^p(X, \tilde{\mu})$ and $1 \le p < +\infty$.

In order to simultaneously treat both discrete and continuous semigroups of Markov operators, we shall introduce the symbol $\mathbb{A}$ to denote either the interval $[0, +\infty[$ or the set $\mathbf{N}$ of all positive integers. The subset $\mathbb{A}$ will be endowed with the usual ordering $\le$ inherited from $\mathbf{R}$. Without no explicit mention, we shall

refer to this ordering when we shall consider converging nets $(x_\tau)^{\leq}_{\tau\in\mathbb{A}}$ in some metric space, whose limit will be denoted by $\lim_{\tau\to\infty} x_\tau$.

We set

$$\rho(X) := \sup\{\rho(x,y) \mid x, y \in X\} \tag{1.4.2}$$

and

$$\mathrm{Lip}(X) := \left\{ f \in \mathscr{C}(X) \mid |f|_{\mathrm{Lip}} := \sup_{\substack{x,y\in X\\ x\neq y}} \frac{|f(x)-f(y)|}{\rho(x,y)} < +\infty \right\}. \tag{1.4.3}$$

Below we state and prove the main result of the section.

Theorem 1.4.2. *Let $(T(\tau))^{\leq}_{\tau\in\mathbb{A}}$ be a net of Markov operators on $\mathscr{C}(X)$ such that*
(i) (Semigroup property) $T(\sigma)\circ T(\tau) = T(\sigma+\tau)$ for every $\sigma, \tau \in \mathbb{A}$.
(ii) (Lipschitz contraction property) For every $\tau \in \mathbb{A}, T(\tau)(\mathrm{Lip}(X)) \subset \mathrm{Lip}(X)$ and

$$|T(\tau)(f)|_{\mathrm{Lip}} \leq c(\tau)|f|_{\mathrm{Lip}} \quad (f \in \mathrm{Lip}(X)),$$

where $c : \mathbb{A} \longrightarrow]0, +\infty[$, with $\lim_{\tau\to\infty} c(\tau) = 0$.
Then
(a) For every $f \in \mathscr{C}(X)$ the net $(T(\tau)(f))^{\leq}_{\tau\in\mathbb{A}}$ converges uniformly on X to a constant function.
(b) There exists a unique probability Borel measure $\tilde{\mu}$ such that $T(\tau) \in \Lambda(\tilde{\mu})$ for every $\tau \in \mathbb{A}$, i.e., $\int_X T(\tau)(f)\,d\tilde{\mu} = \int_X f\,d\tilde{\mu}$ for each $f \in \mathscr{C}(X)$ and $\tau \in \mathbb{A}$. Moreover,

$$\lim_{\tau\to\infty} T(\tau)(f) = \int_X f\,d\tilde{\mu} \quad \textit{uniformly on } X \quad (f \in \mathscr{C}(X)), \tag{1.4.4}$$

as well as

$$\lim_{\tau\to\infty} \widetilde{T}(\tau)(f) = \int_X f\,d\tilde{\mu} \quad \textit{in } L^p(X,\tilde{\mu}) \qquad (f \in L^p(X,\tilde{\mu})) \tag{1.4.5}$$

for every $p \in [1, +\infty[$.
(c) For every $f \in \mathrm{Lip}(X)$ and $\tau \in \mathbb{A}$,

$$\left| T(\tau)(f) - \int_X f\,d\tilde{\mu} \right| \leq c(\tau)\rho(X)|f|_{\mathrm{Lip}}. \tag{1.4.6}$$

Proof. Because of assumption (ii), given $f \in \mathrm{Lip}(X)$ and $\tau \in \mathbb{A}$, for every $x, y \in X$ we get

$$|T(\tau)(f)(x) - T(\tau)(f)(y)| \leq c(\tau)\rho(X)|f|_{\mathrm{Lip}} \tag{1}$$

and hence

$$|T(\tau)(f)(x)\mathbf{1} - T(\tau)(f)| \leq c(\tau)\rho(X)|f|_{\mathrm{Lip}}.$$

Therefore, for $\sigma \in \mathbb{A}$, by recalling that $T(\sigma)$ is a Markov operator and by using the semigroup property (i), we obtain

$$|T(\tau)(f)(x)\mathbf{1} - T(\sigma+\tau)(f)| \le c(\tau)\rho(X)|f|_{\mathrm{Lip}};$$

thus, since $x \in X$ was arbitrarily chosen,

$$|T(\tau)(f) - T(\sigma+\tau)(f)| \le c(\tau)\rho(X)|f|_{\mathrm{Lip}}. \tag{2}$$

The above estimate together with the assumption on the function c show that the family $(T(\tau)(f))^{\le}_{\tau\in\mathbb{A}}$ is a Cauchy net in $\mathscr{C}(X)$ with respect to the uniform norm and hence it is convergent.

Therefore, we may consider the mapping $T : \mathrm{Lip}(X) \longrightarrow \mathscr{C}(X)$ defined by

$$T(f) := \lim_{\tau\to\infty} T(\tau)(f) \quad (f \in \mathrm{Lip}(X)),$$

that is linear, positive and $T(\mathbf{1}) = \mathbf{1}$.

From (1) it follows that, for every $f \in \mathrm{Lip}(X)$ and $x, y \in X, T(f)(x) = T(f)(y)$ so that $T(f)$ is constant. In other words, there exists a positive linear functional $\psi : \mathrm{Lip}(X) \longrightarrow \mathbf{R}$ such that

$$T(f) = \psi(f)\mathbf{1} \quad \text{for every} \quad f \in \mathrm{Lip}(X).$$

The functional ψ extends to a positive linear functional $\varphi : \mathscr{C}(X) \longrightarrow \mathbf{R}$ such that $\varphi(\mathbf{1}) = \psi(\mathbf{1}) = 1$ and hence, by the Riesz representation theorem (see Theorem 1.1.1) there exists a unique probability Borel measure $\tilde{\mu}$ such that $\varphi(f) = \int_X f\,d\tilde{\mu} \quad (f \in \mathscr{C}(X))$. Consequently $\lim_{\tau\to\infty} T(\tau)(f) = \int_X f\,d\tilde{\mu}$ uniformly on X, provided $f \in \mathrm{Lip}(X)$. Since $\mathrm{Lip}(X)$ is dense in $\mathscr{C}(X)$, the same limit relationship extends to $\mathscr{C}(X)$ as well, which shows (1.4.4) and, hence, (1.4.5) (see Proposition 1.4.1, part (2)).

Notice that $T(\tau) \in \Lambda(\tilde{\mu})$ for every $\tau \in \mathbb{A}$ because, given $f \in \mathscr{C}(X)$,

$$\int_X T(\tau)(f)\,d\tilde{\mu} = \lim_{\sigma\to\infty} T(\sigma)(T(\tau)(f)) = \lim_{\sigma\to\infty} T(\sigma+\tau)(f) = \int_X f\,d\tilde{\mu}.$$

Moreover, if $\tilde{\nu}$ is another probability Borel measure and if $T(\tau) \in \Lambda(\tilde{\nu})$ for each $\tau \in \mathbb{A}$, then, from (1.4.4) we obtain

$$\int_X f\,d\tilde{\nu} = \lim_{\tau\to\infty} \int_X T(\tau)(f)\,d\tilde{\nu} = \int_X f\,d\tilde{\mu}$$

for every $f \in \mathscr{C}(X)$, so that $\tilde{\nu} = \tilde{\mu}$.

Finally (1.4.6) follows from (2) letting σ to tend to ∞. □

Remarks 1.4.3.

1. Note that, under the assumptions of Theorem 1.4.2, from (1.4.6) it follows that, for every $r > 0$,

$$\lim_{\tau\to\infty} T(\tau)(f) = \int_X f\,d\tilde{\mu} \tag{1.4.7}$$

uniformly on X and uniformly with respect to $f \in \mathrm{Lip}(X)$, $|f|_{\mathrm{Lip}} \leq r$.
2. Consider the K–functionals ([86, p. 171])

$$K(f,\delta) := \inf_{g\in \mathrm{Lip}(X)} \{\| f-g \|_\infty + \delta |g|_{\mathrm{Lip}}\}$$

$(f \in \mathscr{C}(X), \delta > 0)$
and

$$\tilde{K}(f,\delta) := \inf_{g\in \mathrm{Lip}(X)} \{\| f-g \|_p + \delta |g|_{\mathrm{Lip}}\}$$

$(f \in L^p(X,\tilde{\mu})$, $1 \leq p < +\infty$, $\delta > 0)$.
Then from (1.4.6) it follows that, for every $\tau \in \mathbb{A}$,

$$\left| T(\tau)(f) - \int_X f\, d\tilde{\mu} \right| \leq K(f, c(\tau)\rho(X)) \qquad (f \in \mathscr{C}(X))$$

and

$$\left\| \tilde{T}(\tau)(f) - \int_X f\, d\tilde{\mu} \right\|_p \leq \tilde{K}(f, c(\tau)\delta(X)) \qquad (f \in L^p(X,\tilde{\mu}), 1 \leq p < +\infty).$$

Finally, we point out that, if X is a compact interval of $\mathbf{R}$ or the unit circle $\mathbb{T}$ of $\mathbf{R}^2$, then

$$K(f,\delta) = \frac{1}{2}\tilde{\omega}(f, 2\delta) \leq \omega(f, 2\delta),$$

where $\omega(f,\cdot)$ denotes the usual *modulus of continuity of f* defined by

$$\omega(f,\cdot) := \sup\{|f(x)-f(y)| \mid x,y \in K, \|x-y\|_2 \leq \delta\} \tag{1.4.8}$$

and $\omega(f,\cdot)$ denotes the least concave majorant of $\omega(f,\cdot)$ ([86, Chapter 6, Theorem 2.1 and Chapter 2, Lemma 2.1]).

Below we discuss some consequences of Theorem 1.4.2. Consider a Markov operator $T : \mathscr{C}(X) \longrightarrow \mathscr{C}(X)$ such that

$$T(\mathrm{Lip}(X)) \subset \mathrm{Lip}(X) \tag{1.4.9}$$

and assume that there exists $c \in]0,1[$ such that

$$|T(f)|_{\mathrm{Lip}} \leq c|f|_{\mathrm{Lip}} \tag{1.4.10}$$

for every $f \in \mathrm{Lip}(X)$.

For every $m \in \mathbf{N}$, denote by T^m the iterate of T of order m, defined as in (1.3.11).

Clearly $T^m(\mathrm{Lip}(X)) \subset \mathrm{Lip}(X)$ and

$$|T^m(f)|_{\mathrm{Lip}} \leq c^m |f|_{\mathrm{Lip}} \quad (f \in \mathrm{Lip}(X)). \tag{1.4.11}$$

Therefore, Theorem 1.4.2 applies to $\mathbb{A} = \mathbf{N}, T(m) = T^m$ and $c(m) = c^m$ $(m \in \mathbf{N})$ and we get

Corollary 1.4.4. *Under assumptions (1.4.9) and (1.4.10), there exists a unique probability Borel measure $\tilde{\mu}$ such that $T \in \Lambda(\tilde{\mu})$, i.e., $\int_X T(f)\,d\tilde{\mu} = \int_X f\,d\tilde{\mu}$ for every $f \in \mathscr{C}(X)$. Moreover,*

$$\lim_{m\to\infty} T^m(f) = \int_X f\,d\tilde{\mu} \quad \textit{uniformly on } X \tag{1.4.12}$$

for every $f \in \mathscr{C}(X)$ and, if $p \in [1, +\infty[$ and $f \in L^p(X, \tilde{\mu})$, then

$$\lim_{m\to\infty} \widetilde{T}^m(f) = \int_X f\,d\tilde{\mu} \quad \textit{in } L^p(X, \tilde{\mu}). \tag{1.4.13}$$

Finally, if $f \in \mathrm{Lip}(X)$ and $m \geq 1$, then

$$\left| T^m(f) - \int_X f\,d\tilde{\mu} \right| \leq c^m \rho(X)|f|_{\mathrm{Lip}} \tag{1.4.14}$$

so that the limit (1.4.12) is uniform with respect to $f \in \mathrm{Lip}(X), |f|_{\mathrm{Lip}} \leq r, (r > 0)$.

Remarks 1.4.5.
1. Markov operators admitting only one invariant probability measure are called *uniquely ergodic* ([127, Section 5.1, p. 178]). Under assumptions (1.4.9) and (1.4.10), clearly from (1.4.13) it turns out that for every $f \in L^p(X, \tilde{\mu})$ the sequence $\left(\frac{1}{n}\sum\limits_{k=0}^{n-1} \widetilde{T}^k(f)\right)_{n\geq 1}$ of the Cesàro means converges to $\int\limits_X f\,d\tilde{\mu}$ in $L^p(X, \tilde{\mu})$. On the other hand, by Akcoglu's ergodic theorem ([127, Theorem 2.6, p. 190]), the sequence is $\tilde{\mu}$-a.e. convergent and hence

$$\lim_{n\to\infty} \frac{1}{n}\sum_{k=0}^{n-1} \widetilde{T}^k(f) = \int_X f\,d\tilde{\mu} \quad \tilde{\mu} - \text{a.e..} \tag{1.4.15}$$

2. Estimates of the rate of convergence in (1.4.12) and (1.4.13) can be directly obtained from Remark 1.4.3, 2. For the sake of brevity we omit to explicitly state them.

The problem of checking condition (ii) of Theorem 1.4.2 for a *continuous family* $(T(t))_{t\geq 0}$ of Markov operators on $\mathscr{C}(X)$, (i.e., for a family $(T(t))_{t\geq 0}$ of Markov operators such that, for every $f \in \mathscr{C}(X)$, the mapping $t \geq 0 \mapsto T(t)(f)$ is continuous), seems to be a more delicate task, especially when one does not know an explicit description of the operators $T(t)$ as it generally happens when dealing with the C_0-semigroup generated by some linear operator $A : D(A) \subset \mathscr{C}(X) \longrightarrow \mathscr{C}(X)$.

Below we discuss a simple situation where both Theorem 1.4.2 and Corollary 1.4.4 can be successfully applied. This situation often occurs in the theory of approximation by positive linear operators.

Corollary 1.4.6. *Consider a continuous semigroup $(T(t))_{t\geq 0}$ of Markov operators on $\mathscr{C}(X)$ and assume that there exists a sequence $(L_n)_{n\geq 1}$ of Markov operators*

on $\mathscr{C}(X)$ such that for every $t \geq 0$ there exists a sequence $(k(n))_{n\geq 1}$ of positive integers with $k(n)/n \to t$ and

$$T(t)(f) = \lim_{n\to\infty} L_n^{k(n)}(f) \quad \text{uniformly on } X \tag{1.4.16}$$

for every $f \in \mathscr{C}(X)$.

Furthermore, assume that

(i) There exists $\omega \in \mathbf{R}, \omega < 0$, such that for every $n \geq 1$, $L_n(\mathrm{Lip}(X)) \subset \mathrm{Lip}(X)$ and $|L_n(f)|_{\mathrm{Lip}} \leq \left(1+\frac{\omega}{n}\right)|f|_{\mathrm{Lip}}$ for every $f \in \mathrm{Lip}(X)$.

(ii) There exists a probability Borel measure $\tilde{\mu}$ such that $L_n \in \Lambda(\tilde{\mu})$ for every $n \geq 1$.

Then

(1) For every $n \geq 1$ and $f \in \mathscr{C}(X)$,

$$\lim_{m\to\infty} L_n^m(f) = \int_X f\,d\tilde{\mu} \quad \text{uniformly on } X. \tag{1.4.17}$$

(2) For every $t \geq 0$, $T(t)(\mathrm{Lip}X) \subset \mathrm{Lip}(X)$ and $|T(t)(f)|_{\mathrm{Lip}} \leq \exp(\omega t)|f|_{\mathrm{Lip}}$ ($f \in \mathrm{Lip}(X)$). Moreover, $T(t) \in \Lambda(\tilde{\mu})$ and

$$\lim_{t\to+\infty} T(t)(f) = \int_X f\,d\tilde{\mu} \quad \text{uniformly on } X \tag{1.4.18}$$

for every $f \in \mathscr{C}(X)$.

(3) If $1 \leq p < +\infty$, $n \geq 1$ and $f \in L^p(X,\tilde{\mu})$, then

$$\lim_{m\to\infty} \widetilde{L}_n^m(f) = \int_X f\,d\tilde{\mu} = \lim_{t\to+\infty} \widetilde{T}(t)(f) \quad \text{in } L^p(X,\tilde{\mu}). \tag{1.4.19}$$

Proof. Each operator L_n satisfies conditions (1.4.9) and (1.4.10) so that, by Corollary 1.4.4, there exists a unique probability Borel measure $\tilde{\nu}_n$ such that $L_n \in \Lambda(\tilde{\nu}_n)$. Therefore $\tilde{\nu}_n = \tilde{\mu}$ and hence (1.4.17) follows from (1.4.12).

From (1.4.16) it turns out that, given $t \geq 0, T(t)(\mathrm{Lip}(X)) \subset \mathrm{Lip}(X)$ and

$$|T(t)(f)|_{\mathrm{Lip}} \leq \exp(\omega t)|f|_{\mathrm{Lip}},$$

because $\lim_{n\to\infty}\left(1+\frac{\omega}{n}\right)^{k(n)} = \exp(\omega t)$. Therefore, Theorem 1.4.2 applies and hence there exists a unique probability Borel measure $\tilde{\nu}$ such that $T(t) \in \Lambda(\tilde{\nu})$ for every $t \geq 0$. On the other hand, $T(t) \in \Lambda(\tilde{\mu})$ for every $t \geq 0$ as pointed out in Proposition 1.4.1, part (2), and hence $\tilde{\nu} = \tilde{\mu}$. Accordingly, (1.4.4) implies (1.4.18), and (1.4.19) follows from (1.4.13) and (1.4.5), respectively. □

Remark 1.4.7. Notice that, according to (1.4.6) and (1.4.14), if $f \in \mathrm{Lip}(X)$, then

$$\left|L_n^m(f) - \int_X f\,d\tilde{\mu}\right| \leq \left(1+\frac{\omega}{n}\right)^m \rho(X)|f|_{\mathrm{Lip}} \tag{1.4.20}$$

and

$$\left|T(t)(f) - \int_X f \, d\tilde{\mu}\right| \le \exp(\omega t)\rho(X)|f|_{\mathrm{Lip}} \tag{1.4.21}$$

($n \ge 1, m \ge 1, t \ge 0$). Other estimates for arbitrary functions in $\mathscr{C}(X)$ or in $L^p(X, \tilde{\mu})$ can be obtained by applying Remark 1.4.3, 2 (see also [111]).

1.5 Asymptotic formulae for positive linear operators

Asymptotic formulae play an important role in the analysis of the saturation properties of approximation processes and in the approximation of strongly continuous semigroups by means of iterates of operators. In this section we shall discuss some of them in the setting of convex compact subsets of $\mathbf{R}^d$, $d \geq 1$; moreover, we shall present some results on non compact domains and, in particular, on unbounded intervals. On this topic, for further details, we refer the interested reader to [13], [23], [24] and the references therein.

First of all, let K be a convex compact subset of $\mathbf{R}^d$, $d \geq 1$, whose interior $\mathrm{int}(K)$ is assumed to be non-empty. Given $x \in K$, we denote by $\Psi_x : K \longrightarrow \mathbf{R}^d$ the mapping defined by

$$\Psi_x(y) := y - x \quad (y \in K) \tag{1.5.1}$$

and by $\Phi_x : K \longrightarrow \mathbf{R}$ the function defined by

$$\Phi_x(y) := \|y - x\|_2 \quad (x \in K), \tag{1.5.2}$$

where $\|\cdot\|_2$ stands for the Euclidean norm on $\mathbf{R}^d$.

We recall that the symbol $\mathscr{C}^2(K)$ denotes the space of all real-valued continuous functions $u : K \longrightarrow \mathbf{R}$ which are twice continuously differentiable on $\mathrm{int}(K)$ and whose partial derivatives of order ≤ 2 can be continuously extended to K. Given $u \in \mathscr{C}^2(K)$, then, for every $x, y \in K$,

$$\begin{aligned} u(y) &= u(x) + \sum_{i=1}^{d} \frac{\partial u}{\partial x_i}(x)(y_i - x_i) \\ &+ \frac{1}{2}\sum_{i,j=1}^{d} \frac{\partial^2 u}{\partial x_i \partial x_j}(x)(y_i - x_i)(y_j - x_j) + \omega_u(x,y)\Phi_x^2(y), \end{aligned} \tag{1.5.3}$$

where

$$\omega_u(x,y) := \frac{u(y) - u(x) - \sum\limits_{i=1}^{d} \frac{\partial u}{\partial x_i}(x)(y_i - x_i) - \frac{1}{2}\sum\limits_{i,j=1}^{d} \frac{\partial^2 u}{\partial x_i \partial x_j}(x)(y_i - x_i)(y_j - x_j)}{\|y - x\|_2^2} \tag{1.5.4}$$

if $x \neq y$ and $\omega_u(x,y) := 0$ if $x = y$.

Thus, $\omega_u(x,\cdot)\Phi_x^2 \in \mathscr{C}(K)$. Moreover, by Taylor's formula with integral remainder (see, e.g., [46, §4.3.1]), if $x, y \in \mathrm{int}(K)$,

$$\begin{aligned} &\omega_u(x,y)\Phi_x^2(y) \\ &= \int_0^1 (1-t) \sum_{i,j=1}^{d} \left[\frac{\partial^2 u}{\partial x_i \partial x_j}(x + t(y-x)) - \frac{\partial^2 u}{\partial x_i \partial x_j}(x) \right] (y_i - x_i)(y_j - x_j)\, dt. \end{aligned} \tag{1.5.5}$$

By the continuity of the second partial derivatives and the Lebesgue dominated convergence theorem, the above equality extends to every $x, y \in K$. Therefore, setting

$$M := \sup_{x\in K} \sum_{i,j=1}^{d} \left| \frac{\partial^2 u}{\partial x_i \partial x_j}(x) \right|, \tag{1.5.6}$$

we get

$$|\omega_u(x,y)|\Phi_x^2(y) \le 2M\|y-x\|_2^2 \int_0^1 (1-t)\,dt = M\|y-x\|_2^2$$

and hence

$$|\omega_u(x,y)| \le M \quad (x, y \in K). \tag{1.5.7}$$

Finally, note that the uniform continuity of all the second partial derivatives implies that

$$\lim_{y\to x} \omega_u(x,y) = 0 \qquad \text{uniformly w.r.t. } x \in K. \tag{1.5.8}$$

Indeed, given $\varepsilon > 0$, there exists $\delta > 0$ such that

$$\left| \frac{\partial^2 u}{\partial x_i \partial x_j}(x) - \frac{\partial^2 u}{\partial x_i \partial x_j}(y) \right| \le \frac{2\varepsilon}{d^2}$$

for every $i, j = 1, \dots, d$ and for every $x, y \in K$, $\|x-y\|_2 \le \delta$. Therefore, for any such $x, y \in K$, by using (1.5.5), we get

$$|\omega_u(x,y)\Phi_x^2(y)| \le \varepsilon\|y-x\|_2^2,$$

so that $|\omega_u(x,y)| \le \varepsilon$.

An additional useful estimate of the function ω_u is shown below.

Lemma 1.5.1. *Let $u \in \mathscr{C}^2(K)$. Then for every $\varepsilon > 0$ there exists $\delta > 0$ such that*

$$|\omega_u(x,\cdot)|\Phi_x^2 \le \varepsilon\Phi_x^2 + \frac{M}{\delta^{q-2}}\Phi_x^q$$

for all $x \in K$ and $q \in \mathbf{R}$, $q > 2$, where M is given by (1.5.6).

Proof. Given $\varepsilon > 0$, by (1.5.8) there exists $\delta > 0$ such that $|\omega_u(x,y)|\Phi_x^2(y) \le \varepsilon\|y-x\|_2^2$ for every $\|x-y\|_2 \le \delta$. On the other hand, if $\|x-y\|_2 > \delta$, by (1.5.7)

$$|\omega_u(x,y)|\Phi_x^2(y) \le M\Phi_x^2(y) \le \frac{M}{\delta^{q-2}}\Phi_x^q(y)$$

and hence the result follows. □

The general asymptotic formula which we shall show next is concerned with differential operators of the form

$$A(u)(x) := \frac{1}{2}\sum_{i,j=1}^{d} \alpha_{ij}(x)\frac{\partial^2 u}{\partial x_i \partial x_j}(x) + \sum_{i=1}^{d} \beta_i(x)\frac{\partial u}{\partial x_i}(x) + \gamma(x)u(x) \tag{1.5.9}$$

($u \in \mathscr{C}^2(K)$, $x \in K$), where $\alpha_{ij}, \beta_i, \gamma \in \mathscr{F}(K)$, $i, j = 1, \dots, d$.

Theorem 1.5.2. *Let $\alpha_{ij}, \beta_i, \gamma \in \mathscr{F}(K)$, $i, j = 1, \dots, d$, and consider the differential operator A defined by (1.5.9). Furthermore, consider a divergent sequence $(\varphi(n))_{n\geq 1}$ of positive integers and a sequence $(L_n)_{n\geq 1}$ of positive linear operators from $\mathscr{C}(K)$ into $\mathscr{F}(K)$. Assume that there exists a subset G of K such that*

(i) $\lim_{n\to\infty} \varphi(n)(L_n(\mathbf{1})(x) - 1) - \gamma(x) = 0$ *uniformly w.r.t.* $x \in G$;

(ii) $\lim_{n\to\infty} \varphi(n)L_n(pr_i \circ \Psi_x)(x) - \beta_i(x) = 0$ *uniformly w.r.t.* $x \in G$ *for every* $i = 1, \dots, d$;

(iii) $\lim_{n\to\infty} \varphi(n)L_n((pr_i \circ \Psi_x)(pr_j \circ \Psi_x))(x) - \alpha_{ij}(x) = 0$ *uniformly w.r.t.* $x \in G$, *for every* $i, j = 1, \dots, d$.

If $u \in \mathscr{C}^2(K)$ *and if*

(iv) $\lim_{n\to\infty} \varphi(n)L_n(\omega_u(x, \cdot)\Phi_x^2)(x) = 0$ *uniformly w.r.t.* $x \in G$,

then

$$\lim_{n\to\infty} \varphi(n)(L_n(u) - u) = A(u) \tag{1.5.10}$$

uniformly on G.

In particular, if

(v) $\sup_{\substack{x\in G\\ n\geq 1}} \varphi(n)L_n(\Phi_x^2)(x) < +\infty$

and if

(vi) there exists $q \in \mathbf{R}$, $q > 2$, *such that*

$$\lim_{n\to\infty} \varphi(n)L_n(\Phi_x^q)(x) = 0 \qquad \textit{uniformly w.r.t. } x \in G,$$

then (1.5.10) *holds true for every* $u \in \mathscr{C}^2(K)$.

Proof. Consider $u \in \mathscr{C}^2(K)$ satisfying (iv). Given $x \in G$, from (1.5.3) it follows that

$$\begin{aligned} u &= u(x)\mathbf{1} + \sum_{i=1}^{d} \frac{\partial u}{\partial x_i}(x)(pr_i \circ \Psi_x) \\ &+ \frac{1}{2} \sum_{i,j=1}^{d} \frac{\partial^2 u}{\partial x_i \partial x_j}(x)(pr_i \circ \Psi_x)(pr_j \circ \Psi_x) + \omega_u(x, \cdot)\Phi_x^2 \end{aligned}$$

and hence, for $n \geq 1$,

$$\begin{aligned} L_n(u) &= u(x)L_n(\mathbf{1}) + \sum_{i=1}^{d} \frac{\partial u}{\partial x_i}(x)L_n(pr_i \circ \Psi_x) \\ &+ \frac{1}{2} \sum_{i,j=1}^{d} \frac{\partial^2 u}{\partial x_i \partial x_j}(x)L_n((pr_i \circ \Psi_x)(pr_j \circ \Psi_x)) + L_n(\omega_u(x, \cdot)\Phi_x^2). \end{aligned}$$

Therefore,

$$|\varphi(n)(L_n(u)(x) - u(x)) - Au(x)| \le |\varphi(n)(L_n(\mathbf{1})(x) - 1) - \gamma(x)| \|u\|_\infty$$
$$+ \sum_{i=1}^{d} |\varphi(n)L_n(pr_i \circ \Psi_x)(x) - \beta_i(x)| \left\| \frac{\partial u}{\partial x_i} \right\|_\infty$$
$$+ \frac{1}{2} \sum_{i,j=1}^{d} |\varphi(n)L_n((pr_i \circ \Psi_x)(pr_j \circ \Psi_x))(x) - \alpha_{ij}(x)| \left\| \frac{\partial^2 u}{\partial x_i \partial x_j} \right\|_\infty$$
$$+ \varphi(n)L_n(|\omega_u(x, \cdot)|\Phi_x^2)(x),$$

and hence (1.5.10) follows.

As regards the second part of the statement, it is sufficient to remark that conditions (v) and (vi) imply (iv) by virtue of Lemma 1.5.1. □

Remark 1.5.3. Consider a sequence $(L_n)_{n\ge1}$ of positive linear operators from $\mathscr{C}(K)$ into $\mathscr{F}(K)$ and denote by $(\tilde{\mu}_{n,x})_{n\ge1, x\in K}$ the selection of Borel measures associated with them, i.e.,

$$L_n(f)(x) = \int_K f \, d\tilde{\mu}_{n,x} \quad (f \in \mathscr{C}(K), n \ge 1, x \in K). \tag{1.5.11}$$

Let G be a subset of K and, together with (v), assume that

$$\lim_{n\to\infty} \varphi(n) \int_{K\setminus B(x,\delta)} \Phi_x^2 \, d\tilde{\mu}_{n,x} = 0 \quad \text{uniformly w.r.t. } x \in G \tag{1.5.12}$$

for every $\delta > 0$, where $B(x, \delta)$ denotes the open ball of radius δ centered at x. Then property (iv) holds true for every $u \in \mathscr{C}^2(K)$.

Indeed, given $\varepsilon > 0$, by (1.5.8) there exists $\delta > 0$ such that $|\omega_u(x, y)| \le \frac{\varepsilon}{2N}$ for every $x, y \in K$, $\|x - y\|_2 \le \delta$, where $N := \sup\limits_{\substack{x\in G \\ n\ge1}} \varphi(n)L_n(\Phi_x^2)(x)$.

Choose $\nu \in \mathbf{N}$ such that, for every $n \ge \nu$ and $x \in G$,

$$\varphi(n) \int_{K\setminus B(x,\delta)} \Phi_x^2 \, d\tilde{\mu}_{n,x} \le \frac{\varepsilon}{2(M+1)},$$

where M is given by (1.5.6), so that, on account of (1.5.7),

$$|\varphi(n)L_n(\omega_u(x, \cdot)\Phi_x^2)(x)| \le \varphi(n) \left(\int_{B(x,\delta)} |\omega_u(x, \cdot)|\Phi_x^2 \, d\tilde{\mu}_{n,x} \right.$$
$$\left. + \int_{K\setminus B(x,\delta)} |\omega_u(x, \cdot)|\Phi_x^2 \, d\tilde{\mu}_{n,x} \right)$$
$$\le \frac{\varepsilon}{2N} \varphi(n) \int_K \Phi_x^2 \, d\tilde{\mu}_{n,x} + M\varphi(n) \int_{K\setminus B(x,\delta)} \Phi_x^2 \, d\tilde{\mu}_{n,x} \le \varepsilon.$$

With the help of Theorem 1.5.2 we shall establish several asymptotic formulae for all the approximation processes which will be considered in the subsequent chapters.

The particular case $d = 1$ in Theorem 1.5.2 is worth being stated separately.

Corollary 1.5.4. *Consider a compact real interval* $[a, b]$ *and the differential operator*

$$A(u)(x) := \frac{1}{2}\alpha(x)u''(x) + \beta(x)u'(x) + \gamma(x)u(x) \tag{1.5.13}$$

$(u \in \mathscr{C}^2([a, b]), x \in [a, b])$, *where* $\alpha, \beta, \gamma \in \mathscr{F}([a, b])$.

Furthermore, consider a divergent sequence $(\varphi(n))_{n\geq 1}$ *of positive integers and a sequence* $(L_n)_{n\geq 1}$ *of positive linear operators from* $\mathscr{C}([a, b])$ *into* $\mathscr{F}([a, b])$. *Assume that there exists a subset* J *of* $[a, b]$ *such that*

(i) $\lim_{n\to\infty} \varphi(n)(L_n(\mathbf{1})(x) - 1) - \gamma(x) = 0$ *uniformly w.r.t.* $x \in J$;
(ii) $\lim_{n\to\infty} \varphi(n)L_n(\psi_x)(x) - \beta(x) = 0$ *uniformly w.r.t.* $x \in J$;
(iii) $\lim_{n\to\infty} \varphi(n)L_n(\psi_x^2)(x) - \alpha(x) = 0$ *uniformly w.r.t.* $x \in J$;
(iv) $\sup_{\substack{x\in J\\ n\geq 1}} \varphi(n)L_n(\psi_x^2)(x) < +\infty$;
(v) there exists $q \in \mathbf{R}$, $q > 2$, *such that*

$$\lim_{n\to\infty} \varphi(n)L_n(\psi_x^q)(x) = 0 \qquad \text{uniformly w.r.t. } x \in J,$$

where

$$\psi_x(y) := y - x \quad (x, y \in [a, b]). \tag{1.5.14}$$

Then, for every $u \in \mathscr{C}^2([a, b])$,

$$\lim_{n\to\infty} \varphi(n)(L_n(u) - u) = A(u) \qquad \text{uniformly on } J. \tag{1.5.15}$$

In the sequel we shall consider an arbitrary real interval I, possibly unbounded. We shall denote by $\mathscr{F}(I)$ the space of all real-valued functions defined on I and by $\mathscr{C}(I)$ the space of all real-valued continuous functions on I.

We also set

$$E_2(I) := \left\{ f \in \mathscr{C}(I) \mid \sup_{x\in I} \frac{|f(x)|}{1 + x^2} < +\infty \right\}. \tag{1.5.16}$$

Consider $f \in \mathscr{F}(I)$ and assume that f is twice differentiable at a given point $x_0 \in I$. Then, by the Peano form of the Taylor formula there exists $\omega_f : I \longrightarrow \mathbf{R}$ such that, for every $x \in I$,

$$f(x) = f(x_0) + f'(x_0)(x - x_0) + \frac{f''(x_0)}{2}(x - x_0)^2 + \omega_f(x)\frac{(x - x_0)^2}{2} \tag{1.5.17}$$

and

$$\lim_{x\to x_0} \omega_f(x) = 0. \tag{1.5.18}$$

Therefore, we can write

$$f = f(x_0) + f'(x_0)\psi_{x_0} + \frac{f''(x_0)}{2}\psi_{x_0}^2 + \frac{1}{2}\omega_f\,\psi_{x_0}^2 \tag{1.5.19}$$

(see (1.5.14)).

After these preliminaries we can establish a first local asymptotic formula.

Proposition 1.5.5. *Let $x_0 \in I$ and consider a linear subspace E of $\mathscr{F}(I)$ such that $E_2(I) \subset E$ and $\psi_{x_0}^q \in E$ for some even positive integer $q \geq 4$. Let us consider $\alpha(x_0), \beta(x_0), \gamma(x_0) \in \mathbf{R}$ and let $(\varphi(n))_{n\geq 1}$ be a divergent sequence of positive integers. Consider a sequence $(L_n)_{n\geq 1}$ of positive linear operators from E into $\mathscr{F}(I)$ and assume that*

(i) $\lim_{n\to\infty} \varphi(n)(L_n(1)(x_0) - 1) = \gamma(x_0)$,

(ii) $\lim_{n\to\infty} \varphi(n)L_n(\psi_{x_0})(x_0) = \beta(x_0)$,

(iii) $\lim_{n\to\infty} \varphi(n)L_n(\psi_{x_0}^2)(x_0) = \alpha(x_0)$,

(iv) $\lim_{n\to\infty} \varphi(n)L_n(\psi_{x_0}^q)(x_0) = 0$.

If $f \in E_2(I)$ is twice differentiable at x_0, then

$$\lim_{n\to\infty} \varphi(n)(L_n(f)(x_0) - f(x_0)) = \frac{1}{2}\alpha(x_0)f''(x_0) + \beta(x_0)f'(x_0) + \gamma(x_0)f(x_0).$$

Proof. From (1.5.19) we get

$$\varphi(n)(L_n(f)(x_0) - f(x_0)) = f(x_0)\varphi(n)(L_n(\mathbf{1})(x_0) - 1)$$
$$+ f'(x_0)\varphi(n)L_n(\psi_{x_0})(x_0) + \frac{f''(x_0)}{2}\varphi(n)L_n(\psi_{x_0}^2)(x_0) + \frac{1}{2}\varphi(n)L_n(\omega_f\psi_{x_0}^2)(x_0).$$

Therefore, on account of (i), (ii), (iii), it suffices to show that

$$\lim_{n\to\infty} \varphi(n)L_n(\omega_f\psi_{x_0}^2)(x_0) = 0.$$

To this end, given $\varepsilon > 0$, by (1.5.18) there exists $\delta > 0$ such that

$$|\omega_f(x)| \leq \varepsilon \qquad \text{for every } x \in I,\ |x - x_0| \leq \delta.$$

Setting $M_\delta := \sup_{|x-x_0|\geq\delta} \frac{1+x^2}{(x-x_0)^2}$ and $\|f\|_{E_2} := \sup_{x\in I} \frac{|f(x)|}{1+x^2}$, for $|x - x_0| \geq \delta$ we get, by (1.5.17),

$$|\omega_f(x)| \leq 2\left(\frac{|f(x) - f(x_0)|}{(x - x_0)^2} + \frac{|f'(x_0)|}{|x - x_0|} + \frac{|f''(x_0)|}{2}\right)$$
$$\leq 2\left(M_\delta\|f\|_{E_2} + \frac{|f(x_0)|}{\delta^2} + \frac{|f'(x_0)|}{\delta} + \frac{|f''(x_0)|}{2}\right) =: K_\delta$$
$$\leq K_\delta\frac{(x - x_0)^{q-2}}{\delta^{q-2}}.$$

This reasoning shows, in particular, that ω_f is bounded on I, and hence $\omega_f\psi_{x_0}^2 \in E_2(I)$.

Moreover, for every $x \in I$,

$$|\omega_f(x)\psi_{x_0}^2(x)| \leq \varepsilon\psi_{x_0}^2(x) + \frac{K_\delta}{\delta^{q-2}}\psi_{x_0}^q(x);$$

therefore, for any $n \geq 1$,

$$\varphi(n)|L_n(\omega_f \psi_{x_0}^2)(x_0)| \leq \varepsilon\varphi(n)L_n(\psi_{x_0}^2)(x_0) + \frac{K_\delta}{\delta^{q-2}}\varphi(n)L_n(\psi_{x_0}^q)(x_0),$$

and hence the result follows on account of (iv). □

Our next result concerns asymptotic formulae which hold true uniformly on compact subintervals of I.

We need, however, to state several preliminaries. Denote by $\mathscr{C}^2(\text{int}(I))$ the space of all real functions which are twice-continuously differentiable on the interior $\text{int}(I)$ of the interval I, and with $\mathscr{C}_b^2(\text{int}(I))$ the subspace of $\mathscr{C}^2(\text{int}(I))$ whose elements have bounded second-order derivatives.

Moreover, if $f \in \mathscr{C}(I) \cap \mathscr{C}_b^2(\text{int}(I))$, then, setting $M := \sup_{x\in \text{int}(I)} |f''(x)|$ and choosing $x_0 \in \text{int}(I)$, since

$$f(x) = \int_{x_0}^{x}\left(\int_{x_0}^{t} f''(s)\,ds\right)dt + f'(x_0)(x-x_0) + f(x_0) \qquad (x \in \text{int}(I)),$$

it follows that

$$|f(x)| \leq \frac{M}{2}(x-x_0)^2 + |f'(x_0)|\,|x-x_0| + |f(x_0)| \tag{1.5.20}$$

for every $x \in \text{int}(I)$ and hence, by continuity, for every $x \in I$.

Therefore $f \in E_2(I)$. Furthermore, by the Lagrange form of the Taylor formula there exists $\omega_f : \text{int}(I) \times I \longrightarrow \mathbf{R}$ such that, for every $x \in \text{int}(I)$ and $y \in I$,

$$f(y) = f(x) + f'(x)(y-x) + \frac{f''(x)}{2}(y-x)^2 + \omega_f(x,y)(y-x)^2 \tag{1.5.21}$$

and

$$\omega_f(x,y) = \frac{f''(\xi) - f''(x)}{2} \tag{1.5.22}$$

for some ξ lying into the interior of the interval having x and y as endpoints.

Therefore

$$|\omega_f(x,y)| \leq M \qquad \text{for every } x \in \text{int}(I), y \in I. \tag{1.5.23}$$

We proceed to show that, if $J \subset \text{int}(I)$ is a compact subinterval, then

$$\lim_{y\to x} \omega_f(x,y) = 0 \qquad \text{uniformly w.r.t. } x \in J. \tag{1.5.24}$$

Indeed, assume $J = [a,b]$ and fix $\varepsilon > 0$. Let $\delta_1 > 0$ such that $[a-\delta_1, b+\delta_1] \subset \text{int}(I)$; since f'' is uniformly continuous on $[a-\delta_1, b+\delta_1]$ there exists $0 < \delta \leq \delta_1$ such that $|f''(x) - f''(y)| \leq 2\varepsilon$ for every $x, y \in [a-\delta_1, b+\delta_1]$, $|x-y| \leq \delta$.

Therefore, if $x \in J$ and $y \in I$, $|x-y| \leq \delta$, necessarily $y \in [a-\delta_1, b+\delta_1]$ and hence $|f''(x) - f''(y)| \leq 2\varepsilon$, i.e., $|\omega_f(x,y)| \leq \varepsilon$.

After these preliminaries we state another result concerning asymptotic formulae with respect to a given weight which naturally occurs in several concrete situations.

Theorem 1.5.6. *Let E be a linear subspace of $\mathscr{F}(I)$ such that $E_2(I) \subset E$ and $\psi_x^q \in E$ for some even positive integer $q \geq 4$ and for every $x \in I$. Let us consider $\alpha, \beta, \gamma, w : \operatorname{int}(I) \longrightarrow \mathbf{R}$ and let $(\varphi(n))_{n\geq 1}$ be a divergent sequence of positive integers. Consider a sequence $(L_n)_{n\geq 1}$ of positive linear operators from E into $\mathscr{F}(I)$ and fix a compact subinterval J of $\operatorname{int}(I)$.*

Assume that

(i) $\lim_{n\to\infty} w(x)x^k[\varphi(n)(L_n(1)(x) - 1) - \gamma(x)] = 0$ *uniformly w.r.t.* $x \in J$, *k=0,2.*

(ii) $\lim_{n\to\infty} w(x)x^k[\varphi(n)L_n(\psi_x)(x) - \beta(x)] = 0$ *uniformly w.r.t.* $x \in J$, *k=0,1.*

(iii) $\lim_{n\to\infty} w(x)[\varphi(n)L_n(\psi_x^2)(x) - \alpha(x)] = 0$ *uniformly w.r.t.* $x \in J$.

(iv) $\sup_{\substack{n\geq 1\\ x\in I}} |w(x)|\varphi(n)L_n(\psi_x^2)(x) < +\infty$.

(v) $\lim_{n\to\infty} w(x)\varphi(n)L_n(\psi_x^q)(x) = 0$ *uniformly w.r.t.* $x \in J$.

Then, for every $f \in \mathscr{C}(I) \cap \mathscr{C}_b^2((\operatorname{int}(I)))$,

$$\lim_{n\to\infty} w(x)[\varphi(n)(L_n(f)(x) - f(x)) - A(f)(x)] = 0 \tag{1.5.25}$$

uniformly w.r.t. $x \in J$, *where*

$$A(f)(x) := \frac{1}{2}\alpha(x)f''(x) + \beta(x)f'(x) + \gamma(x)f(x) \qquad (x \in J). \tag{1.5.26}$$

In particular, if $\inf_{x\in I} w(x) > 0$, *then*

$$\lim_{n\to\infty} \varphi(n)(L_n(f) - f) = A(f) \qquad \text{uniformly on } J.$$

Proof. First of all, note that the inequality $|x| \leq 1 + x^2$ $(x \in \mathbf{R})$ implies that assumption (i) holds true for $k = 1$, too. Moreover, (1.5.21) can be rewritten, for a given $x \in \operatorname{int}(I)$, as

$$f = f(x) + f'(x)\psi_x + \frac{f''(x)}{2}\psi_x^2 + \omega_f(x,\cdot)\psi_x^2$$

and hence

$$\begin{aligned}
&|w(x)[\varphi(n)(L_n(f)(x) - f(x)) - Af(x)]| \\
&\leq |w(x)f(x)[\varphi(n)(L_n(\mathbf{1})(x) - 1) - \gamma(x)]| \\
&+ |w(x)f'(x)[\varphi(n)L_n(\psi_x)(x) - \beta(x)]| \\
&+ \left|w(x)\frac{f''(x)}{2}[\varphi(n)L_n(\psi_x^2)(x) - \alpha(x)]\right| \\
&+ |w(x)\varphi(n)L_n(\omega_f(x,\cdot)\psi_x^2)(x)|.
\end{aligned}$$

Choosing $x_0 \in \operatorname{int}(I)$, we get

$$|f'(x)| \leq \left|\int_{x_0}^{x} f''(t)dt\right| + |f'(x_0)| \leq M|x - x_0| + |f'(x_0)| \tag{1}$$

and, as in (1.5.20),

$$|f(x)| \leq \frac{M}{2}(x-x_0)^2 + |f'(x_0)|\,|x-x_0| + |f(x_0)|$$

where, as before, $M := \sup_{x\in \mathrm{int}(I)} |f''(x)|$.

Therefore, on account of assumptions (i)-(iii), it suffices to show that

$$\lim_{n\to\infty} w(x)\varphi(n)L_n(\omega_f(x,\cdot)\psi_x^2)(x) = 0 \qquad \text{uniformly w.r.t. } x \in J.$$

Set, indeed, $K := \sup_{\substack{n\geq 1 \\ x\in J}} |w(x)|\varphi(n)L_n(\psi_x^2)(x)$, and choose $\varepsilon > 0$. By (1.5.24) and (v), there exists $\delta > 0$ such that

$$|\omega_f(x,y)| \leq \frac{\varepsilon}{2(K+1)} \qquad \text{for every } x \in J, y \in I, |x-y| \leq \delta,$$

and there exists $\nu \in \mathbf{N}$ such that for every $n \geq \nu$ and $x \in J$

$$|w(x)\varphi(n)L_n(\psi_x^q)(x)| \leq \frac{\varepsilon}{2(M+1)}\delta^{q-2}.$$

If $x \in J$, we have (see also (1.5.23)) $|\omega_f(x,\cdot)\psi_x^2| \leq \frac{\varepsilon}{2(K+1)}\psi_x^2 + \frac{M}{\delta^{q-2}}\psi_x^q$ and hence, for $n \geq \nu$,

$$|w(x)|\varphi(n)L_n(\omega_f(x,\cdot)\psi_x^2)(x) \leq \frac{\varepsilon}{2(K+1)}|w(x)|\varphi(n)L_n(\psi_x^2)(x)$$
$$+ \frac{M}{\delta^{q-2}}|w(x)|\varphi(n)L_n(\psi_x^q)(x) \leq \varepsilon. \qquad \square$$

For more details about asymptotic formulae on unbounded intervals we refer the interested reader to [24].

1.6 Moduli of smoothness and degree of approximation by positive linear operators

We end this chapter by collecting some properties concerning the classical moduli of continuity and smoothness and some extensions to the more general setting of compact convex subsets of (possible infinite-dimensional) locally convex Hausdorff spaces. Moreover, we present some results which will help us to determine some quantitative error estimates for the (pointwise or uniform) approximation by positive linear operators.

Let K be a convex subset of $\mathbf{R}^d$. For every $f \in \mathscr{C}_b(K)$, $k \geq 1$ and $h > 0$, we define the *k-th difference $\Delta_h^k f(x)$ of f with step h at a point x* as

$$\Delta_h^k f(x) := \begin{cases} \sum_{l=0}^{k} (-1)^{k-l} \binom{k}{l} f(x+lh) & \text{if } x, x+kh \in K; \\ 0 & \text{otherwise.} \end{cases} \tag{1.6.1}$$

Then, according to (1.4.8), the first modulus of continuity of f with argument $\delta > 0$ is defined as

$$\omega(f,\delta) := \sup\{|f(x) - f(y)| \mid x, y \in K, \|x-y\|_2 \leq \delta\}$$

and, for every $k > 1$, the *k-th modulus of smoothness of f with respect to δ* is defined as

$$\omega_k(f,\delta) := \sup\{|\Delta_h^k f(x)| \mid x, x+kh \in K, |h| \leq \delta\}. \tag{1.6.2}$$

In the following proposition, whose proof is similar to that of [18, Lemma 5.5.1], we state some properties of the moduli of smoothness.

Proposition 1.6.1. *Let K be a convex compact subset of $\mathbf{R}^d$ and fix $f, g \in \mathscr{C}(K)$ and $k, j \geq 1$. Then*
(1) $\omega_k(f,\delta_1) \leq \omega_k(f,\delta_2)$, provided that $0 < \delta_1 \leq \delta_2$.
(2) $\omega_k(f+g,\delta) \leq \omega_k(f,\delta) + \omega_k(g,\delta)$ for every $\delta > 0$.
(3) $\lim_{\delta \to 0^+} \omega_k(f,\delta) = 0$.
(4) $\omega_{k+j}(f,\delta) \leq 2^j \omega_k(f,\delta)$ for every $\delta > 0$.
(5) $\omega_k(f,n\delta) \leq n^k \omega(f,\delta)$ for every $\delta > 0$ and $n \geq 1$.
(6) $\omega_k(f,\lambda\delta) \leq (1+[\lambda])^k \omega_k(f,\delta)$ for every $\delta > 0$ and $\lambda > 0$; here $[\lambda]$ denotes the integer part of λ.

Below we present some general estimates of the degree of approximation by means of positive linear operators in terms of the first and the second moduli of smoothness, both in one-dimensional and in multidimensional settings.

We begin by stating the following result (for a proof, see [158, Theorem 2.2.1]) in the context of bounded intervals of $\mathbf{R}$; nevertheless, we point out that it holds true for arbitrary intervals of $\mathbf{R}$.

Theorem 1.6.2. *Consider a positive linear operator* $L : \mathscr{C}([a,b]) \longrightarrow \mathscr{F}([a,b])$. *Then, for every* $\delta > 0$, *for every* $f \in \mathscr{C}([a,b])$ *and for every* $x \in [a,b]$,

$$\begin{aligned}|L(f)(x) - f(x)| &\le |L(\mathbf{1})(x) - 1||f(x)| + \frac{1}{\delta}|L(\psi_x)(x)|\omega(f,\delta)\\ &+ \left[L(\mathbf{1})(x) + \frac{1}{2\delta^2}L(\psi_x^2)(x)\right]\omega_2(f,\delta),\end{aligned} \tag{1.6.3}$$

where, according to (1.5.14),

$$\psi_x(t) := t - x \quad (t \in I).$$

Finally, we state, under suitable assumptions, a general estimate (in the multidimensional setting) of the rate of approximation by means of positive linear operators in terms of the second modulus of smoothness in $\mathbf{R}^d$ $(d \ge 1)$.

Let K be a convex compact subset of $\mathbf{R}^d$ $(d \ge 1)$ and consider a positive linear operator $L : \mathscr{C}(K) \longrightarrow \mathscr{C}(K)$; set

$$e := \sum_{i=1}^{d} pr_i^2 \tag{1.6.4}$$

(see (1.1.21)) and

$$\begin{aligned}\lambda_\infty := \max\{&\|L(\mathbf{1}) - \mathbf{1}\|_\infty, \|L(pr_1) - pr_1\|_\infty, \dots,\\ &\|L(pr_d) - pr_d\|_\infty, \|L(e) - e\|_\infty\}.\end{aligned} \tag{1.6.5}$$

The following result can be found in [52, Theorem 2'] and it holds true under the more general assumption that K is the closure of a bounded region of $\mathbf{R}^d$ satisfying the cone property (see, e.g., [1, p. 66]). We explicitly point out that, by means of [139, pp. 14-15], every convex compact subset K of $\mathbf{R}^d$ shares the cone property.

Theorem 1.6.3. *Let* K *be a convex compact subset of* $\mathbf{R}^d$ *and consider a positive linear operator* $L : \mathscr{C}(K) \longrightarrow \mathscr{C}(K)$. *Then, for every* $f \in \mathscr{C}(K)$,

$$\|L(f) - f\|_\infty \le \max\{\|L\| + 1, C\}(\lambda_\infty \|f\|_\infty + \omega_2(f, \lambda_\infty^{1/2})), \tag{1.6.6}$$

where C *is an absolute constant, depending on* d *and* K, *only.*

In what follows, we briefly describe a suitable modulus of continuity, first introduced in [152], which generalizes the classical first modulus of smoothness $\omega(f,\delta)$ to the case of approximation processes on spaces of continuous functions defined on convex compact subsets of (possibly infinite dimensional) locally convex spaces.

Let X be a locally convex space and let K be a compact convex subset of X; we denote by the symbol X' the dual space of X and we consider the space

$$L(K) := \{\varphi_{|K} \mid \varphi \in X'\}. \tag{1.6.7}$$

If $m \geq 1$, $h_1, \dots, h_m \in L(K)$ and $\delta > 0$, we set

$$H(h_1, \dots, h_m, \delta) := \left\{ (x, y) \in K \times K \mid \sum_{j=1}^{m} (h_j(x) - h_j(y))^2 \leq \delta^2 \right\}. \tag{1.6.8}$$

Fix $f \in \mathscr{B}(K)$; the *modulus of continuity of f with respect to $h_1, \dots, h_m$* is defined as

$$\omega(f; h_1, \dots, h_m, \delta) := \sup\{|f(x) - f(y)| \mid (x, y) \in H(h_1, \dots, h_m, \delta)\}. \tag{1.6.9}$$

Moreover, we define the *total modulus of continuity of f* as

$$\Omega(f, \delta) := \inf \Big\{ \omega(f; h_1, \dots, h_m, \delta) \mid m \geq 1, \; h_1, \dots, h_m \in L(K) \text{ and } \Big\| \sum_{j=1}^{m} h_j^2 \Big\|_\infty = 1 \Big\}. \tag{1.6.10}$$

It is easy to see that

$$\Omega(f, \delta) = \inf \Big\{ \omega(f; h_1, \dots, h_m, 1) \mid m \geq 1, \; h_1, \dots, h_m \in L(K) \text{ and } \Big\| \sum_{j=1}^{m} h_j^2 \Big\|_\infty = \frac{1}{\delta^2} \Big\}.$$

If $X = \mathbf{R}^d$, $d \geq 1$, there is a simple relationship between $\Omega(f, \delta)$ and the (first) modulus of continuity $\omega(f, \delta)$ defined by (1.4.8); indeed,

$$\omega(f; pr_1, \dots, pr_d, \delta) = \omega(f, \delta), \tag{1.6.11}$$

so that, setting $r(K) := \max\{\|x\|_2 \mid x \in K\}$,

$$\Omega(f, \delta) \leq \omega(f, \delta r(K)), \tag{1.6.12}$$

the last inequality being an equality if $d = 1$.

Moreover, the following result holds (for a proof, see [18, Lemma 5.1.3]).

Proposition 1.6.4. *Fix $m \geq 1$, $h_1, \dots, h_m \in L(K)$ and $f \in \mathscr{C}(K)$. Then*
(1) $\omega(f; h_1, \dots, h_m, \delta_1) \leq \omega(f; h_1, \dots, h_m, \delta_2)$ provided that $0 < \delta_1 \leq \delta_2$. Accordingly, if $0 < \delta_1 \leq \delta_2$, then $\Omega(f, \delta_1) \leq \Omega(f, \delta_2)$.
(2) $\lim_{\delta \to 0^+} \Omega_k(f, \delta) = 0$.
(3) $\omega(f; h_1, \dots, h_m, \lambda\delta) \leq [\lambda]\omega(f; h_1, \dots, h_m, \delta)$ for every $\delta > 0$ and $\lambda > 0$; here $[\lambda]$ denotes the integer part of λ. Accordingly, $\Omega(f, \lambda\delta) \leq [\lambda]\Omega(f, \delta)$.

Now, let L be a positive linear operator from $\mathscr{C}(K)$ into $\mathscr{B}(K)$ and, for a given $f \in \mathscr{C}(K)$ and $x \in K$, set

$$\mu(x, L, f) := L((f - f(x)\mathbf{1})^2)(x). \tag{1.6.13}$$

Then, under suitable assumptions, it is possible to estimate the pointwise error in approximating f by $L(f)$ by means of $\mu(x, L, f)$, as the following result shows (see, [18, Proposition 5.1.4]).

Proposition 1.6.5. *Let $L : \mathscr{C}(K) \longrightarrow \mathscr{B}(K)$ be a positive linear operator and consider $f \in \mathscr{C}(K)$ and $x \in K$. Then, for every $m \geq 1$, $h_1, \dots, h_m \in L(K)$ and $\delta > 0$,*

$$|L(f)(x) - f(x)L(\mathbf{1})(x)| \leq \left(L(\mathbf{1})(x) + \frac{1}{\delta^2} \sum_{j=1}^{m} \mu(x, L, h_j) \right) \omega(f; h_1, \dots, h_m, \delta). \tag{1.6.14}$$

In particular,

$$|L(f)(x) - f(x)L(\mathbf{1})(x)| \leq \left(L(\mathbf{1})(x) + \frac{\tau(\delta, x)}{\delta^2} \right) \Omega(f, \delta), \tag{1.6.15}$$

where

$$\tau(\delta, x) := \sup \left\{ \sum_{j=1}^{m} \mu(x, L, h_j) \mid m \geq 1, h_1, \dots, h_m \in L(K) \text{ and } \left\| \sum_{j=1}^{m} h_j^2 \right\|_\infty = \frac{1}{\delta^2} \right\}. \tag{1.6.16}$$

If, in addition, $L(h) = h$ for every $h \in A(K)$, then

$$|L(f)(x) - f(x)| \leq \left(1 + \frac{1}{\delta^2} \left(L\left(\sum_{j=1}^{m} h_j^2 \right)(x) - \sum_{j=1}^{m} h_j^2(x) \right) \right) \omega(f; h_1, \dots, h_m, \delta) \tag{1.6.17}$$

and

$$|L(f)(x) - f(x)| \leq \left(1 + \frac{\sigma(\delta, x)}{\delta^2} \right) \Omega(f, \delta), \tag{1.6.18}$$

where

$$\sigma(\delta, x) := \sup \left\{ L\left(\sum_{j=1}^{m} h_j^2 \right)(x) - \sum_{j=1}^{m} h_j^2(x) \mid m \geq 1, h_1, \dots, h_m \in L(K) \text{ and } \left\| \sum_{j=1}^{m} h_j^2 \right\|_\infty = \frac{1}{\delta^2} \right\}. \tag{1.6.19}$$

We close this section by discussing some results concerning the preservation of global smoothness of functions under the action of bounded operators. With the help of these results, in the subsequent chapters we shall stress under which conditions the involved approximation process preserves Hölder continuity.

Consider a compact metric space (X, ρ). We shall refer to the subspace $\mathrm{Lip}(X)$ defined by (1.4.3) and to the relevant seminorm $|\cdot|_{\mathrm{Lip}}$.

As in (1.4.8), also in the setting of metric spaces we can define a modulus of continuity, which is said to be the *ρ-modulus of continuity of f with respect to δ*, by

$$\omega_\rho(f,\delta) := \sup\{|f(x)-f(y)| \mid x,y \in X,\ \rho(x,y) \le \delta\} \tag{1.6.20}$$

$(f \in \mathscr{C}(X), \delta > 0)$.

Furthermore, for any $M \ge 0$ and $0 < \alpha \le 1$, we denote by

$$\mathrm{Lip}(M,\alpha) := \{f \in \mathscr{C}(X) \mid |f(x)-f(y)| \le M\rho(x,y)^\alpha \text{ for every } x,y \in X\} \tag{1.6.21}$$

the space of all *Hölder continuous functions with exponent α and constant M*.

In particular, $\mathrm{Lip}(M,1)$ is said to be the space of all *Lipschitz continuous functions with constant M*.

For the sequel it is necessary to assume that

$$\omega_\rho(f,t\delta) \le (1+t)\omega_\rho(f,\delta) \tag{1.6.22}$$

for every $f \in \mathscr{C}(X)$, $\delta, t > 0$.

This assumption is satisfied whenever X is a subset of a metrizable locally convex space E and ρ is induced by a metric ρ_0 on E such that

$$\rho_0(x+z,y+z) = \rho_0(x,y) \text{ and } \rho_0(\alpha x,0) \le \alpha\rho_0(x,0) \tag{1.6.23}$$

for every $x,y,z \in E$ and $\alpha \in [0,1]$ (see [155]). In particular, (1.6.22) is fulfilled for every compact subset of a normed space. Other examples are those metric spaces having a coefficient of deformation equal to one ([119]).

Theorem 1.6.6. *Under assumption* (1.6.22) *let T be a non-null bounded operator from $\mathscr{C}(X)$ into $\mathscr{C}(X)$. Assume that $T(\mathrm{Lip}(X)) \subset \mathrm{Lip}(X)$ and that there exists $c \ge 0$ such that $|T(f)|_{\mathrm{Lip}} \le c|f|_{\mathrm{Lip}}$ for every $f \in \mathrm{Lip}(X)$. Then*

(i) For each $f \in \mathscr{C}(X)$ and $\delta > 0$,

$$\omega_\rho(T(f),\delta) \le (\|T\|+c)\omega_\rho(f,\delta).$$

(ii) For each $M \ge 0$ and $0 < \alpha \le 1$,

$$T(\mathrm{Lip}(M,\alpha)) \subset \mathrm{Lip}\left(Mc^\alpha\|T\|^{1-\alpha},\alpha\right).$$

Therefore, if $c \le 1$ and $\|T\| \le 1$, then

$$T(\mathrm{Lip}(M,\alpha)) \subset \mathrm{Lip}(M,\alpha).$$

1.7 Notes and comments

Radon measures are a powerful analytical tool and they play a central role in this book. For more details the reader is referred to [59], [50] and [77].

The notion of Choquet boundary was introduced by Bauer ([47]) in connection with the study of the abstract Dirichlet problems. In terms of it Bishop and de Leeuw ([55]) extended Choquet's integral representation theory to arbitrary compact Hausdorff spaces. Choquet boundaries are also strongly connected with Korovkin-type approximation theorems.

For more deepenings on this subject we refer to [5], [18, Section 2.6 and the relevant final notes], [77, Vol. II, Section 29], [135].

The notion of Choquet simplex was introduced by Choquet ([76]) in connection with the problem of the uniqueness of representing maximal measures for convex compact subsets. For more details we refer to [77, Vol. II, Section 28] or to [135, Section 6.8].

Bauer simplices are related to the abstract Dirichlet problem on $\partial_e X$, namely statement (c) of Theorem 1.1.12. This last theorem includes results due to Bauer ([47]) (see also [5, Theorems II.4.1 and II.4.2]) and to Rogalski ([175]) (see also [7]).

As a general reference to the Korovkin-type approximation theory we mention Altomare-Campiti ([18]) and Altomare ([11]) and the references therein. Within these monographs the reader can find several connections not only with classical approximation theory but also with functional analysis, real analysis and partial differential equations.

Furthermore, in the Appendix D of [18] a subject classification of the Korovkin-type approximation theory is presented. The classification shows the main lines along which this theory has been developed until 1994. More recent contributions are indicated in [11].

All the results in Section 1.3 are taken from [12]. In this article the reader can find several applications.

Corollary 1.3.6 generalizes a result obtained by Altomare ([8], [18, Theorem 3.3.3], [11, Theorem 10.3]) in the particular case $S = T$, T being a Markov projection.

Theorem 1.3.8 has been obtained by Raşa ([171]) in the particular case $X = [0,1]$, $\Phi(x) = x^2, (0 \le x \le 1)$ and $T : \mathscr{C}([0,1]) \longrightarrow \mathscr{C}([0,1])$ is the natural Markov projection defined by

$$T(f)(x) := (1-x)f(0) + xf(1)$$

$(0 \le x \le 1, f \in \mathscr{C}([0,1]))$.

The implication (e)⇒(a) of Theorem 1.3.9 has been first established by Gavrea-Ivan ([104, Theorem 3.1]; see also [103, Theorem 1], [171, Theorem 2.2] and [105, Theorem 1]. In all these papers the reader can also find several applications.

We also mention that the study of the asymptotic behaviour of iterates of Markov operators is important not only in approximation theory but also in ergodic theory and, in particular, for ergodic theorems (see. e.g., [127]).

In Section 1.4 we present some other results in this respect. However, as shown in Corollary 1.4.6, these results can be also useful to investigate the asymptotic behaviour of strongly continuous semigroups of operators which, among other things, gives useful information about the behaviour for large time of the solutions to the Cauchy problems governed by them (see, e.g., [92], [204] and the references therein).

All the results of Section 1.4 are taken from a recent paper by Altomare and Raşa ([36]). In that paper the authors also discuss an application concerning Bernstein-Durrmeyer operators with Jacobi weights.

Asymptotic formulae are intimately connected with the saturation phenomenon for positive approximation processes as well as with the approximation of semigroups of operators by iterates of operators.

The most celebrated of such formulae was established by Voronovskaja ([207]) for the sequence of Bernstein operators. Later on, Mamedov ([137]) generalized it to arbitrary sequences of positive linear operators and his result is, in fact, Corollary 1.5.4. Further refinements have been obtained by Sikkema ([194]) and Khan ([124]). The extension to multidimensional convex compact domains, namely Theorem 1.5.2, is due to Dahmen and Micchelli ([80]).

More recently, asymptotic formulae in weighted continuous function spaces on real intervals have been investigated, among others, by Altomare and Amiar ([13]) and Altomare and Diomede ([24]).

In particular, in [13] the reader can also find an application concerning the Central Limit Theorem, one of the most fundamental results of probability theory.

Recently, Altomare ([10]) and Altomare and Diomede ([24]) provided a characterization of those continuous functions for which a pointwise asymptotic formula holds true. The characterization is expressed in terms of smoothness of the given functions together with a suitable behaviour of the first and the second derivatives near the end points of the underlying interval.

Finally, in [23] general asymptotic formulae for convex (not necessarily compact) domains of Banach spaces are discussed as well.

For some references concerning the connections between asymptotic formulae and saturation theorems or approximation problems for C_0-semigroups of opera-

tors we refer, e.g., to [13] and to the final notes of Chapter 4.

Pointwise, uniform and L_p-estimates of rates of convergence by positive linear operators can be found in several articles and books. Without any claim of completeness, for more deepenings we refer to Anastassiou and Gal ([38]), Berens and De Vore ([52]), Butzer and Nessel ([62]), Dai and Xu ([81]), De Vore ([85]), De Vore and Lorentz ([86]), Ditzian ([88]), Ditzian and Totik ([89]), Gavrea, Gonska, Păltănea and Tachev ([102]), Lorentz, Golitschek and Makovoz ([133]), Păltănea ([158]), Sendov and Popov ([189]).

Theorem 1.6.6 is due to Anastassiou, Cottin and Gonska ([37]) (see also [18, Corollary 6.1.20]).

2 C_0-semigroups of operators and linear evolution equations

It is difficult to argue that the main aim of the mathematician should be to solve problems. The problems are basically a way, the ultimate goal remaining the creation of powerful tools to lead to the simplification and unification of mathematics.

Gustave Choquet

In this second preliminary chapter we sketch the main features of the theory of C_0-semigroups of operators and, in particular, we stress its strong relationship with different questions arising from functional analysis, probability theory, partial differential equations (evolution equations) and other fields.

In what follows, we collect some properties of C_0-semigroups and the main generation theorems for them. Moreover, we present some results about the approximation of semigroups which will be very useful in the subsequent chapters. Further, we recall some preliminaries about Feller and Markov semigroups, also in relation with the theory of Markov processes; finally, we prove how the theory of Feller semigroups may be fruitfully employed in order to solve diffusion problems connected with elliptic second-order differential operators in one- and multi-dimensional settings.

2.1 C_0-semigroups of operators and abstract Cauchy problems

This section is devoted to review the main features of the theory of C_0-semigroups; in particular, we recall the basic definitions and the properties of C_0-semigroups and their generators. Moreover, we review the main generations theorems and we stress the deep relationship between the study of certain differential problems and the theory of C_0-semigroups.

For the reader's convenience, we prefer to split up this paragraph into two subsections.

2.1.1 C_0-semigroups and their generators

Unless otherwise stated, from now on we denote by $\mathbf{K}$ either the field $\mathbf{R}$ of real numbers or the field $\mathbf{C}$ of complex numbers.

Let $(E, \|\cdot\|)$ be a Banach space over $\mathbf{K}$ and denote by $\mathscr{L}(E)$ the space of all bounded linear operators on E; $\mathscr{L}(E)$ turns out to be a Banach space, once

endowed with the *operator norm* $\|\cdot\|$ which we recall is defined by

$$\|B\| := \sup_{\substack{f\in E \\ \|f\|\le 1}} \|B(f)\| \qquad (B \in \mathscr{L}(E)).$$

Definition 2.1.1. *A family* $(T(t))_{t\ge 0}$ *in* $\mathscr{L}(E)$ *is said to be a* semigroup of bounded linear operators on E *if*
(i) $T(0) = I$.
(ii) $T(t+s) = T(s)T(t)$ *for every* $s, t \ge 0$,
where I *stands for the identity operator on* E *and the symbol* $T(s)T(t)$ *means the composition* $T(s) \circ T(t)$.

A semigroup $(T(t))_{t\ge 0}$ *is said to be a* strongly continuous *or a* C_0-semigroup *if, for every* $t_0 \ge 0$ *and* $f \in E$,

$$\lim_{t\to t_0^+} T(t)(f) = T(t_0)(f) \qquad \text{in } (E, \|\cdot\|).$$

A semigroup $(T(t))_{t\ge 0}$ *is said to be a* uniformly continuous *semigroup if, for every* $t_0 \ge 0$,

$$\lim_{t\to t_0^+} T(t) = T(t_0) \qquad \text{in } \mathscr{L}(E).$$

Finally, if $(E, \|\cdot\|)$ *is a Banach lattice with respect to a certain ordering* $\le$, *then a semigroup* $(T(t))_{t\ge 0}$ *on* E *is said to be a* positive *semigroup if* $T(t)$ *is a positive operator (i.e.,* $T(t)(f) \ge 0$ *for every* $f \ge 0$*) for every* $t \ge 0$.

Remarks 2.1.2.
1. Clearly, every uniformly continuous semigroup is also a C_0-semigroup, but the converse does not hold true, unless E is finite-dimensional.

From now on, we limit ourselves to the study of C_0-semigroups, since, as we shall see in the next examples, the uniform continuity is too strong a condition, so that it is fulfilled only in very special situations (see next Example 2.1.4, 1 and Example 2.1.6, 1).
2. A semigroup $(T(t))_{t\ge 0}$ is a C_0-semigroup if and only if for every $f \in E$ the *orbit map*

$$\xi_f : t \ge 0 \longmapsto \xi_f(t) := T(t)(f) \tag{2.1.1}$$

is continuous from $\mathbf{R}^+$ into $(E, \|\cdot\|)$.

In what follows we recall a useful characterization of C_0-semigroups (for a proof, see, e.g., [93, Chapter I, Proposition 5.3]).

Proposition 2.1.3. *Let* $(T(t))_{t\ge 0}$ *be a semigroup on a Banach space* $(E, \|\cdot\|)$. *Then the following statements are equivalent:*
(a) $(T(t))_{t\ge 0}$ *is a* C_0*-semigroup.*
(b) $\lim_{t\to 0^+} T(t)(f) = f$ *for every* $f \in E$.

(c) *There exist $\delta > 0$, $M \geq 1$ and a dense subset D of E such that*
   (i) $\|T(t)\| \leq M$ *for every* $t \in [0, \delta]$;
   (ii) $\lim_{t \to 0^+} T(t)(f) = f$ *for every* $f \in D$.

Moreover, if $(T(t))_{t \geq 0}$ is a C_0-semigroup, there always exist $\omega \in \mathbf{R}$ and $M \geq 1$ such that

$$\|T(t)\| \leq M e^{\omega t} \qquad \text{for every } t \geq 0 \tag{2.1.2}$$

(see [93, Chapter I, Proposition 5.5]).

The *growth bound* ω_0 of the semigroup $(T(t))_{t \geq 0}$ is defined as

$$\begin{aligned}\omega_0 := \inf\{\omega \in \mathbf{R} \mid \ & \text{there exists } M \geq 1 \text{ such that } \|T(t)\| \leq M e^{\omega t} \\ & \text{for every } t \geq 0\}.\end{aligned} \tag{2.1.3}$$

Thus, $\omega_0 \in \mathbf{R} \cup \{-\infty\}$.

In particular, a C_0-semigroup is said to be a *contractive* semigroup if $M = 1$ and $\omega_0 = 0$, i.e., $\|T(t)\| \leq 1$ for every $t \geq 0$.

Below we present some examples of C_0-semigroups.

Examples 2.1.4.

1. *Let $A \in \mathscr{L}(E)$ and, for any $n \geq 0$, set*

$$A^n := \begin{cases} I & \text{if } n = 0; \\ A \circ A^{n-1} & \text{if } n \geq 1. \end{cases}$$

Moreover, for every $t \geq 0$, set

$$T(t) = e^{tA} := \sum_{n=0}^{\infty} \frac{t^n A^n}{n!}. \tag{2.1.4}$$

Then $(T(t))_{t \geq 0}$ is a uniformly continuous semigroup on E and $\|T(t)\| \leq e^{t\|A\|}$ for every $t \geq 0$.

2. *Set*

$$(E, \|\cdot\|) := \begin{cases} (L^p(\mathbf{R}), \|\cdot\|_p), & 1 \leq p < +\infty; \\ (\mathscr{C}_0(\mathbf{R}), \|\cdot\|_\infty); \\ (\mathscr{U}\mathscr{C}_b(\mathbf{R}), \|\cdot\|_\infty). \end{cases}$$

Here the symbol $\mathscr{C}_0(\mathbf{R})$ stands for the space of all continuous functions on $\mathbf{R}$ which vanish at infinity and $\mathscr{U}\mathscr{C}_b(\mathbf{R})$ is the space of all uniformly continuous and bounded functions on $\mathbf{R}$.

For every $t \geq 0$, consider the bounded linear operator $T(t)$ on E defined by setting, for every $f \in E$ and $x \in \mathbf{R}$,

$$T(t)(f)(x) := f(x + t).$$

Then $(T(t))_{t\geq 0}$ is a positive C_0-semigroup on E and it is called the (left) translation *semigroup.*

3. Let $(E, \|\cdot\|)$ and $(F, \|\cdot\|')$ be Banach spaces and consider an isometric isomorphism $\Phi : F \longrightarrow E$ and a C_0-semigroup $(T(t))_{t\geq 0}$ on E.

For every $t \geq 0$, set

$$S(t) := \Phi^{-1}T(t)\Phi.$$

Then $(S(t))_{t\geq 0}$ is a C_0-semigroup on F, which is said to be similar *to $(T(t))_{t\geq 0}$; in particular, for every $t \geq 0$, $\|S(t)\|' = \|T(t)\|$.*

4. Let $(T(t))_{t\geq 0}$ be a C_0-semigroup on E and let ω_0 be its growth bound (see (2.1.3)). Fix $\mu \in \mathbf{K}$ and $\alpha > 0$; for every $t \geq 0$, set

$$S(t) := e^{\mu t}T(\alpha t).$$

Then $(S(t))_{t\geq 0}$ is a C_0-semigroup on E, which is called the rescaled semigroup *of $(T(t))_{t\geq 0}$ with parameters μ and α.*

In particular, if $\mu = -\omega_0$ (resp., $\mu < -\omega_0$) and $\alpha = 1$, then the rescaled semigroup $(S(t))_{t\geq 0}$ has growth bound equal to (or less than) 0.

Before presenting some aspects of the theory of C_0-semigroups, we recall some preliminaries on operator theory.

Given a linear operator $(A, D(A))$ on a Banach space $(E, \|\cdot\|)$, on the domain $D(A)$ of A we may consider the norm $\|\cdot\|_A$, defined by setting, for every $u \in D(A)$,

$$\|u\|_A := \|u\| + \|A(u)\|;$$

such a norm is usually referred to as the *graph norm.*
A linear operator $(A, D(A))$ is said to be *closed* if the normed space $(D(A), \|\cdot\|_A)$ turns out to be a Banach space.

More explicitly, this means that for every sequence $(f_n)_{n\geq 1}$ in $D(A)$ such that the limits $\lim_{n\to\infty} f_n = f \in E$ and $\lim_{n\to\infty} A(f_n) = g \in E$ exist, it follows that $f \in D(A)$ and $A(f) = g$; in other words, the graph $\{(f, A(f)) \mid f \in D(A)\}$ is closed in $E \times E$.

If $A \in \mathscr{L}(E)$, then A is obviously a closed operator; on the other hand, a closed operator $(A, D(A))$ is bounded if and only if $D(A)$ is closed in E.

Given a linear operator $(A, D(A))$ on E, a linear operator $(B, D(B))$ is said to be an *extension* of $(A, D(A))$ if $D(A) \subset D(B)$ and $B = A$ on $D(A)$.

A linear operator $(A, D(A))$ is *closable* if it admits a closed extension.

Equivalently, a linear operator $(A, D(A))$ is closable if and only if for any sequence $(f_n)_{n\geq 1}$ in $D(A)$ such that $\lim_{n\to\infty} f_n = 0$ and $(A(f_n))_{n\geq 1}$ is convergent in E, it follows that $\lim_{n\to\infty} A(f_n) = 0$.

If $(A, D(A))$ is a closable operator, then the smallest closed extension is called the *closure* of $(A, D(A))$ and it is denoted by $(\overline{A}, D(\overline{A}))$. It is not difficult to show

that

$$D(\overline{A}) := \{f \in E \mid \text{ there exists } (f_n)_{n\geq 1} \text{ in } D(A) \text{ such that } \lim_{n\to\infty} f_n = f$$
$$\text{and } (A(f_n))_{n\geq 1} \text{ is convergent}\}$$

and, for every $f \in D(\overline{A})$,

$$\overline{A}(f) := \lim_{n\to\infty} A(f_n),$$

where $(f_n)_{n\geq 1}$ is an arbitrary sequence in $D(A)$ such that $\lim_{n\to\infty} f_n = f$ and $(A(f_n))_{n\geq 1}$ is convergent in E.

Of course, if $(A, D(A))$ is a closed operator, then it is also closable and its closure $(\overline{A}, D(\overline{A}))$ coincides with $(A, D(A))$.

Finally, if $(A, D(A))$ is a bounded linear operator, then it is closable; in particular, $D(\overline{A}) = \overline{D(A)}$ and $\overline{A}$ is the natural extension of A to $\overline{D(A)}$.

A *core* for a closed operator $(A, D(A))$ is a linear subspace D_0 of $D(A)$ which is dense in $D(A)$ with respect to the graph norm.

Remark 2.1.5. If $(A, D(A))$ is a closed operator, $\lambda \in \mathbf{K}$ and the operator $\lambda I - A$ is invertible, then, by the open mapping theorem, the inverse operator $(\lambda I - A)^{-1}$ is continuous from $(E, \|\cdot\|)$ into $(D(A), \|\cdot\|_A)$ and, hence, $(\lambda I - A)^{-1} \in \mathscr{L}(E)$. If this is the case, a subspace D_0 of $D(A)$ is a core for $(A, D(A))$ if and only if $(\lambda I - A)(D_0)$ is dense in E.

Finally, we recall some definitions from the spectral theory.

Let $(A, D(A))$ be a closed operator on a Banach space $(E, \|\cdot\|)$. We shall denote by $\rho(A)$ the *resolvent set* of A, i.e.,

$$\rho(A) := \{\lambda \in \mathbf{K} \mid \lambda I - A \text{ is invertible}\},$$

and by $\sigma(A)$ the *spectrum* of A, namely

$$\sigma(A) := \mathbf{K} \setminus \rho(A).$$

In particular, if $\lambda \in \rho(A)$, we denote by $R(\lambda, A)$ the operator $(\lambda I - A)^{-1}$; $R(\lambda, A)$ is said to be the *resolvent* of A.

Clearly, by the open mapping theorem, we have that $R(\lambda, A) \in \mathscr{L}(E)$ for every $\lambda \in \rho(A)$. Moreover, the following identity, which is also called the *resolvent identity*, holds true: for every $\lambda, \mu \in \rho(A)$,

$$R(\lambda, A) - R(\mu, A) = (\mu - \lambda)R(\lambda, A)R(\mu, A). \tag{2.1.5}$$

Coming back to the setting of C_0-semigroups, we now recall the definition of the generator of a C_0-semigroup and its main properties.

Given a C_0-semigroup $(T(t))_{t\geq 0}$ on a Banach space $(E, \|\cdot\|)$, we may consider the linear operator $(A, D(A))$ on E defined as follows:

$$D(A) := \left\{ f \in E \mid \text{ there exists } \lim_{t\to 0^+} \frac{T(t)(f) - f}{t} \in E \right\} \tag{2.1.6}$$

and, for every $f \in D(A)$,

$$A(f) := \lim_{t \to 0^+} \frac{T(t)(f) - f}{t}. \tag{2.1.7}$$

The operator $(A, D(A))$ is said to be the *generator* of the semigroup $(T(t))_{t \geq 0}$.

Below, we present some examples of generators of C_0-semigroups; other examples may be found, e.g., in [93, Chapter II, Section 2].

Examples 2.1.6.

1. Let $A \in \mathscr{L}(E)$ and let $(T(t))_{t \geq 0}$ be the uniformly continuous semigroup presented in (2.1.4), i.e., $T(t) = e^{tA}$ for every $t \geq 0$; then A is the generator of $(T(t))_{t \geq 0}$.

On the other hand, if $(T(t))_{t \geq 0}$ is a C_0-semigroup and its generator $A \in \mathscr{L}(E)$, then necessarily $T(t) = e^{tA}$ for every $t \geq 0$ (for a proof, see [93, Chapter II, Corollary 1.5]).

In other words, every uniformly continuous semigroup $(T(t))_{t \geq 0}$ is necessarily of the form $T(t) = e^{tA}$ $(t \geq 0)$, for some $A \in \mathscr{L}(E)$.

2. Consider the Banach space $(\mathscr{UC}_b(\mathbf{R}), \|\cdot\|_\infty)$ and let $(T(t))_{t \geq 0}$ be the corresponding translation semigroup (see Example 2.1.4, 2). Then the generator $(A, D(A))$ of $(T(t))_{t \geq 0}$ is given by

$$D(A) := \{u \in \mathscr{UC}_b(\mathbf{R}) \mid u \text{ is differentiable and } u' \in \mathscr{UC}_b(\mathbf{R})\}$$

and, for every $u \in D(A)$,

$$A(u) = u'.$$

Further, if we consider the translation semigroup $(T(t))_{t \geq 0}$ on the Banach space $(L^p(\mathbf{R}), \|\cdot\|_p)$, $1 \leq p < +\infty$, then its generator $(A, D(A))$ is defined as follows:

$$D(A) := \{u \in L^p(\mathbf{R}) \mid u \text{ is absolutely continuous and } u' \in L^p(\mathbf{R})\}$$

and, for every $u \in D(A)$,

$$A(u) = u'.$$

3. Let $(E, \|\cdot\|)$ and $(F, \|\cdot\|')$ be Banach spaces and consider an isometric isomorphism $\Phi : F \longrightarrow E$. If $(T(t))_{t \geq 0}$ is a C_0-semigroup on E with generator $(A, D(A))$, then the generator $(B, D(B))$ of the similar semigroup $(S(t))_{t \geq 0}$ on F (see Example 2.1.4, 3) is defined as follows: $D(B) := \{u \in F \mid \Phi(u) \in D(A)\}$ and $B := \Phi^{-1} A \Phi$.

Moreover, we have that $\sigma(A) = \sigma(B)$ and $R(\lambda, B) = \Phi^{-1} R(\lambda, A) \Phi$ for every $\lambda \in \rho(A) = \rho(B)$.

4. If $(S(t))_{t \geq 0}$ is the rescaled semigroup of the C_0-semigroup $(T(t))_{t \geq 0}$ with parameters $\mu \in \mathbf{K}$ and $\alpha > 0$ (see Example 2.1.4, 4), then the generator of $(S(t))_{t \geq 0}$ is the operator $(\alpha A + \mu I, D(A))$, where $(A, D(A))$ is the generator of $(T(t))_{t \geq 0}$.

In what follows, we recall the main properties of the generator $(A, D(A))$ of a C_0-semigroup $(T(t))_{t\geq 0}$; for a proof, we refer the interested reader, e.g., to [93, Chapter II, Lemma 1.3, Theorem 1.4, Proposition 1.7].

Proposition 2.1.7. *Let $(T(t))_{t\geq 0}$ be a C_0-semigroup on a Banach space $(E, \|\cdot\|)$ and let $(A, D(A))$ be its generator. Then*

(a) $D(A)$ is dense in E, A is a closed operator and it determines the semigroup uniquely.

(b) $T(t)(D(A)) \subset D(A)$ for every $t \geq 0$.

(c) For every $f \in E$ the orbit map $\xi_f : \mathbf{R}^+ \longrightarrow E$ defined by (2.1.1) is differentiable in $\mathbf{R}^+$ if and only if $f \in D(A)$; moreover, for every $f \in D(A)$ and $t \geq 0$,

$$\xi'(t) = \frac{d}{dt}T(t)(f) = A(T(t)(f)) = T(t)(A(f)).$$

(d) For every $t \geq 0$ and $f \in E$, $\int_0^t T(s)(f)\,ds \in D(A)$ and

$$A\left(\int_0^t T(s)(f)\,ds\right) = T(t)(f) - f.$$

(e) For every $t \geq 0$ and $f \in D(A)$, one has

$$\int_0^t T(s)(A(f))\,ds = A\left(\int_0^t T(s)(f)\,ds\right) = T(t)(f) - f.$$

(f) A subspace D_0 of $D(A)$ which is dense in E and invariant under the semigroup $(T(t))_{t\geq 0}$ (i.e., $T(t)(D_0) \subset D_0$ for every $t \geq 0$) is a core for $(A, D(A))$.

(g) $\rho(A) \neq \emptyset$ and, if $(B, D(B))$ is a closed operator and if there exists a linear subspace $D_0 \subset D(A) \cap D(B)$ which is a core for $(A, D(A))$ and $A = B$ on D_0, then $(B, D(B)) = (A, D(A))$.

Since, by means of statement (g) in the previuous result, if $(A, D(A))$ is the generator of a strongly continuous semigroup $(T(t))_{t\geq 0}$, necessarily $\rho(A) \neq \emptyset$, then the *spectral bound* $s(A)$ of A defined by

$$s(A) := \sup\{\mathrm{Re}(\lambda) \mid \lambda \in \sigma(A)\} \tag{2.1.8}$$

is always finite.

Furthermore, if ω_0 is the growth bound of $(T(t))_{t\geq 0}$ (see (2.1.3)), then $s(A) \leq \omega_0 < +\infty$ (see [93, Chapter IV, Proposition 2.2]).

Finally, we recall the main properties of the resolvent $R(\lambda, A)$ of the generator $(A, D(A))$ of a C_0-semigroup $(T(t))_{t\geq 0}$ (see, e.g., [93, Chapter II, Theorem 1.10 and Corollary 1.11, Chapter III, Corollary 5.5]).

Proposition 2.1.8. *Let $(A, D(A))$ be the generator of a C_0-semigroup $(T(t))_{t\geq 0}$ such that, for any $t \geq 0$, $\|T(t)\| \leq Me^{\omega t}$, for some $M \geq 1$ and $\omega \in \mathbf{R}$. Then*

(a) *If* $R(\lambda)(f) := \int_0^\infty e^{-\lambda s}T(s)(f)\,ds$ *exists for some* $\lambda \in \mathbf{C}$ *and every* $f \in E$*, then* $\lambda \in \rho(A)$ *and, for every* $f \in E$*,*

$$R(\lambda, A)(f) = \int_0^\infty e^{-\lambda s}T(s)(f)\,ds. \tag{2.1.9}$$

(b) *If* $\lambda \in \mathbf{C}$ *and* $\mathrm{Re}(\lambda) > \omega$*, then* $\lambda \in \rho(A)$ *and* $R(\lambda, A)$ *is given by* (2.1.9)*. Moreover,* $\|R(\lambda, A)\| \le \frac{M}{\mathrm{Re}(\lambda)-\omega}$.

(c) *If* $n \ge 1$*,* $\lambda \in \mathbf{C}$ *and* $\mathrm{Re}(\lambda) > \omega$*, then* $R(\lambda, A)$ *is differentiable* n*-times with respect to* λ *and, for every* $f \in E$*,*

$$\begin{aligned} R(\lambda, A)^n(f) &= \frac{(-1)^{n-1}}{(n-1)!}\frac{d^{n-1}}{d\lambda^{n-1}}R(\lambda, A)(f) \\ &= \frac{1}{(n-1)!}\int_0^{+\infty} s^{n-1}e^{-\lambda s}T(s)(f)\,ds. \end{aligned}$$

In particular,

$$\|R(\lambda, A)^n\| \le \frac{M}{(Re(\lambda)-\omega)^n}.$$

(d) *For every* $f \in D(A)$ *and* $\lambda \in \mathbf{C}$*,* $\mathrm{Re}(\lambda) > \omega$*,*

$$R(\lambda, A)(A(f)) = AR(\lambda, A)(f) = \int_0^{+\infty} e^{-\lambda s}T(s)(A(f))\,ds.$$

(e) *For every* $f \in E$ *and* $t \ge 0$*,*

$$T(t)(f) = \lim_{n\to\infty}\left(\frac{n}{t}R\left(\frac{n}{t}, A\right)\right)^n(f),$$

the above limit being uniform for t *varying in compact intervals.*

2.1.2 Generation theorems and abstract Cauchy problems

One of the most fundamental problems of semigroups theory concerns the characterization of those closed operators which are the generators of C_0-semigroups; in what follows, we present some of the main results in this respect.

The first generation theorem, due independently to Hille and Yoshida, was originally proved for contractive semigroups (for a proof, see, e.g., [115], [209], [93, Chapter II, Theorem 3.5 and Corollary 3.6]), but it may be also stated in the following form.

Theorem 2.1.9. *Let* $\omega \in \mathbf{R}$ *and let* $(A, D(A))$ *be a linear operator on a Banach space* $(E, \|\cdot\|)$*. Then the following statements are equivalent:*

(a) $(A, D(A))$ *generates a* C_0*-semigroup* $(T(t))_{t\ge 0}$ *such that* $\|T(t)\| \le e^{\omega t}$ *for every* $t \ge 0$*.*

(b) $(A, D(A))$ *is closed, densely defined,* $]\omega, +\infty[\subset \rho(A)$ *and, for every* $\lambda > \omega$*, one has* $\|R(\lambda, A)\| \leq \frac{1}{\lambda-\omega}$.

(c) $(A, D(A))$ *is closed, densely defined and, for every* $\lambda \in \mathbf{C}$, $\mathrm{Re}(\lambda) > \omega$*, one has* $\lambda \in \rho(A)$ *and* $\|R(\lambda, A)\| \leq \frac{1}{\mathrm{Re}(\lambda)-\omega}$.

With the next result, which goes back to Feller (see [99]), Miyadera ([146]) and Phillips ([160]) (see, also, [93, Chapter II, Theorem 3.8], [159, Chapter 1, Theorem 5.2]), we present a generation result for arbitrary C_0-semigroups.

Theorem 2.1.10. *Let* $\omega \in \mathbf{R}$, $M \geq 1$ *and let* $(A, D(A))$ *be a linear operator on a Banach space* $(E, \|\cdot\|)$*. Then the following statements are equivalent:*

(a) $(A, D(A))$ *generates a* C_0*-semigroup* $(T(t))_{t\geq 0}$ *such that* $\|T(t)\| \leq Me^{\omega t}$ *for every* $t \geq 0$.

(b) $(A, D(A))$ *is closed, densely defined,* $]\omega, +\infty[\subset \rho(A)$ *and, for every* $\lambda > \omega$*, one has* $\|R(\lambda, A)^n\| \leq \frac{M}{(\lambda-\omega)^n}$ *for any* $n \geq 1$.

(c) $(A, D(A))$ *is closed, densely defined, for every* $\lambda \in \mathbf{C}$, $\mathrm{Re}(\lambda) > \omega$*, one has* $\lambda \in \rho(A)$ *and* $\|R(\lambda, A)^n\| \leq \frac{M}{(\mathrm{Re}(\lambda)-\omega)^n}$ *for any* $n \geq 1$.

Theorems 2.1.9 and 2.1.10 in general require an explicit knowledge of the resolvent operator or, at least, explicit estimates of the norm of (the powers of) the resolvent itself. To overcome such a difficulty, other generation theorems have been introduced.

In order to present these results, we first recall the following definition, due to Phillips (see [161]).

Let $(E, \|\cdot\|)$ be a Banach space. For every $f \in E$, set

$$\mathfrak{U}(f) := \{\varphi \in E' \mid \varphi(f) = \|f\|^2 = \|\varphi\|^2\},$$

where the symbol E' stands for the dual of E.

By the Hahn-Banach theorem, $\mathfrak{U}(f) \neq \emptyset$ for every $f \in E$.

A linear operator $(A, D(A))$ is said to be *dissipative* if for every $f \in D(A)$ there exists $\varphi \in \mathfrak{U}(f)$ such that $\mathrm{Re}(\varphi(A(f))) \leq 0$.

Equivalently (see, e.g., [159, Chapter 1, Theorem 4.2]), this also means that

$$\|(\lambda I - A)(f)\| \geq \lambda\|f\|$$

for every $f \in D(A)$ and $\lambda > 0$.

We are now ready to state the following theorem, which was proved by Lumer and Phillips (see [136], [159, Chapter 1, Theorem 4.3]).

Theorem 2.1.11. *Let* $(A, D(A))$ *be a linear operator on a Banach space* $(E, \|\cdot\|)$ *such that* $D(A)$ *is dense in* E*. Then the following statements are equivalent:*

(a) $(A, D(A))$ *is the generator of a contractive* C_0*-semigroup* $(T(t))_{t\geq 0}$ *on* E.

(b) $(A, D(A))$ *is a dissipative operator and there exists* $\lambda_0 > 0$ *such that* $(\lambda_0 I - A)(D(A)) = E$.

Moreover, if statement (a) or, equivalently, statement (b) holds true, then $(\lambda I - A)(D(A)) = E$ *for every* $\lambda > 0$ *and* $\mathrm{Re}(\varphi(A(f))) \leq 0$ *for every* $f \in D(A)$ *and* $\varphi \in \mathfrak{U}(f)$.

Corollary 2.1.12. *If* $(A, D(A))$ *is a dissipative operator on a Banach space* $(E, \|\cdot\|)$ *such that* $D(A)$ *is dense in* E, *then the following statements are equivalent:*
(a) $(A, D(A))$ *is closable and its closure* $(\overline{A}, D(\overline{A}))$ *is the generator of a contractive* C_0*-semigroup* $(T(t))_{t\geq 0}$ *on* E.
(b) There exists $\lambda_0 > 0$ *such that the range* $(\lambda_0 I - A)(D(A))$ *is dense in* E.

It is sometimes useful to look at an operator as sum of two or more simpler terms. In such a view, the following generation result holds (see [93, Chapter III, Theorem 1.3], [106, Theorem 1.6.5]).

Theorem 2.1.13. *Let* $(A, D(A))$ *be the generator of a* C_0*-semigroup* $(T(t))_{t\geq 0}$ *such that* $\|T(t)\| \leq Me^{\omega t}$ *for every* $t \geq 0$ *and for some* $M \geq 1$ *and* $\omega \in \mathbf{R}$. *Moreover, consider* $B \in \mathscr{L}(E)$. *Then the operator* $(A + B, D(A))$ *defined as*

$$(A + B)(u) = A(u) + B(u),$$

for every $u \in D(A)$, *generates a* C_0*-semigroup* $(S(t))_{t\geq 0}$ *such that, for every* $t \geq 0$,

$$\|S(t)\| \leq Me^{(\omega + M\|B\|)t}. \tag{2.1.10}$$

Moreover, setting, for every $t \geq 0$ *and* $n \geq 0$, $S_0(t) := T(t)$ *and*

$$S_{n+1}(t)(f) := \int_0^t T(t-s)BS_n(s)(f)\,ds \qquad (f \in E),$$

we have

$$S(t) = \sum_{n=0}^{\infty} S_n(t), \tag{2.1.11}$$

the above series being absolutely convergent and uniformly convergent on compact subintervals of $[0, +\infty[$.

The operator B is said to be a *bounded additive perturbation* of the operator $(A, D(A))$. Other perturbation results may be found, e.g., in [93, Chapter III, Sections 1-3], [106, Section 1.6].

We end this section by stressing the strong relationship between the semigroups theory and the solvability of certain (abstract) differential equations.

To this end, we recall the following fundamental definition.

Let $(E, \|\cdot\|)$ be a Banach space and consider a linear operator $(A, D(A))$ on E. The initial value problem

$$\begin{cases} \dfrac{du}{dt}(t) = A(u(t)) & t \geq 0; \\[2ex] u(0) = u_0 & u_0 \in D(A) \end{cases} \tag{2.1.12}$$

is said to be the *abstract Cauchy problem associated with* $(A, D(A))$ *and to the initial value* u_0 (in short, (ACP)).

A function $u : \mathbf{R}^+ \longrightarrow E$ is said to be a *classical solution* of (ACP) if
(i) $u(t) \in D(A)$ for every $t \geq 0$.
(ii) u is continuously differentiable with respect to E.
(iii) u satisfies (2.1.12).

The notion of (ACP) has a deep connection with the study of mathematical models describing the evolution in time of some physical phenomena.

In fact, the evolution in time of a physical system may be described by some initial value problems, which involve (eventually partial) differential equations and certain initial and/or lateral conditions. The differential initial value problem may be translated, in turn, into a suitable (ACP) associated with an operator $(A, D(A))$ on a certain Banach space $(E, \|\cdot\|)$. In such a view, the initial conditions in the initial value problem are usually included in the domain $D(A)$ of the operator A and, in particular, the search for solutions to an initial value problem is equivalent to the search for classical solutions to the corresponding (ACP).

The following example explains better the above mentioned method; other ones may be found, e.g., in [106].

Example 2.1.14. *Let* $\alpha, \beta \in \mathscr{C}(]0,1[)$, *fix* $u_0 \in \mathscr{C}([0,1]) \cap \mathscr{C}^2(]0,1[)$ *such that* $\lim\limits_{\substack{x\to 0^+ \\ x \to 1^-}} u_0''(x) = 0$ *and consider the following initial value problem:*

$$\begin{cases} \dfrac{\partial u}{\partial t}(x,t) = \alpha(x)\dfrac{\partial^2 u}{\partial x^2}(x,t) + \beta(x)\dfrac{\partial u}{\partial x}(x,t) & x \in]0,1[,\ t \geq 0; \\ u(x,0) = u_0(x) & x \in [0,1]; \\ u(\cdot,t) \in \mathscr{C}([0,1]) & t \geq 0; \\ \lim\limits_{\substack{x\to 0^+ \\ x\to 1^-}} \alpha(x)\dfrac{\partial^2 u}{\partial x^2}(x,t) + \beta(x)\dfrac{\partial u}{\partial x}(x,t) = 0 & t \geq 0. \end{cases} \tag{2.1.13}$$

For suitable choices of the functions α *and* β, *such an initial value problem is of concern in the description of several mathematical models, arising from biology, population dynamics and mathematical finance (see Subsections 2.3.3 and 2.3.4).*

A (classical) solution to (2.1.13) *is a function* $v : [0,1] \times [0,+\infty[\longrightarrow \mathbf{R}$ *such that*
(i) There exists $\frac{\partial v}{\partial t}$ *and it is continuous on* $[0,1] \times [0,+\infty[$.
(ii) There exist $\frac{\partial v}{\partial x}$ *and* $\frac{\partial^2 v}{\partial x^2}$ *as continuous functions on* $]0,1[\times [0,+\infty[$.
(iii) v *satisfies* (2.1.13).
In order to determine the solution to (2.1.13), *it may be useful to transform problem* (2.1.13) *into a suitable (ACP). Consider the Banach space* $\mathscr{C}([0,1])$ *and*

the operator $(A, D(A))$ *defined by setting*

$$A(u)(x) := \begin{cases} \alpha(x)u''(x) + \beta(x)u'(x) & \text{if } x \in]0,1[; \\ 0 & \text{if } x = 0 \text{ or } x = 1, \end{cases}$$

for every $u \in D(A)$, *where*

$$D(A) := \left\{ u \in \mathscr{C}([0,1]) \cap \mathscr{C}^2(]0,1[) \mid \lim_{\substack{x \to 0^+ \\ x \to 1^-}} \alpha(x)u''(x) + \beta(x)u'(x) = 0 \right\}.$$

Then, if we consider the (ACP)

$$\begin{cases} \dfrac{du}{dt}(t) = A(u(t)) & t \geq 0; \\ u(0) = u_0 & u_0 \in D(A), \end{cases}$$

it can be proved that a function $u : [0, +\infty[\longrightarrow \mathscr{C}([0,1])$ *is a classical solution to (ACP) if and only if the function* $v : [0,1] \times [0, +\infty[\longrightarrow \mathbf{R}$ *defined by setting, for every* $x \in [0,1]$ *and* $t \geq 0$,

$$v(x,t) := u(t)(x)$$

is solution to (2.1.13).

The last example shows that, in order to determine the (classical) solutions to some differential initial value problems, it may be useful to know how to determine the (classical) solutions to the corresponding (ACP) and this can be achieved if the operator $(A, D(A))$ is the generator of a C_0-semigroup. In fact, the following result holds (see, e.g., [93, Chapter II, Theorem 6.7]).

Theorem 2.1.15. *Let* $(A, D(A))$ *be a closed operator on a Banach space* $(E, \|\cdot\|)$. *Then* $(A, D(A))$ *is the generator of a* C_0*-semigroup* $(T(t))_{t\geq 0}$ *on* E *if and only if* $\rho(A) \neq \emptyset$ *and for every* $u_0 \in D(A)$ *there exists a unique classical solution to the abstract Cauchy problem* (2.1.12) *associated with* $(A, D(A))$ *and with initial value* u_0. *If this is the case, for every* $u_0 \in D(A)$ *the (unique) solution to (ACP) is given by*

$$u(t) = T(t)(u_0) \qquad (t \geq 0). \tag{2.1.14}$$

Remarks 2.1.16.

1. The solution (2.1.14) is continuously dependent on the initial datum.

In fact, if $u(\cdot) = T(t)(u_0)$ and $v(\cdot) = T(t)(v_0)$ are the solutions of (2.1.12) with initial data u_0 and v_0, respectively, then

$$\|u(t) - v(t)\| \leq \|T(t)\|\|u_0 - v_0\| \leq Me^{\omega t}\|u_0 - v_0\|,$$

for every $t \geq 0$, some $M \geq 1$ and $\omega \in \mathbf{R}$.

2. The operator $(A, D(A))$ presented in Example 2.1.14 is closed and, as we

shall see in the Subsection 2.3.3, under suitable assumptions on the coefficients α and β, it is the generator of a C_0-semigroup $(T(t))_{t\geq 0}$; hence the initial value problem (2.1.13) admits a unique (classical) solution $v : [0,1] \times [0,+\infty[\longrightarrow \mathbf{R}$ such that, for every $x \in [0,1]$ and $t \geq 0$,

$$v(x,t) = T(t)(u_0)(x).$$

3. In general, we say that an abstract Cauchy problem associated with the operator $(A, D(A))$ is *well posed* if $\rho(A) \neq \emptyset$ and for every $u_0 \in D(A)$ there exists a unique classical solution to the abstract Cauchy problem (2.1.12) associated with $(A, D(A))$ and with initial value u_0.

Hence, Theorem 2.1.15 may be re-stated by saying that a closed operator $(A, D(A))$ is the generator of a C_0-semigroup if and only if the abstract Cauchy problem associated with $(A, D(A))$ is well posed.

We point out that there are other possible definitions of the well-posedness of an abstract Cauchy problem. For more details about the subject we refer the interested reader to [93, Chapter II, Section 6].

Let $f : [0,+\infty[\longrightarrow E$ and consider the (more general) *inhomogeneous abstract Cauchy problem*

$$\begin{cases} \dfrac{du}{dt}(t) = A(u(t)) + f(t) & t \geq 0; \\ \\ u(0) = u_0 & u_0 \in D(A). \end{cases} \tag{2.1.15}$$

(for short, (IACP)).

As before, a function $u : \mathbf{R}^+ \longrightarrow E$ is said to be a *classical solution* of (IACP) if

(i) $u(t) \in D(A)$ for every $t \geq 0$.

(ii) u is continuously differentiable with respect to E.

(iii) u satisfies (2.1.15).

If $(A, D(A))$ is the generator of a C_0-semigroup $(T(t))_{t\geq 0}$, then by means of Theorem 2.1.15, the (homogeneous) abstract Cauchy problem associated with $(A, D(A))$ admits a unique (classical) solution $u : [0,+\infty[\longrightarrow D(A)$, given by $u(t) := T(t)(u_0)$ $(t \geq 0)$.

Under suitable assumptions on the function f, this is also a sufficient condition for (2.1.15) to have a unique classical solution. Moreover, such a solution may be represented in terms of the given semigroup $(T(t))_{t\geq 0}$, as the following result shows (for a proof, see [106, Chapter 2. Section 1.3]).

Theorem 2.1.17. *Let $(A, D(A))$ be the generator of a C_0-semigroup $(T(t))_{t\geq 0}$ on E. If either*

(i) f takes values in $D(A)$ and f and $A(f)$ are continuous functions with respect to E

or

(ii) f is continuously differentiable with respect to E,
then, for every $u_0 \in D(A)$, the inhomogeneous abstract Cauchy problem (2.1.15) *associated with $(A, D(A))$ and with initial datum u_0 has a unique classical solution $u : [0, +\infty[\longrightarrow D(A)$, given by*

$$u(t) = T(t)(u_0) + \int_0^t T(t-s) f(s)\, ds \qquad (t \geq 0). \tag{2.1.16}$$

2.2 Approximation of C_0-semigroups

Given a linear operator $(A, D(A))$ on some Banach space $(E, \|\cdot\|)$, Theorems 2.1.9, 2.1.10 and 2.1.11 determine necessary and sufficient conditions in order that $(A, D(A))$ is the generator of a C_0-semigroup $(T(t))_{t\geq 0}$. However, those theorems do not provide an explicit representation for the semigroup $(T(t))_{t\geq 0}$ and, consequently, for the solutions to the (ACP) associated with $(A, D(A))$ (see Theorem 2.1.15). In particular, the above-mentioned generation theorems are not constructive and they do not give any information about the semigroup itself and, hence, about the solutions of the (ACP) associated with its generator.

However, some other generation results are available which allow, not only to determine sufficient conditions in order that a suitable operator is the generator of a C_0-semigroup, but also to approximate that semigroup by means of iterates of linear operators, so that we may infer some preservation properties of the semigroup itself by studying the corresponding ones held by the relevant approximating operators.

The first result in this direction is due to Trotter (see [202], [159, Chapter 3, Section 3.4]).

Theorem 2.2.1. *Let $(E, \|\cdot\|)$ be a Banach space and let $(L_n)_{n\geq 1}$ be a sequence of bounded linear operators on E. Moreover, consider a sequence $(\rho(n))_{n\geq 1}$ of positive real numbers such that $\lim_{n\to\infty} \rho(n) = 0$. Suppose that there exist $M \geq 1$ and $\omega \in \mathbf{R}$ such that*

$$\|L_n^k\| \leq M e^{\omega\rho(n)k} \tag{2.2.1}$$

for every $k, n \geq 1$.

Let $(A, D(A))$ be the linear operator on E defined as

$$A(f) := \lim_{n\to\infty} \frac{L_n(f) - f}{\rho(n)}, \tag{2.2.2}$$

for every $f \in D(A)$, where

$$D(A) := \left\{ g \in E \mid \text{ there exists } \lim_{n\to\infty} \frac{L_n(g) - g}{\rho(n)} \right\}. \tag{2.2.3}$$

Assume that

(a) $D(A)$ is dense in E;

(b) the range $(\lambda I - A)(D(A))$ is dense in E for some $\lambda > \omega$.

Then $(A, D(A))$ is closable and its closure is the generator of a C_0-semigroup $(T(t))_{t\geq 0}$, such that for every $t \geq 0$ and for every sequence $(k(n))_{n\geq 1}$ of positive integers satisfying $\lim_{n\to\infty} k(n)\rho(n) = t$, one has

$$T(t)(f) = \lim_{n\to\infty} L_n^{k(n)}(f) \tag{2.2.4}$$

for every $f \in E$.

In particular, $\|T(t)\| \leq M e^{\omega t}$ for every $t \geq 0$.

Remark 2.2.2. By choosing, for every $n \geq 1$, $k(n) := [t/\rho(n)]$, where the symbol $[t/\rho(n)]$ stands for the integer part of $t/\rho(n)$, we get

$$T(t)(f) = \lim_{n\to\infty} L_n^{[t/\rho(n)]}(f) \tag{2.2.5}$$

for every $f \in E$, and the convergence is uniform in t varying on compacts subintervals of $[0, +\infty[$.

As a consequence of Theorem 2.2.1, the following result holds.

Corollary 2.2.3. *Let $(L_n)_{n\geq 1}$ be a sequence of bounded linear operators on E and let $(\rho(n))_{n\geq 1}$ be a sequence of positive real numbers tending to 0 as $n \to \infty$. Suppose that there exist $M \geq 1$ and $\omega \in \mathbf{R}$ such that*

$$\|L_n^k\| \leq Me^{\omega\rho(n)k} \tag{2.2.6}$$

for every $k, n \geq 1$.

Let (A_0, D_0) be a linear operator defined on a dense subspace D_0 of E and assume that
(i) $(\lambda I - A_0)(D_0)$ is dense in E for some $\lambda > \omega$;
(ii) $\lim\limits_{n\to\infty} \frac{L_n(u)-u}{\rho(n)} = A_0(u)$ for every $u \in D_0$.
Then (A_0, D_0) is closable and its closure is the generator of a C_0-semigroup $(T(t))_{t\geq 0}$ on E such that
(1) $\|T(t)\| \leq Me^{\omega t}$ for every $t \geq 0$;
(2) $T(t)(f) = \lim\limits_{n\to\infty} L_n^{k(n)}(f)$ for every $f \in E$, $t \geq 0$ and for every sequence $(k(n))_{n\geq 1}$ of positive integers satisfying $\lim\limits_{n\to\infty} k(n)\rho(n) = t$.

Proof. Let $(B, D(B))$ be the linear operator defined by

$$B(u) := \lim_{n\to\infty} \frac{L_n(u) - u}{\rho(n)}$$

for every $u \in D(B) := \left\{v \in E \mid \text{ there exists } \lim\limits_{n\to\infty} \frac{L_n(v)-v}{\rho(n)} \in E\right\}$.

Since $D_0 \subset D(B)$ and $B = A_0$ on D_0, $D(B)$ is dense in E along with $(\lambda I - B)(D(B))$. Therefore, by Trotter's theorem, $(B, D(B))$ is closable and its closure $(\overline{B}, D(\overline{B}))$ generates a C_0-semigroup $(T(t))_{t\geq 0}$ satisfying (1) and (2).

On the other hand, (A_0, D_0) is closable because it possesses a closable extension, namely $(B, D(B))$. Moreover, denoted by $(A, D(A))$ its closure, clearly $D(A) \subset D(\overline{B})$ and $\overline{B} = A$ on $D(A)$. From (i) it follows that $(\lambda I - \overline{B})(D_0) = (\lambda I - A_0)(D_0)$ is dense in E; hence, D_0 is a core for $(\overline{B}, D(\overline{B}))$ and, obviously, for $(A, D(A))$ as well. Therefore, by Proposition 2.1.7, part (g), $(A, D(A)) = (\overline{B}, D(\overline{B}))$ and the result follows. □

Remarks 2.2.4.

1. According to the preceding proof, if $u, v \in E$ and if $\lim\limits_{n\to\infty} \frac{L_n(u)-u}{\rho(n)} = v$, then $u \in D(A)$ and $A(u) = v$.

In particular, if $\lim_{n\to\infty} \frac{L_n(u)-u}{\rho(n)} = 0$, then $u \in D(A)$ and $A(u) = 0$.

2. Let $(A, D(A))$ be the generator of a C_0-semigroup $(T(t))_{t\geq 0}$ such that $\|T(t)\| \leq Me^{\omega t}$ $(t \geq 0)$, for some $M \geq 1$ and $\omega \in \mathbf{R}$. Moreover, consider a sequence $(L_n)_{n\geq 1}$ of bounded linear operators on E and a sequence $(\rho(n))_{n\geq 1}$ of positive real numbers such that $\lim_{n\to\infty} \rho(n) = 0$ and assume that
(i) $\|L_n^k\| \leq Me^{\omega\rho(n)k}$ for every $n, k \geq 1$,
(ii) $\lim_{n\to\infty} \frac{L_n(u)-u}{\rho(n)} = A(u)$ for every $u \in D_0$, D_0 being a core for $(A, D(A))$.
Then, as a consequence of Corollary 2.2.3, for every $f \in E$,

$$T(t)(f) = \lim_{n\to\infty} L_n^{k(n)}(f),$$

where $t \geq 0$ and $(k(n))_{n\geq 1}$ is a sequence of positive integers such that $\lim_{n\to\infty} k(n)\rho(n) = t$.

We proceed now to present other results about the approximation of semigroups, which generalize similar ones contained in [61, Corollary 1.2.8] and which allow to approximate a C_0-semigroup, just by knowing its behaviour on suitable knots.

Let $(E, \|\cdot\|)$ be a Banach space and consider a continuous mapping $T : [0,1] \longrightarrow \mathscr{L}(E)$. For every $n \geq 1$ and $x \in [0,1]$, consider the linear operator $B_n(T,x) \in \mathscr{L}(E)$ defined by

$$B_n(T,x) := \sum_{j=0}^{n} \binom{n}{j} x^j (1-x)^{n-j} T\left(\frac{j}{n}\right). \tag{2.2.7}$$

Then the following result holds (see [61, Proposition 1.2.9]).

Proposition 2.2.5. *For every $n \geq 1$, $f \in E$ and $x \in [0,1]$,*

$$\|B_n(T,x)(f) - T(x)(f)\| \leq \frac{3}{2}\omega\left(T(\cdot)(f), \frac{1}{\sqrt{n}}\right), \tag{2.2.8}$$

where, similarly to (1.6.20), for every $\delta > 0$ and $f \in E$,

$$\omega(T(\cdot)(f), \delta) := \sup\{\|T(x)(f) - T(y)(f)\| \mid x, y \in [0,1],\ |x-y| \leq \delta\}. \tag{2.2.9}$$

In particular, for every $f \in E$,

$$\lim_{n\to\infty} B_n(T,x)(f) = T(x)(f) \quad \textit{uniformly on } [0,1]. \tag{2.2.10}$$

Fix now $t \geq 0$ and consider a sequence $(\rho(n))_{n\geq 1}$ of positive real numbers such that $\lim_{n\to\infty} \rho(n) = 0$ and a sequence $(k(n))_{n\geq 1}$ of positive integers such that $\lim_{n\to\infty} \rho(n)k(n) = t$. Then there exists $M \geq 1$ such that $\frac{\rho(n)k(n)}{t} \leq M$ for every $n \geq 1$.

Fix a continuous mapping $T : [0, M] \longrightarrow \mathscr{L}(E)$ and, for every $n \geq 1$ and $x \in [0,1]$, set

$$B_n^*(T,x) := \sum_{j=0}^{k(n)} \binom{k(n)}{j} x^j (1-x)^{k(n)-j} T\left(\frac{j\rho(n)}{t}\right) \in \mathscr{L}(E). \tag{2.2.11}$$

Then

$$B_n^*(T,x) = B_{k(n)}(T_n,x), \tag{2.2.12}$$

where $B_{k(n)}$ is defined by (2.2.7) and $T_n : [0,1] \longrightarrow \mathscr{L}(E)$ is defined by

$$T_n(x) := T\left(x\frac{k(n)\rho(n)}{t}\right) \quad (0 \le x \le 1). \tag{2.2.13}$$

After these preliminaries, we state the following result.

Theorem 2.2.6. *For every $f \in E$,*

$$\lim_{n\to\infty} B_n^*(T,x)(f) = T(x)(f) \quad \textit{uniformly w.r.t. } x \in [0,1]. \tag{2.2.14}$$

Moreover,

$$\int_0^1 T(x)(f)\,dx = \lim_{n\to\infty} \frac{1}{k(n)+1} \sum_{j=0}^{k(n)} T\left(\frac{j\rho(n)}{t}\right)(f). \tag{2.2.15}$$

Finally, if $(T(s))_{s\ge 0}$ is a C_0-semigroup, then, for every $f \in E$,

$$T(x)(f) = \lim_{n\to\infty} \left[xT\left(\frac{\rho(n)}{t}\right) + (1-x)I\right]^{k(n)}(f) \tag{2.2.16}$$

uniformly w.r.t. $x \in [0,1]$ and

$$\int_0^1 T(x)(f)\,dx = \lim_{n\to\infty} \frac{1}{k(n)+1} \sum_{j=0}^{k(n)} T\left(\frac{\rho(n)}{t}\right)^j (f). \tag{2.2.17}$$

Proof. For a given $f \in E$, since the mapping $x \in [0,1] \mapsto T(x)(f)$ is continuous and, hence, uniformly continuous, we have that

$$\lim_{n\to\infty} T_n(x)(f) = T(x)(f) \tag{1}$$

uniformly on $[0,1]$.

Moreover, it is easy to prove that

$$\omega(T_n(\cdot)(f),\delta) \le \omega(T(\cdot)(f), M\delta) \tag{2}$$

for every $\delta > 0$ and $n \ge 1$(see (2.2.9)).

Consequently, taking (2.2.8), (2.2.12) and (2) into account, we get that, for every $n \ge 1$, $x \in [0,1]$ and $f \in E$,

$$\begin{aligned}
&\|B_n^*(T,x)(f) - T(x)(f)\| \\
&\le \|B_{k(n)}(T_n,x)(f) - T_n(x)(f)\| + \|T_n(x)(f) - T(x)(f)\| \\
&\le \frac{3}{2}\omega\left(T(\cdot)(f), \frac{M}{\sqrt{n}}\right) + \|T_n(x)(f) - T(x)(f)\|
\end{aligned}$$

and, from this and (1), (2.2.14) holds true.

Moreover, if $(T(s))_{s\geq 0}$ is a C_0-semigroup, then (2.2.16) follows at once from (2.2.14), taking the identity

$$\left[xT\left(\frac{\rho(n)}{t}\right)+(1-x)I\right]^{k(n)}=\sum_{j=0}^{k(n)}\binom{k(n)}{j}x^j(1-x)^{k(n)-j}T\left(\frac{j\rho(n)}{t}\right)$$

into account.

Further, from (2.2.14), we get

$$\begin{aligned}\int_0^1 T(x)(f)\,dx &= \lim_{n\to\infty}\int_0^1 B_n^*(T(x))(f)\,dx\\ &= \lim_{n\to\infty}\sum_{j=0}^{k(n)}\binom{k(n)}{j}T\left(\frac{j\rho(n)}{t}\right)(f)\int_0^1 x^j(1-x)^{k(n)-j}\,dx,\end{aligned}$$

so that (2.2.15) and (2.2.17) hold true. □

Remark 2.2.7. Formula (2.2.16) generalizes a result by Kendall obtained for $\rho(n)=\frac{1}{n}$ and $k(n)=n$ $(n\geq 1)$ and $t=1$ (see, also [61, Corollary 1.2.8]).

Turning back to the main subject of this paragraph, in concrete situations it may be not easy to prove assumption (b) in Theorem 2.2.1. For this reason, it would be useful to determine other generation and approximation results which do not require such a condition.

One of them is a theorem due to Schnabl (see [187, Satz 4]). Since the article [187] is not very easy to find, for the convenience of the reader we prefer to present here the details of the proof.

We begin with the following lemma (see [187], [153]).

Lemma 2.2.8. *Let $(E,\|\cdot\|)$ be a Banach space, $F\in\mathscr{L}(E)$ and $(C_n)_{n\geq 1}$ be a sequence of bounded linear operators on E such that $\lim_{n\to\infty}\|C_n\|=0$. Moreover, fix $t\geq 0$ and consider a sequence $(\rho(n))_{n\geq 1}$ of positive real numbers such that $\lim_{n\to\infty}\rho(n)=0$ and a sequence $(k(n))_{n\geq 1}$ of natural numbers such that $\lim_{n\to\infty}\rho(n)k(n)=t$. Then*

$$\lim_{n\to\infty}(I+\rho(n)F+\rho(n)C_n)^{k(n)}=e^{tF} \tag{2.2.18}$$

and

$$\lim_{n\to\infty}\frac{1}{k(n)+1}\sum_{j=0}^{k(n)}(I+\rho(n)F+\rho(n)C_n)^j=\int_0^1 e^{t\beta F}\,d\beta, \tag{2.2.19}$$

with respect to the operator norm on $\mathscr{L}(E)$, where $e^{\alpha F}:=\sum_{k=0}^{\infty}\frac{\alpha^k F^k}{k!}$ $(\alpha\geq 0)$.

Proof. First of all, we prove that, given $C,D\in\mathscr{L}(E)$ and $s\in\mathbf{N}$,

$$\|(I+C+D)^s-e^{sD}\|\leq s\|C\|e^{s(\|D\|+\|C\|)}+e^{s\|D\|}-(1+\|D\|)^s. \tag{1}$$

To this end, we preliminarily show that, for every $k \geq 1$,

$$\|(D+C)^k - D^k\| \leq k\|C\|(\|D\| + \|C\|)^{k-1}. \tag{2}$$

Indeed,

$$(D+C)^k - D^k = C((D+C)^{k-1} + (D+C)^{k-2}D + \dots + (D+C)D^{k-2} + D^{k-1}).$$

Hence

$$\begin{aligned}
&\|(D+C)^k - D^k\| \\
&\leq \|C\|\|(D+C)^{k-1} + (D+C)^{k-2}D + \dots + (D+C)D^{k-2} + D^{k-1}\| \\
&\leq \|C\| \left((\|D\| + \|C\|)^{k-1} + (\|C\| + \|D\|)(\|D\| + \|C\|)^{k-2} + \dots \right.\\
&\left. +(\|C\| + \|D\|)^{k-1}\right) = k\|C\|(\|D\| + \|C\|)^{k-1}.
\end{aligned}$$

Since

$$\begin{aligned}
(I+D+C)^s - e^{sD} &= \sum_{j=0}^{s} \binom{s}{j} (D+C)^j - \sum_{j=0}^{\infty} \frac{(sD)^j}{j!} \\
&= \sum_{j=0}^{s} \binom{s}{j} ((D+C)^j - D^j) + \sum_{j=0}^{s} \left(\binom{s}{j} - \frac{s^j}{j!}\right) D^j - \sum_{j=k+1}^{\infty} \frac{(sD)^j}{j!},
\end{aligned}$$

taking (2) into account, we infer that

$$\begin{aligned}
&\|(I+D+C)^s - e^{sD}\| \\
&\leq \sum_{j=0}^{s} \binom{s}{j} \|(D+C)^j - D^j\| + \sum_{j=0}^{s} \left(\frac{s^j}{j!} - \binom{s}{j}\right) \|D\|^j + \sum_{j=k+1}^{\infty} \frac{(s\|D\|)^j}{j!} \\
&\leq \sum_{j=1}^{s} \binom{s}{j} j\|C\|(\|C\| + \|D\|)^{j-1} + e^{s\|D\|} - (1 + \|D\|)^s \\
&\leq s\|C\| \sum_{j=0}^{s-1} \frac{(s-1)^j}{j!} (\|C\| + \|D\|)^j + e^{s\|D\|} - (1 + \|D\|)^s \\
&\leq s\|C\| e^{s(\|D\|+\|C\|)} + e^{s\|D\|} - (1 + \|D\|)^s,
\end{aligned}$$

so that (1) holds true.

Now, putting, for every $n \geq 1$, $s = k(n)$, $D = \rho(n)F$, $C = \rho(n)C_n$, we get

$$\begin{aligned}
&\|(I + \rho(n)F + \rho(n)C_n)^{k(n)} - e^{tF}\| \\
&\leq \|(I + \rho(n)F + \rho(n)C_n)^{k(n)} - e^{k(n)\rho(n)F}\| + \|e^{k(n)\rho(n)F} - e^{tF}\| \\
&\leq \rho(n)k(n)\|C_n\| e^{\rho(n)k(n)(\|F\|+\|C_n\|)} + e^{\rho(n)k(n)\|F\|} - (1 + \rho(n)\|F\|)^{k(n)} \\
&+ \|e^{k(n)\rho(n)F} - e^{tF}\|.
\end{aligned}$$

Letting $n \to \infty$, since $\lim_{n\to\infty} \|C_n\| = 0$ and $\lim_{n\to\infty} k(n)\rho(n) = t$, we get (2.2.18).

To show (2.2.19), fix $n \geq 1$; by applying again (1) for $C = \rho(n)C_n$ and $D = \rho(n)F$, for every $j = 0, \dots, k(n)$,

$$\|(I + \rho(n)F + \rho(n)C_n)^j - e^{j\rho(n)F}\|$$
$$\leq j\rho(n)\|C_n\|e^{j\rho(n)(\|F\|+\|C_n\|)} + e^{j\rho(n)\|F\|} - (1 + \rho(n)\|F\|)^j.$$

Hence,

$$\left\|\frac{1}{k(n)+1}\sum_{j=0}^{k(n)}\left((I + \rho(n)F + \rho(n)C_n)^j - e^{j\rho(n)F}\right)\right\|$$
$$\leq \frac{\rho(n)\|C_n\|}{k(n)+1}\sum_{j=0}^{k(n)} je^{j\rho(n)(\|F\|+\|C_n\|)}$$
$$+ \frac{1}{k(n)+1}\sum_{j=0}^{k(n)}\left(e^{j\rho(n)\|F\|} - (1 + \rho(n)\|F\|)^j\right)$$
$$\leq k(n)\rho(n)\|C_n\|e^{k(n)\rho(n)(\|F\|+\|C_n\|)} + e^{k(n)\rho(n)\|F\|} - (1 + \rho(n)\|F\|)^{k(n)},$$

where the last inequality holds true since the function $f(j) := e^{j\rho(n)\|F\|} - (1 + \rho(n)\|F\|)^j$ $(j \in \mathbf{N})$ is increasing.

Accordingly,

$$\left\|\frac{1}{k(n)+1}\sum_{j=0}^{k(n)}(I + \rho(n)F + \rho(n)C_n)^j - \int_0^1 e^{t\beta F}\,d\beta\right\|$$
$$\leq \left\|\frac{1}{k(n)+1}\sum_{j=0}^{k(n)}\left((I + \rho(n)F + \rho(n)C_n)^j - e^{j\rho(n)F}\right)\right\|$$
$$+ \left\|\frac{1}{k(n)+1}\sum_{j=0}^{k(n)} e^{j\rho(n)F} - \int_0^1 e^{t\beta F}\,d\beta\right\|$$
$$\leq k(n)\rho(n)\|C_n\|e^{k(n)\rho(n)(\|F\|+\|C_n\|)} + e^{k(n)\rho(n)\|F\|} - (1 + \rho(n)\|F\|)^{k(n)}$$
$$+ \left\|\frac{1}{k(n)+1}\sum_{j=0}^{k(n)} e^{j\rho(n)F} - \int_0^1 e^{t\beta F}\,d\beta\right\|$$

Letting $n \to \infty$, since $\lim_{n\to\infty}\|C_n\| = 0$, $\lim_{n\to\infty} k(n)\rho(n) = t$ and since, by applying (2.2.17) to $T(t) := e^{tF}$, we have that $\lim_{n\to\infty}\frac{1}{k(n)+1}\sum_{j=0}^{k(n)} e^{j\rho(n)F} = \int_0^1 e^{t\beta F}\,d\beta$, (2.2.19) holds true and this completes the proof. □

We are now in a position to state the following fundamental result, due to Schnabl (see [187, Satz 4]).

Theorem 2.2.9. *Let $(E, \|\cdot\|)$ be a Banach space, consider a sequence $(L_n)_{n\geq 1}$ of linear contractions (i.e., $\|L_n\| \leq 1$ for every $n \geq 1$) on E and let $(\rho(n))_{n\geq 1}$ be a*

sequence of positive real numbers such that $\lim_{n\to\infty} \rho(n) = 0$. *Consider, as in* (2.2.2) *and* (2.2.3), *the linear operator* $(A, D(A))$ *defined by*

$$A(f) := \lim_{n\to\infty} \frac{L_n(f) - f}{\rho(n)},$$

for every $f \in D(A)$, *where*

$$D(A) := \left\{ g \in E \mid \text{ there exists } \lim_{n\to\infty} \frac{L_n(g) - g}{\rho(n)} \right\}.$$

Moreover, assume that there exists a family $(E_i)_{i\in I}$ *of finite dimensional subspaces of* $D(A)$ *which are invariant under each* L_n (*i.e.*, $L_n(E_i) \subset E_i$ *for every* $n \geq 1$ *and* $i \in I$) *and whose union* $\bigcup_{i\in I} E_i$ *is dense in* E.

Then $(A, D(A))$ *is closable and its closure* $(\overline{A}, D(\overline{A}))$ *is the generator of a* C_0-*semigroup* $(T(t))_{t\geq 0}$. *Moreover, if* $t \geq 0$ *and if* $(k(n))_{n\geq 1}$ *is a sequence of positive integers satisfying* $\lim_{n\to\infty} k(n)\rho(n) = t$, *then, for every* $f \in E$,

$$T(t)(f) = \lim_{n\to\infty} L_n^{k(n)}(f) \tag{2.2.20}$$

and

$$\int_0^1 T(tu)(f)\, du = \lim_{n\to\infty} \frac{1}{k(n)+1} \sum_{j=0}^{k(n)} L_n^j(f). \tag{2.2.21}$$

Proof. Consider $t \geq 0$ and a sequence $(k(n))_{n\geq 1}$ of positive integers satisfying $\lim_{n\to\infty} k(n)\rho(n) = t$. Moreover, fix $i \in I$ and consider the Banach space $(E_i, \|\cdot\|_i)$, where $\|\cdot\|_i$ stands for the norm induced on E_i by the norm $\|\cdot\|$ on E. Denote by A_i the restriction of A to E_i and by $L_{n,i}$ the restriction of each L_n $(n \geq 1)$ to E_i. We remark that $A_i \in \mathscr{L}(E_i)$ since $L_n(E_i) \subset E_i$ and E_i is a finite dimensional (and, hence, closed) subspace of $(E, \|\cdot\|)$.

Moreover, for every $n \geq 1$ denote by $F_{n,i}$ the bounded linear operator $\frac{L_{n,i}(g)-g}{\rho(n)} - A_i$ on E_i. For every $h \in E_i$, we get that $\lim_{n\to\infty} F_{n,i}(h) = 0$ and hence $\lim_{n\to\infty} \|F_{n,i}\| = 0$, since E_i is finite-dimensional.

By applying Lemma 2.2.8 with $C_n = F_{n,i}$ and $F = A_i$, for every $t \geq 0$ and for every sequence $(k(n))_{n\geq 1}$ of positive integers such that $\lim_{n\to\infty} k(n)\rho(n) = t$, we have

$$\lim_{n\to\infty} L_{n,i}^{k(n)} = e^{tA_i}$$

and

$$\lim_{n\to\infty} \frac{1}{k(n)+1} \sum_{j=0}^{k(n)} L_{n,i}^j = \int_0^1 e^{tuA_i}\, du,$$

with respect to the operator norm on $\mathscr{L}(E)$. In particular, since the L_n's are contractions, it follows that the operators e^{tA_i} and $\int_0^1 e^{tuA_i}\, du$ are contractions for every $t \geq 0$.

We can then consider the linear operator $L(t) : \bigcup_{i\in I} E_i \longrightarrow E$ $(t \geq 0)$ such that, for every $f \in \bigcup_{i\in I} E_i$, $L(t)(f) := e^{tA_i}(f)$, where $f \in E_i$ for some $i \in I$. We notice that $L(t)$ is well defined, because, for $j \neq i$, $A_j = A_i = A$ on $E_i \cap E_j$. Clearly, $L(t)$ is a contraction, too.

Moreover, since $\bigcup_{i\in I} E_i$ is dense in E, for every $t \geq 0$ the operator $L(t)$ admits a norm preserving extension $T(t)$ to E. Reasoning in a similar way, we may define the operator $\int_0^1 T(tu)\,du$ which is the norm preserving extension of $\int_0^1 L(tu)\,du$ to E.

The family $(T(t))_{t\geq 0}$ is a semigroup, since each $(e^{tA_i})_{t\geq 0}$ is a semigroup for every $i \in I$ (see Example 2.1.4, 1).

Fix now $f \in \bigcup_{i\in I} E_i$; then there exists $i \in I$ such that $f \in E_i$ and $\lim_{t\to 0^+} T(t)(f) = \lim_{t\to 0^+} e^{tA_i}(f) = f$. Since $\bigcup_{i\in I} E_i$ is dense in E and $(T(t))_{t\geq 0}$ is contractive, from Proposition 2.1.3 we infer that $(T(t))_{t\geq 0}$ is a C_0-semigroup. Clearly, it satisfies (2.2.20) and (2.2.21).

Finally, we have to prove that $(A, D(A))$ is closable and its closure is the generator of $(T(t))_{t\geq 0}$.

Let us denote by $(B, D(B))$ the generator of $(T(t))_{t\geq 0}$. We first show that $(B, D(B))$ is an extension of $(A, D(A))$.

To this end, fix $f \in D(A)$ and set $g := A(f)$. Then, setting $z_n := \left\|\frac{L_n(f)-f}{\rho(n)} - g\right\|$ for every $n \geq 1$, we have that $\lim_{n\to\infty} z_n = 0$.

Moreover, since $\|L_n\| \leq 1$, for every $j = 1, \dots, k(n)$, we get

$$\left\|\frac{L_n^j(f) - L_n^{j-1}(f)}{\rho(n)} - L_n^{j-1}(g)\right\| \leq z_n.$$

Therefore,

$$\begin{aligned}&\left\|\frac{L_n^{k(n)}(f) - f}{\rho(n)} - \sum_{j=1}^{k(n)} L_n^{j-1}(g)\right\| \\ &\leq \sum_{j=1}^{k(n)} \left\|\frac{L_n^j(f) - L_n^{j-1}(f)}{\rho(n)} - L_n^{j-1}(g)\right\| \leq k(n) z_n\end{aligned}$$

and hence

$$\lim_{n\to\infty} \frac{1}{k(n)} \left\|\frac{L_n^{k(n)}(f) - f}{\rho(n)} - \sum_{j=1}^{k(n)} L_n^{j-1}(g)\right\| = 0. \tag{1}$$

Moreover,

$$\left\|\frac{T(t)(f) - f}{t} - \int_0^1 T(tu)(g)\,du\right\|$$

$$\leq \left\| \frac{T(t)(f)-f}{t} - \frac{L_n^{k(n)}(f)-f}{\rho(n)k(n)} \right\| + \frac{1}{k(n)} \left\| \frac{L_n^{k(n)}(f)-f}{\rho(n)} - \sum_{j=1}^{k(n)} L_n^{j-1}(g) \right\|$$
$$+ \left\| \frac{1}{k(n)} \sum_{j=1}^{k(n)} L_n^{j-1}(g) - \int_0^1 T(tu)(g)\,du \right\|.$$

Accordingly, since (2.2.20), (2.2.21) and (1) hold true, for $t > 0$,

$$\frac{T(t)(f)-f}{t} = \int_0^1 T(tu)(g)\,du = \frac{1}{t}\int_0^t T(u)(g)\,du.$$

Then, for $t \to 0^+$, we get that $f \in D(B)$ and $B(f) = g$ and hence $D(A) \subset D(B)$ and $B = A$ on $D(A)$. Consequently, $(A, D(A))$ is a closable operator and $(B, D(B))$ is an extension for $(\overline{A}, D(\overline{A}))$, too.

We prove now that $(\overline{A}, D(\overline{A})) = (B, D(B))$. To this end, we preliminarily note that, if $f \in E_i$ for some $i \in I$, then $L_n^k \in E_i$ for every $n, k \geq 1$ so that, given $t \geq 0$ and considered a sequence $(k(n))_{n\geq 1}$ of positive integers such that $\lim_{n\to\infty} \rho(n)k(n) = t$, we obtain that $T(t)(f) = \lim_{n\to\infty} L_n^{k(n)}(f) \in E_i$ since E_i is closed.

This reasoning shows that $T(t)\left(\bigcup_{i\in I} E_i\right) \subset \bigcup_{i\in I} E_i$ for every $t \geq 0$ and hence $\bigcup_{i\in I} E_i$ is a core for $(B, D(B))$ (see Proposition 2.1.7, (f)).

Now, the result follows from statement (g) of Proposition 2.1.7. □

Remark 2.2.10. Actually, the assumptions stated in Theorem 2.2.9 are stronger than those of [187, Satz 4] and were considered in such a form in [18, Theorem 1.6.8]. In particular, in his result Schnabl does not require that the involved finite dimensional subspaces which are invariant under the sequence $(L_n)_{n\geq 1}$ are contained in $D(A)$ (see (2.2.2) and (2.2.3)). However, this assumption seems to be essential in order to show that the generator of the semigroup $(T(t))_{t\geq 0}$ is the closure of the limit operator $(A, D(A))$, which is one of the crucial points of Theorem 2.2.9.

Finally, the following result holds.

Corollary 2.2.11. *Let $(L_n)_{n\geq 1}$ be a sequence of linear contractions on E and let $(\rho(n))_{n\geq 1}$ be a sequence of positive real numbers such that $\lim_{n\to\infty} \rho(n) = 0$. Let (A_0, D_0) be a linear operator defined on a subspace D_0 of E and assume that*

(i) there exists a family $(E_i)_{i\in I}$ of finite dimensional subspaces of D_0 which are invariant under each L_n and whose union $\bigcup_{i\in I} E_i$ is dense in E.

(ii) $\lim_{n\to\infty} \frac{L_n(u)-u}{\rho(n)} = A_0(u)$ for every $u \in D_0$.

Then (A_0, D_0) is closable and its closure $(A, D(A))$ is the generator of a contraction C_0-semigroup $(T(t))_{t\geq 0}$ on E satisfying (2.2.20) and (2.2.21). Furthermore, $\bigcup_{i\in I} E_i$ is a core for $(A, D(A))$.

Proof. Let $(B, D(B))$ be the linear operator defined by

$$B(u) := \lim_{n\to\infty} \frac{L_n(u) - u}{\rho(n)}$$

for every $u \in D(B) := \left\{ v \in E \mid \text{ there exists } \lim_{n\to\infty} \frac{L_n(v)-v}{\rho(n)} \in E \right\}$.

Because of Theorem 2.2.9, $(B, D(B))$ is closable and its closure $(\overline{B}, D(\overline{B}))$ generates a contraction C_0-semigroup $(T(t))_{t\geq 0}$ on E satisfying (2.2.20) and (2.2.21). Since $(B, D(B))$ is a closed extension of (A_0, D_0), then (A_0, D_0) is closable as well and, denoting its closure by $(A, D(A))$, we get $D(A) \subset D(\overline{B})$ and $\overline{B} = A$ on $D(A)$.

Note that each E_i, $i \in I$, is closed and invariant under every L_n, $n \geq 1$. From (2.2.20) it follows that $T(t)(E_i) \subset E_i$ for every $t \geq 0$ and hence $E_\infty := \bigcup_{i\in I} E_i$ is invariant under $(T(t))_{t\geq 0}$. By Proposition 2.1.7, (f), E_∞ is a core for $(\overline{B}, D(\overline{B}))$. In particular, it turns out that D_0 is a core for $(\overline{B}, D(\overline{B}))$. On the other hand, D_0 is a core for $(A, D(A))$, so that $(A, D(A)) = (\overline{B}, D(\overline{B}))$ and now the proof is complete. □

Remarks 2.2.12.

1. The above proof shows that, if $u, v \in E$ and if $\lim_{n\to\infty} \frac{L_n(u)-u}{\rho(n)} = v$, then $u \in D(A)$ and $A(u) = v$.

In particular, if $\lim_{n\to\infty} \frac{L_n(u)-u}{\rho(n)} = 0$, then $u \in D(A)$ and $A(u) = 0$.

2. Let $(A, D(A))$ be the generator of a contraction C_0-semigroup $(T(t))_{t\geq 0}$ and consider a sequence $(L_n)_{n\geq 1}$ of linear contractions on E. Assume that

(i) there exists a family $(E_i)_{i\in I}$ of finite dimensional subspaces of $D(A)$ which are invariant under each L_n and whose union $\bigcup_{i\in I} E_i$ is dense in E;

(ii) there exists a null sequence $(\rho(n))_{n\geq 1}$ of positive numbers such that

$$\lim_{n\to\infty} \frac{L_n(u) - u}{\rho(n)} = A(u)$$

for every $u \in D_0$, D_0 being a core for $(A, D(A))$.

Then, by Corollary 2.2.11, it follows that, for every $f \in E$, $t \geq 0$ and for every sequence $(k(n))_{n\geq 1}$ of positive integers such that $\lim_{n\to\infty} k(n)\rho(n) = t$,

$$T(t)(f) = \lim_{n\to\infty} L_n^{k(n)}(f).$$

2.3 Feller and Markov semigroups of operators

As we have pointed out in Section 2.1, C_0-semigroups may be fruitfully employed in order to solve some differential problems connected with the evolution in time of particular physical deterministic phenomena.

Anyway, classical physics does not always describe with accuracy all the natural phenomena, so that new probabilistic theories, such as stochastic analysis or Markov processes, were developed in the last century, in order to provide more effective mathematical models.

A precious link between these two different approaches is represented by the so-called Feller semigroups. In what follows, we sketch the main features of the theory of Feller semigroups and of the Markov processes; moreover, we outline the many applications of such theories to differential problems in one- and multidimensional settings.

2.3.1 Basic properties

In what follows, we are interested in studying particular C_0-semigroups mainly acting on the space $\mathscr{C}(X)$, X being a compact Hausdorff space.

Definition 2.3.1. *Let X be a compact Hausdorff space. A C_0-semigroup $(T(t))_{t\geq 0}$ on $\mathscr{C}(X)$ is said to be a* positive semigroup *if $T(t)$ is a positive operator, for every $t \geq 0$.*

Any contractive positive semigroup $(T(t))_{t\geq 0}$ on $\mathscr{C}(X)$ is said to be a Feller semigroup.

If $(T(t))_{t\geq 0}$ is a positive semigroup such that $T(t)(\mathbf{1}) = \mathbf{1}$ for every $t \geq 0$ (and, hence, $\|T(t)\| = 1$ for every $t \geq 0$), then $(T(t))_{t\geq 0}$ is said to be a Markov semigroup.

The following characterization of Markov semigroups is very useful.

Proposition 2.3.2. *Let X be a compact Hausdorff space and consider a positive C_0-semigroup $(T(t))_{t\geq 0}$ on $\mathscr{C}(X)$, with generator $(A, D(A))$. Then the following statements are equivalent:*
(i) $(T(t))_{t\geq 0}$ is a Markov semigroup.
(ii) $\mathbf{1} \in D(A)$ and $A(\mathbf{1}) = 0$.

Proof. The implication (i)⇒(ii) being straightforward, we pass on proving the reverse implication.

Consider the mapping $v : [0, +\infty[\longrightarrow \mathscr{C}(X)$ defined by setting $v(t) = \mathbf{1}$ for every $t \geq 0$. Then v is a solution to the (ACP)

$$\begin{cases} \dfrac{du}{dt}(t) = A(u(t)) \quad t \geq 0; \\ \\ u(0) = \mathbf{1}. \end{cases}$$

On the other hand, taking Theorem 2.1.15 into account, the above mentioned (ACP) admits a unique solution given by $u(t) = T(t)(\mathbf{1})$ for every $t \geq 0$, and this completes the proof. □

We discuss now a generation result for Feller semigroups. To this end, we need some preliminaries (see, e.g., [195]).

Definition 2.3.3. *Let X be a compact Hausdorff space and consider a linear operator $(A, D(A))$ on $\mathscr{C}(X)$.*

We say that $(A, D(A))$ satisfies the positive maximum principle *if, for every $u \in D(A)$ such that $\sup_{x\in X} u(x) > 0$ and for every $x_0 \in X$ such that $\sup_{x\in X} u(x) = u(x_0)$, one has $A(u)(x_0) \leq 0$.*

If this is the case, then necessarily $(A, D(A))$ is dissipative (see Subsection 2.1), i.e.,

$$\|\lambda u - A(u)\| \geq \lambda\|u\| \qquad \text{for all } u \in D(A) \text{ and } \lambda > 0.$$

The following result holds (see [98], [58]).

Theorem 2.3.4. *Let X be a compact Hausdorff space and let $(A, D(A))$ be a linear operator on $\mathscr{C}(X)$. The following statements are equivalent:*

(a) $(A, D(A))$ is closable and its closure generates a Feller semigroup $(T(t))_{t\geq 0}$ on $\mathscr{C}(X)$ (resp., $(A, D(A))$ generates a Feller semigroup $(T(t))_{t\geq 0}$ on $\mathscr{C}(X)$).

(b) $D(A)$ is dense in $\mathscr{C}(X)$, $(A, D(A))$ satisfies the positive maximum principle and there exists $\lambda > 0$ such that the range $(\lambda I - A)(D(A))$ is dense in $\mathscr{C}(X)$ (resp., $(\lambda I - A)(D(A)) = \mathscr{C}(X)$).

Moreover, if statement (a) or, equivalently, statement (b) holds true, then $(T(t))_{t\geq 0}$ is a Markov semigroup if and only if $\mathbf{1} \in D(A)$ and $A(\mathbf{1}) = 0$.

We proceed now to present a characterization of the domains of the generators of Feller semigroups in terms of pointwise asymptotic formulae. For an extension of the following result to the case when X is a locally compact Hausdorff space we refer the interested reader to [10].

The next result is well-known but we present the (short) proof for the convenience of the reader.

Lemma 2.3.5. *Let $(A, D(A))$ be the generator of a C_0-semigroup on a Banach space E. Then $(A, D(A))$ does not admit (non-trivial) dissipative extensions.*

Proof. Let $(B, D(B))$ be a dissipative extension of $(A, D(A))$, i.e., $D(A) \subset D(B)$ and $B = A$ on $D(A)$. Given a sufficiently large $\lambda > 0$ and $u \in D(B)$, since $\lambda I - A$ is bijective (here the symbol I stands for the identity operator), there exists $v \in D(A) \subset D(B)$, such that $\lambda v - A(v) = \lambda u - B(u)$.

Since $\lambda I - B$ is injective on $D(B)$, it follows that $u = v \in D(A)$. Therefore $D(B) = D(A)$. □

Theorem 2.3.6. *Let $(A, D(A))$ be the generator of a C_0-semigroup on $\mathscr{C}(X)$ and consider a net $(L_i)_{i\in I}^{\le}$ of positive linear contractions on $\mathscr{C}(X)$ and a net $(\varphi(i))_{i\in I}^{\le}$ of positive real numbers such that $\lim_{i\in I^{\le}} \varphi(i) = +\infty$ and $\lim_{i\in I^{\le}} \varphi(i)(L_i(u) - u) = A(u)$ pointwise on X for every $u \in D(A)$.*

Then $D(A)$ coincides with the subspace of all functions $u \in \mathscr{C}(X)$ for which there exists $v \in \mathscr{C}(X)$ such that $\lim_{i\in I^{\le}} \varphi(i)(L_i(u) - u) = v$ pointwise on X.

In particular, if $u \in \mathscr{C}(X)$ and $\lim_{i\in I^{\le}} \varphi(i)(L_i(u) - u) = 0$ pointwise on X, then $u \in D(A)$ and $A(u) = 0$.

Proof. Denote by $D(B)$ the subspace of all functions $u \in \mathscr{C}(X)$ for which there exists $v \in \mathscr{C}(X)$ such that $\lim_{i\in I^{\le}} \varphi(i)(L_i(u) - u) = v$ pointwise on X and consider the linear operator

$$B(u) := \lim_{i\in I^{\le}} \varphi(i)(L_i(u) - u) \qquad (u \in D(B)).$$

By assumption, $D(A) \subset D(B)$ and $B = A$ on $D(A)$. On account of Lemma 2.3.5 to get the result it is sufficient to show that $(B, D(B))$ is dissipative.

In fact, we shall show that $(B, D(B))$ verifies the positive maximum principle and, to this end, fix $u \in D(B)$ and $x_0 \in X$ satisfying $\sup_{x\in X} u(x) = u(x_0) > 0$.

Then, for every $i \in I$,

$$L_i(u) \le u(x_0) L_i(1) \le u(x_0).$$

In particular, $B(u)(x_0) = \lim_{i\in I^{\le}} \varphi(i)(L_i(u)(x_0) - u(x_0)) \le 0$ and this completes the proof. □

Corollary 2.3.7. *Let $(A, D(A))$ be the generator of a Feller semigroup $(T(t))_{t\ge 0}$ on $\mathscr{C}(X)$ and fix $u \in \mathscr{C}(X)$. Assume that*

(i) there exist a net $(t(i))_{i\in I}^{\le}$ in $]0, +\infty[$ converging to 0 and $v \in \mathscr{C}(X)$ such that $\lim_{i\in I^{\le}} \frac{T(t(i))(u) - u}{t(i)} = v$ pointwise on X.

Then $u \in D(A)$ (and $A(u) = v$).

In particular, if $\lim_{i\in I^{\le}} \frac{T(t(i))(u) - u}{t(i)} = 0$ pointwise on X, then $u \in D(A)$ and $A(u) = 0$, i.e., $T(t)(u) = u$ for every $t \ge 0$.

Proof. It is sufficient to apply Theorem 2.3.6 to the nets $L_i := T(t(i))$ and $\varphi(i) := 1/t(i)$ $(i \in I)$. □

We finally point out that the previous assumption (i) is satisfied, for instance, if there exists the limit $\lim_{t\to 0^+} \frac{T(t)(u) - u}{t} = v \in \mathscr{C}(X)$ pointwise on X or, under

the additional hypothesis that X is metrizable, if the family $\left(\frac{T(t)(u)-u}{t}\right)_{t\geq 0}$ is equicontinuous and pointwise bounded on X (apply Ascoli-Arzelà theorem).

We pass now to recall some perturbation results for Feller semigroups. The first one concerns additive perturbations (see [195, Theorem 9.3.3 and Corollary 9.3.4]).

Theorem 2.3.8. *Let $(A, D(A))$ be the generator of a Feller semigroup on $\mathscr{C}(X)$ and let $B \in \mathscr{L}(\mathscr{C}(X))$. If either $(A+B, D(A))$ or B satisfies the positive maximum principle, then $(A+B, D(A))$ is the generator of a Feller semigroup on $\mathscr{C}(X)$.*

Remark 2.3.9. As an example of a bounded operator satisfying the positive maximum principle, consider $q \in \mathscr{C}(X)$, $q \leq 0$, and set $B(f) := qf$ for every $f \in \mathscr{C}(X)$; then it is easy to show that $B \in \mathscr{L}(\mathscr{C}(X))$ and it satisfies the positive maximum principle.

Moreover, we have the following result (for a proof, see [179]).

Theorem 2.3.10. *Let $(A, D(A))$ be the generator of a Feller semigroup on $\mathscr{C}(X)$. Moreover, assume that $\mathbf{1} \in D(A)$ and fix $c \in \mathbf{R}$ such that $c \leq -A(\mathbf{1})(x)$ for every $x \in X$. Then $(A + cI, D(A))$ is the generator of a Feller semigroup on $\mathscr{C}(X)$.*

Finally, we present a result concerning multiplicative perturbations (see [40, Chapter B-II, Theorem 1.20], [90]).

Theorem 2.3.11. *Let $(A, D(A))$ be the generator of a Feller semigroup on $\mathscr{C}(X)$ and consider $q \in \mathscr{C}(X)$ such that $q(x) > 0$ for every $x \in X$. Then $(qA, D(A))$ generates a Feller semigroup.*

Occasionally, throughout the book we shall also deal with positive C_0-semigroups on Banach lattices of continuous functions defined on a non-compact space. Here we briefly outline some relevant results. For more details we refer, e.g., to [16], [17], [31], [30], [58], [203].

Given a locally compact Hausdorff space X, we shall denote by $\mathscr{C}_b(X)$ the space of all real-valued bounded continuous functions on X endowed with the natural pointwise order and the sup-norm with respect to which it becomes a Banach lattice. We shall also denote by $\mathscr{K}(X)$ the subspace of $\mathscr{C}_b(X)$ of all continuous real-valued functions on X having compact support. The closure $\overline{\mathscr{K}(X)}$ (with respect to the sup-norm) of $\mathscr{K}(X)$ will be denoted by $\mathscr{C}_0(X)$. If X is non-compact, then $\mathscr{C}_0(X)$ consists of all continuous functions $f \in \mathscr{C}(X)$ which vanish at infinity, i.e., for any $\varepsilon > 0$ a compact subset K of X may be found satisfying $|f(x)| \leq \varepsilon$ for every $x \in X \setminus K$.

The space $\mathscr{C}_0(X)$ is a Banach sublattice of $\mathscr{C}_b(X)$.

Similarly as in the compact case, a linear operator $T : \mathscr{C}_0(X) \longrightarrow \mathscr{C}_0(X)$ is said to be positive if $T(f) \geq 0$ for every $f \geq 0$.

Every positive linear operator on $\mathscr{C}_0(X)$ is continuous (see [6, Theorem 12.3]).

A C_0-semigroup of positive linear operators on $\mathscr{C}_0(X)$ will be briefly referred to as a positive C_0-semigroup. A *Feller semigroup on* $\mathscr{C}_0(X)$ is a positive and contractive C_0-semigroup on $\mathscr{C}_0(X)$.

The next result generalizes Theorem 2.3.4 for non-compact spaces.

Theorem 2.3.12. *Let* $A : D(A) \subset \mathscr{C}_0(X) \longrightarrow \mathscr{C}_0(X)$ *be a densely defined linear operator and* $\omega \in \mathbf{R}$. *The following statement are equivalent:*

(a) $(A, D(A))$ *is closable and its closure is the generator of a positive* C_0*-semigroup* $(T(t))_{t\geq 0}$ *on* $\mathscr{C}_0(X)$ *satisfying* $\|T(t)\| \leq \exp \omega t$ *for every* $t \geq 0$.

(b) (i) There exists $\lambda > \omega$ *such that* $(\lambda I - A)(D(A))$ *is dense in* $\mathscr{C}_0(X)$.

(ii) If $u \in D(A)$ *and* $\sup\limits_{x\in X} u(x) > 0$, *then for every* $x_0 \in X$ *such that* $u(x_0) = \sup\limits_{x\in X} u(x)$, *one gets* $A(u)(x_0) \leq \omega u(x_0)$.

A result similar to Theorem 2.3.11 reads as follows (cfr. [16, Corollary 2.7]).

Theorem 2.3.13. *Consider a linear operator* $(A, D(A))$ *on* $\mathscr{C}_0(X)$ *which is the generator of a positive* C_0*-semigroup* $(T(t))_{t\geq 0}$ *satisfying* $\|T(t)\| \leq \exp \omega t$ *for every* $t \geq 0$ *and some* $\omega \in \mathbf{R}$. *Let* $m \in \mathscr{C}_b(X)$ *and assume that* $m(x) > 0$ *for every* $x \in X$. *Then* $(mA, D(A))$ *is closable and its closure generates a positive* C_0*-semigroup* $(S(t))_{t\geq 0}$ *on* $\mathscr{C}_0(X)$ *such that* $\|T(t)\| \leq \exp \omega_1 t$ *for every* $t \geq 0$, *where* $\omega_1 := \omega^+ \max\{\|m\|_\infty, 1\}$, ω^+ *being the positive part of* ω, *i.e.,* $\omega^+ := \max\{\omega, 0\}$.

In the next subsection we discuss some applications of the theory of Feller semigroups.

2.3.2 Markov Processes

First of all, we review the deep relationship between Feller semigroups and Markov processes; for more details, we refer the interested reader, for example, to [48], [56], [195], [199].

We restrict ourselves to consider compact spaces only. For locally compact spaces we refer, e.g., to the above-mentioned references.

So, let X be a compact metric space endowed with some distance ρ. Moreover, let ∞ be a point not belonging to X and consider the one-point compactification X_∞ of X.

Let $(\Omega, \mathscr{U})$ be a measurable space, $(P^x)_{x\in X_\infty}$ a family of probability Borel measures on $\mathscr{U}$ and $(Z_t)_{0\leq t\leq +\infty}$ a family of random variables from Ω into X_∞.

Then the quadruple $\left(\Omega, \mathscr{U}, (P^x)_{x\in X_\infty}, (Z_t)_{0\leq t\leq +\infty}\right)$ is said to be a *Markov process with state space* X_∞ if

(i) Z_t is $(\mathscr{U}, B_{X_\infty})$-measurable for every $t \geq 0$; here B_{X_∞} denotes the Borel σ-algebra on X_∞;

(ii) $Z_{+\infty}(\omega) = \infty$ for every $\omega \in \Omega$ and if $Z_t(\omega) = \infty$ for some $\omega \in \Omega$ and $t \geq 0$, then $Z_s(\omega) = \infty$ for every $s \geq t$;

(iii) the function $x \in X \mapsto P^x(A)$ is Borel measurable for every $A \in \mathscr{U}$;

(iv) $P^x\{Z_{t+s} \in B \mid \mathscr{U}_s\} = P^{Z_s}\{Z_t \in B\}$ P^x-a.e., for every $s, t \geq 0, x \in X_\infty, B \in B_{X_\infty}$.

Here, by the symbol $\{Z_t \in B\}$ we denote the set $Z_t^{-1}(B)$, $\mathscr{U}_s$ denotes the σ-algebra generated by the family $(Z_u)_{0 \leq u \leq s}$, with $P^{Z_s}\{Z_t \in B\}$ we indicate the random variable $\omega \in \Omega \mapsto P^{Z_s(\omega)}\{Z_t \in B\}$ and $P^x\{Z_{t+s} \in B \mid \mathscr{U}_s\}$ is the *conditional probability of* $\{Z_{t+s} \in B\}$ *given* $\mathscr{U}_s$, namely, the almost surely uniquely determined $\mathscr{U}_s$-measurable and positive function on Ω such that, for every $A \in \mathscr{U}_s$,

$$\int_A P^x\{Z_{t+s} \in B \mid \mathscr{U}_s\}\, dP^x = P^x(A \cap \{Z_{t+s} \in B\}).$$

A Markov process is said to be *normal* if

$$P^x\{Z_0 = x\} = 1 \qquad \text{for every } x \in X_\infty. \tag{2.3.1}$$

Intuitively, we may interpret these mathematical conditions as follows: imagining a particle which moves in X_∞ after a random experiment $\omega \in \Omega$, then $Z_t(\omega)$ represents the position of the particle at time $t \geq 0$ after the experiment ω. Moreover, for every $A \in \mathscr{U}$, $P^x(A)$ represents the probability that the particle lies in A if it starts at position $x \in X_\infty$ and $P^x\{Z_t \in B\}$ is the probability that a particle, starting at position $x \in X_\infty$, at time t will be found in B.

In such a view, condition (iv) means that, under the experiment ω, the future behaviour of a particle does not depend on what has happened during the interval $[0, t]$ and, hence, knowing its history up to the time t is the same as considering a particle starting afresh at the position $Z_t(\omega)$.

Given a Markov process $\left(\Omega, \mathscr{U}, (P^x)_{x \in X_\infty}, (Z_t)_{0 \leq t \leq +\infty}\right)$, for every $\omega \in \Omega$ the map $t \in [0, +\infty] \mapsto Z_t(\omega) \in X_\infty$ is said to be a *path* of the process.

Moreover, the random variable $\varsigma : \Omega \longrightarrow \mathbf{R} \cup \{+\infty\}$ defined by setting, for every $\omega \in \Omega$,

$$\varsigma(\omega) := \inf\{t \in [0, +\infty] \mid Z_t(\omega) = \infty\} \tag{2.3.2}$$

is said to be the *lifetime* of the process. In such a view, condition (ii) in the definition of a Markov process means that, once a particle reaches the point ∞, it sticks there forever.

A Markov process is *right-continuous* if, for every $x \in X$,

$$P^x\{\omega \in \Omega \mid \text{ the path } t \mapsto Z_t(\omega) \text{ is right-continuous on } [0, +\infty[\} = 1 \tag{2.3.3}$$

and it is said to be *continuous* if, for every $x \in X$,

$$P^x\{\omega \in \Omega \mid \text{ the path } t \mapsto Z_t(\omega) \text{ is continuous on } [0, \varsigma(\omega)[\} = 1. \tag{2.3.4}$$

We now proceed to introduce the definition of Markov transition functions which is deeply connected with the theory of Markov processes.

Definition 2.3.14. *A family* $(P_t)_{t\geq 0}$ *of functions* $P_t : X \times B_X \longrightarrow \mathbf{R}$ $(t \geq 0)$ *is said to be a* Markov transition function *if*
(i) $P_t(x,\cdot)$ *is a Borel measure on* X *such that* $P_t(x,X) = 1$ *for every* $t \geq 0$ *and* $x \in X$;
(ii) $P_t(\cdot, B)$ *is a positive Borel measurable function for every* $t \geq 0$ *and* $B \in B_X$;
(iii) $P_{t+s}(x,B) = \int_X P_s(\cdot, B)\, dP_t(x,\cdot)$ *for every* $s,t \geq 0$, $x \in X$ *and* $B \in B_X$.
Condition (iii) is also known as Chapman-Kolmogorov *equation.*

A Markov transition function $(P_t)_{t\geq 0}$ *is said to be* normal *if* $P_0(x,\{x\}) = 1$ *for every* $x \in X$.

There is a strong relationship between normal Markov transition functions and normal Markov processes, as the following result shows. For a proof, we refer the interested reader to [48, Theorem 12.4.5 and Theorem 12.4.6].

Theorem 2.3.15. *Let* $\left(\Omega, \mathcal{U}, (P^x)_{x\in X_\infty}, (Z_t)_{0\leq t\leq +\infty}\right)$ *be a normal Markov process with state space* X_∞ *and, for every* $t \geq 0$, $x \in X$ *and* $B \in B_X$, *set*

$$P_t(x,B) := P^x\{Z_t \in B\}. \tag{2.3.5}$$

Then the family $(P_t)_{t\geq 0}$ *is a normal Markov transition function on* X.

Conversely, if $(P_t)_{t\geq 0}$ *is a normal Markov transition function on* X, *then there exists a normal Markov process* $\left(\Omega, \mathcal{U}, (P^x)_{x\in X_\infty}, (Z_t)_{0\leq t\leq +\infty}\right)$ *such that* (2.3.5) *holds true.*

A Markov transition function $(P_t)_{t\geq 0}$ is called *uniformly stochastically continuous* on X if, for every $\varepsilon > 0$,

$$\lim_{t\to 0^+} \sup_{x\in X} (1 - P_t(x, B(x,\varepsilon)) = 0, \tag{2.3.6}$$

where $B(x,\varepsilon) := \{y \in X \mid d(x,y) < \varepsilon\}$.

For uniformly stochastically continuous transition functions the following result holds (see [91]).

Theorem 2.3.16. *If* $(P_t)_{t\geq 0}$ *is a normal uniformly stochastically continuous transition function such that, for every* $\varepsilon > 0$ *and every compact subset* K *of* X,

$$\lim_{t\to 0^+} \sup_{x\in K} P_t(x, X \setminus B(x,\varepsilon)) = 0, \tag{2.3.7}$$

then there exists a right-continuous normal Markov process $\big(\Omega, \mathcal{U}, (P^x)_{x\in X_\infty}$, $(Z_t)_{0\leq t\leq +\infty}\big)$ *with state space* X_∞, *whose paths have left-hand limits on* $[0, \varsigma(\omega)[$ *(see* (2.3.2)*) for* P_x*-almost all* $\omega \in \Omega$ *and satisfying* (2.3.5).

Moreover, if, for every $\varepsilon > 0$,

$$\lim_{t\to 0^+} \frac{1}{t} \sup_{x\in X} P_t(x, X \setminus B(x,\varepsilon)) = 0, \tag{2.3.8}$$

then the Markov process $\left(\Omega, \mathcal{U}, (P^x)_{x\in X_\infty}, (Z_t)_{0\le t\le+\infty}\right)$ *has continuous paths on* $[0, \varsigma(\omega)[$ *for* P_x*-almost all* $\omega \in \Omega$.

Markov transition functions are objects of interest also from the point of view of semigroup theory.

More precisely, consider the space

$$B_0(X) := \{f : X \longrightarrow \mathbf{R} \mid f \text{ is Borel measurable and bounded}\}, \tag{2.3.9}$$

equipped with the sup-norm.

If $(P_t)_{t\ge0}$ is a normal Markov transition function on X, then for every $t \ge 0$ we may consider the positive linear operator $T(t) : B_0(X) \longrightarrow B_0(X)$ defined by setting, for every $f \in B_0(X)$ and $x \in X$,

$$T(t)(f)(x) := \int_X f\, dP_t(x, \cdot) = \int_\Omega f \circ Z_t\, dP^x, \tag{2.3.10}$$

where $\left(\Omega, \mathcal{U}, (P^x)_{x\in X_\infty}, (Z_t)_{0\le t\le+\infty}\right)$ is the normal Markov process associated with $(P_t)_{t\ge0}$ via Theorem 2.3.15.

Then, from condition (iii) in Definition 2.3.14, it follows that $(T(t))_{t\ge0}$ is a contraction semigroup on $B_0(X)$, and it is called the *transition semigroup* associated with $(P_t)_{t\ge0}$.

A Markov transition function $(P_t)_{t\ge0}$ is said to be a *Feller transition function* if $T(t)(\mathscr{C}(X)) \subset \mathscr{C}(X)$ for every $t \ge 0$ and, in this case, we keep to denote by $T(t)$ the restriction of the operator $T(t)$ to $\mathscr{C}(X)$.

We point out that in general the semigroup $(T(t))_{t\ge0}$ is not a C_0-semigroup (and, hence, a Feller semigroup). However, the following fundamental result, which furnishes a characterization of Feller semigroups in terms of Markov transition functions, holds (see [199, Theorem 3.1 and Theorem 3.2]).

Theorem 2.3.17. *If* $(P_t)_{t\ge0}$ *is a normal uniformly stochastically continuous Feller transition function satisfying* (2.3.7), *then the semigroup* $(T(t))_{t\ge0}$ *is a Feller semigroup on* $\mathscr{C}(X)$.

Conversely, every Feller semigroup $(T(t))_{t\ge0}$ *on* $\mathscr{C}(X)$ *is the transition semigroup of a normal uniformly stochastically continuous Feller transition function* $(P_t)_{t\ge0}$ *on* X *satisfying* (2.3.7) *and such that* (2.3.10) *holds true, or, equivalently, of a right-continuous normal Markov process with state space* X_∞, *whose paths have left-hand limits on* $[0, \varsigma(\omega)[$ *for* P_x*-almost all* $\omega \in \Omega$.

In particular, if $(T(t))_{t\ge0}$ *is a Markov semigroup, then the Markov process may be chosen in such a way that its state space is* X.

2.3.3 Second-order differential operators on real intervals and Feller theory

In this subsection we present one of the most effective tools in order to determine whether a second-order differential operator defined on suitable domains of con-

tinuous functions on a real interval generates a C_0-semigroup. The main results are due to Feller (see [98]) and they constitute the so-called *Feller theory.*

Consider an interval J of $\mathbf{R}$ and set

$$r_1 := \inf J \quad \text{and} \quad r_2 := \sup J.$$

Set $\tilde{J} := [r_1, r_2] \subset \mathbf{R} \cup \{-\infty, +\infty\}$ and consider the space

$$\mathscr{C}(\tilde{J}) := \left\{ f \in \mathscr{C}(J) \mid \text{ there exists } \lim_{x \to r_i} f(x) \in \mathbf{R} \text{ whenever } r_i \notin J, i = 1, 2 \right\}. \tag{2.3.11}$$

The main aim is to establish under what conditions suitable second-order differential operators generate Markov semigroups on $\mathscr{C}(\tilde{J})$.

More precisely, fix $\alpha, \beta \in \mathscr{C}(\text{int}(J))$ such that $\alpha(x) > 0$ for every $x \in \text{int}(J)$ and consider the linear operator

$$A(u)(x) := \begin{cases} \alpha(x)u''(x) + \beta(x)u'(x) & \text{if } x \in \text{int}(J); \\ \\ \lim_{t \to r_i} \alpha(t)u''(t) + \beta(t)u'(t) & \text{if } x = r_i, \ i = 1, 2. \end{cases} \tag{2.3.12}$$

on one of the linear spaces defined by setting

$$D_M(A) := \left\{ u \in \mathscr{C}(\tilde{J}) \cap \mathscr{C}^2(\text{int}(J)) \mid \text{ there exist } \lim_{\substack{x \to r_1^+ \\ x \to r_2^-}} A(u)(x) \in \mathbf{R} \right\}, \tag{2.3.13}$$

$$D_V(A) := \left\{ u \in D_M(A) \mid \lim_{\substack{x \to r_1^+ \\ x \to r_2^-}} A(u)(x) = 0 \right\} \tag{2.3.14}$$

$$D_{VM}(A) := \left\{ u \in D_M(A) \mid \lim_{x \to r_1 +} A(u)(x) = 0 \right\} \tag{2.3.15}$$

and

$$D_{MV}(A) := \left\{ u \in D_M(A) \mid \lim_{x \to r_2^-} A(u)(x) = 0 \right\}. \tag{2.3.16}$$

$D_M(A)$ is often referred to as the *maximal domain* for the operator A; $D_V(A)$, instead, is referred to as *Ventcel domain* for A, whereas $D_{VM}(A)$ and $D_{MV}(A)$ are called *mixed domains* for A.

It is a natural question to understand under what conditions on α and β the operator $(A, D_M(A))$ (resp., $(A, D_V(A)), (A, D_{VM}(A)), (A, D_{MV}(A))$) generates a C_0-semigroup.

To answer this question, an important role is played by the so-called *Feller theory,* which relates this problem to some integrability properties of the functions α and β with respect to the end-points r_1 and r_2 of J (see [98], [93, Chapter VI, Section 4.c]).

More precisely, fix $x_0 \in J$ and, for every $x \in J$, set

$$W(x) := \exp\left(-\int_{x_0}^{x} \frac{\beta(t)}{\alpha(t)}\,dt\right), \tag{2.3.17}$$

$$Q(x) := \frac{1}{\alpha(x)W(x)} \int_{x_0}^{x} W(t)\,dt \quad \text{and} \quad R(x) := W(x) \int_{x_0}^{x} \frac{1}{\alpha(t)W(t)}\,dt. \tag{2.3.18}$$

W is often referred to as the *Wronskian* of the operator A.

We say that

r_1 is a *regular* end-point if $Q \in L^1(r_1, x_0)$ and $R \in L^1(r_1, x_0)$;

r_1 is an *entrance* end-point if $Q \in L^1(r_1, x_0)$ and $R \notin L^1(r_1, x_0)$;

r_1 is an *exit* end-point if $Q \notin L^1(r_1, x_0)$ and $R \in L^1(r_1, x_0)$;

r_1 is a *natural* end-point if $Q \notin L^1(r_1, x_0)$ and $R \notin L^1(r_1, x_0)$.

Similar definitions can be given for the end-point r_2.

Remark 2.3.18. If $R \in L^1(r_1, x_0)$, then $W \in L^1(r_1, x_0)$. Analogously, if $Q \in L^1(r_1, x_0)$ then $(\alpha W)^{-1} \in L^1(x_0, r_1)$.

Consequently, if $(\alpha W)^{-1} \notin L^1(x_0, r_1)$, then r_1 cannot be an entrance or a regular end-point; in the same way, if $W \notin L^1(r_1, x_0)$, then r_1 cannot be an exit or a regular end-point.

Moreover, if $W \in L^1(r_1, x_0)$ and $(\alpha W)^{-1} \in L^1(r_1, x_0)$, then $R, Q \in L^1(r_1, x_0)$ and, consequently, r_1 is a regular end-point.

Similar results hold true also for the end-point r_2 (for more details, see [93, Remark 4.10, p. 394]).

After these preliminaries, we are ready to present the above-mentioned generation result (see [98], [78], [201]).

Theorem 2.3.19. *The operator $(A, D_M(A))$ generates a Markov semigroup on $\mathscr{C}(\tilde{J})$ if and only if r_1 and r_2 are either entrance or natural end-points.*

Moreover, the operator $(A, D_V(A))$ generates a Markov semigroup on $\mathscr{C}(\tilde{J})$ if and only if r_1 and r_2 are not entrance end-points.

The operator $(A, D_{VM}(A))$ generates a Markov semigroup on $\mathscr{C}(\tilde{J})$ if and only if r_1 is not an entrance end-point and and r_2 is either an entrance or a natural end-point.

Finally, the operator $(A, D_{MV}(A))$ generates a Markov semigroup on $\mathscr{C}(\tilde{J})$ if and only if r_1 is either an entrance or a natural end-point and r_2 is not an entrance end-point.

Remark 2.3.20. According to Feller himself ([98]), in the light of Theorems 2.3.17 and 2.3.19, the probabilistic interpretation of the previous classification of the end-points is as follows: an end-point r_i $(i = 1, 2)$ is of regular or exit type if the probability that the process reaches r_i in a finite time is strictly positive and it is natural if the probability that the process reaches r_i in a finite time is 0.

Ventcel's boundary condition (2.3.14) means that, once the process reaches one of the end-points r_i, it stops there forever.

In the special case where J is not closed, we are also interested in stating some generation results in the Banach lattice

$$\mathscr{C}_0(J) := \left\{ f \in \mathscr{C}(J) \mid \lim_{x \to r_i} f(x) = 0 \text{ whenever } r_i \notin J, \ i = 1, 2 \right\}$$

and for the differential operator

$$A_0 := A|_{D_V(A_0)} : D_V(A_0) \longrightarrow \mathscr{C}_0(J) \tag{2.3.19}$$

defined on

$$\begin{aligned} D_V(A_0) &= D_V(A) \cap \mathscr{C}_0(J) \\ &= \left\{ u \in \mathscr{C}_0(J) \cap \mathscr{C}(\mathrm{int}(J)) \mid \lim_{\substack{x \to r_1^+ \\ x \to r_2^-}} \alpha(x)u''(x) + \beta(x)u'(x) = 0 \right\}. \end{aligned} \tag{2.3.20}$$

By means of Theorem 2.3.19 and Theorem 2.3.12 (with $\omega = 0$) it is not difficult to show the next result (for more details see [17, Proposition 2.2]).

Corollary 2.3.21. *If the interval J is not closed and the end-points r_1 and r_2 are not entrance points, then $(A_0, D_V(A_0))$ is the generator of a Feller semigroup on $\mathscr{C}_0(J)$.*

In sections 4.5 and 5.7 we shall discuss in more details several applications of Theorem 2.3.19.

For an extension of Theorem 2.3.19 to the setting of weighted spaces of continuous functions the interested reader is referred to [16], [17], [29], [30], [31], [32], [33], [143].

2.3.4 Multidimensional second-order differential operators and Markov semigroups

In this subsection we present some examples of multi-dimensional second-order differential operators which, under suitable assumptions, generate Markov semigroups.

First of all, we focus our attention on some elliptic operators acting on the d-dimensional simplex.

Let K_d be the canonical d-dimensional simplex and denote by $v_0, \dots, v_d$ its vertices (see (1.1.45) and (1.1.46)).

Fix $\beta_1, \dots, \beta_d \in \mathscr{C}(K_d)$ and consider the second-order differential operator

$$L_{K_d}(u)(x) := \sum_{i,j=1}^{d} \frac{x_i(\delta_{ij} - x_j)}{2} \frac{\partial^2 u}{\partial x_i \partial x_j}(x) + \sum_{i=1}^{d} \beta_i(x) \frac{\partial u}{\partial x_i}(x) \tag{2.3.21}$$

($u \in \mathscr{C}^2(K_d), x = (x_1, \dots, x_d) \in K_d$), where δ_{ij} is the Kronecker symbol.

Such operators are often referred to as *Fleming-Viot* operators.

They appear in the theory of Fleming-Viot processes which are involved in the description of a stochastic process associated with a diffusion approximation of a gene frequency model in population genetics. In particular, the functions $\beta_1, \dots, \beta_d$ are those coefficients which, in the model, take phenomena such as mutation, migration and selection into account ([95], [96], [97], [101]).

The operators (2.3.21) have been object of investigation by several authors (see, for example, [3], [4], [71], [66], [190], [191], [192], [193] and the references quoted therein).

The difficulties in studying the operators (2.3.21) are determined by the fact that the boundary ∂K_d of K_d is non-smooth, due to the presence of sides and corners; moreover, L_{K_d} is an elliptic second-order operator which degenerates on ∂K_d.

Nevertheless, under suitable assumptions on the coefficients $\beta_1, \dots, \beta_d$, Ethier (see [95], [96, Theorem 2.8]) proved that $(V, \mathscr{C}^2(K_d))$ is closable and its closure is the generator of a Markov semigroup.

Theorem 2.3.22. *Let $(L_{K_d}, \mathscr{C}^2(K_d))$ be the differential operator defined by* (2.3.21) *and assume that $\beta_1, \dots, \beta_d$ are Lipschitz-continuous and, moreover, that*

$$\langle \beta(x), \nabla \det(x_i(\delta_{ij} - x_j)) \rangle \geq 0 \quad \textit{for every } x = (x_1, \dots, x_d) \in \partial K_d, \tag{2.3.22}$$

where $\beta := (\beta_1, \dots, \beta_d)$ and the symbols $\langle \cdot, \cdot \rangle$ and ∇ stand, respectively, for the scalar product in $\mathbf{R}^d$ and for the usual gradient of a function.

Then $(L_{K_d}, \mathscr{C}^2(K_d))$ is closable and its closure $(B_{K_d}, D(B_{K_d}))$ is the generator of a Markov semigroup on $\mathscr{C}(K_d)$. Moreover, $\mathscr{C}^2(K_d))$ is a core for $(B_{K_d}, D(B_{K_d}))$.

Remark 2.3.23. In [66, p. 56] Campiti and Raşa remarked that condition (2.3.22) may be stated in an easier form; more precisely, (2.3.22) holds true if and only if for every $x = (x_1, \dots, x_d) \in \partial K_d$

$$x_h = 0 \Rightarrow \beta_h(x_1, \dots, x_d) \geq 0, \qquad h = 0, 1, \dots d, \tag{2.3.23}$$

where $\beta_0 := -\sum_{i=1}^{d} \beta_i$ and $x_0 := 1 - \sum_{i=1}^{d} x_i$.

It means that on every face of the simplex the drift vector β is directed inside the simplex, i.e., phenomena such as migration, selection and mutation give rise to an inside directed drift term.

In particular, Theorem 2.3.22 applies to the diffusion problem associated with a second-order differential operator of the form (2.3.21) related to the so-called *Wright-Fisher model* (see, for more details, [96, Chapter 10]).

That model describes the evolution of the genetic pool of a population of diploid organisms, assuming that the generations are not overlapping, and taking into account factors as natural selection and mutation.

According to the Wright-Fisher model, the genetic information is devolved upon the genes, which are arranged on the chromosomes in linear order at certain positions called *loci*. The different alternatives of the types of genes that can occur at a given locus are named *alleles*.

Thus, if there are $d+1$ feasible alleles $A_0, \dots, A_d$ and, for every $i = 1, \dots, d$, x_i represents the frequency of the allele A_i, then $\sum_{i=1}^{d} x_i \leq 1$ (so that the d-dimensional simplex is the natural framework for such problems), and the frequency x_0 of the allele A_0 is given by $1 - \sum_{i=1}^{d} x_i$.

The above mentioned stochastic model (for more details, see [96], [192], [193]) is typically approximated by a diffusion process connected with the second-order differential operator defined by setting, for every $u \in \mathscr{C}^2(K_d)$ and $x = (x_1, \dots, x_d) \in K_d$,

$$V_1(u)(x) := \sum_{i,j=1}^{d} \frac{x_i(\delta_{ij} - x_j)}{2} \frac{\partial^2 u}{\partial x_i \partial x_j}(x) + \sum_{i=1}^{d} \frac{\mu_i - \nu x_i}{2} \frac{\partial u}{\partial x_i}(x) \tag{2.3.24}$$

with $\mu_i \geq 0$ for every $i = 1, \dots, d$ and $\nu \geq \sum_{i=1}^{d} \mu_i$, that is to a second-order differential operator of the form (2.3.21), where, for every $x = (x_1, \dots, x_d) \in K_d$,

$$\beta_i(x_1, \dots, x_d) := \frac{\mu_i - \nu x_i}{2} \qquad i = 1, \dots, d$$

and

$$\beta_0(x_1, \dots, x_d) := \frac{1}{2}\left(-\sum_{i=1}^{d} \mu_i + \nu \sum_{i=1}^{d} x_i\right)$$

(see (2.3.23)).

Clearly, V_1 satisfies condition (2.3.23), so that Theorem 2.3.22 applies; consequently $(V_1, \mathscr{C}^2(K_d))$ is closable and its closure generates a Markov semigroup $(T(t))_{t\geq 0}$ on $\mathscr{C}(K_d)$.

In [66], among other things, the asymptotic behaviour of the semigroup $(T(t))_{t\geq 0}$ was studied, whereas in [19] the authors presented an approximation

of such semigroup by means of powers of suitable positive linear operators (see Section 5.8 and, in particular, formula (5.8.2) and Corollary 5.8.3).

Moreover, if we choose $\mu := (\mu_1, \dots, \mu_{d+1}) \in]-1, +\infty[^{d+1}$ and we consider

$$\beta_i(x) := \mu_i + 1 - \sum_{j=1}^{d+1} (\mu_j + 1) x_i \qquad (x = (x_1, \dots, x_d) \in K_d), \tag{2.3.25}$$

then the corresponding Fleming-Viot operator

$$V_\mu(u)(x) := \sum_{i,j=1}^{d} x_i(\delta_{ij} - x_j) \frac{\partial^2 u}{\partial x_i \partial x_j}(x) + \sum_{i=1}^{d} \beta_i(x) \frac{\partial u}{\partial x_i}(x)$$

also appears in some models from population genetics (see [190; 191]); it can be proved that the closure of $(V_\mu, \mathscr{C}^2(K_d))$ in $L^p_{w_\mu}(K_d)$ is the generator of a positive C_0-semigroup (see [3], [71]). Here $L^p_{w_\mu}(K_d)$ stands for the space of all measurable functions $f : K_d \longrightarrow \mathbf{R}$ such that

$$\int_{K_d} |f(x)|^p w_\mu(x)\, dx < +\infty,$$

where $w_\mu(x) := (1 - x_1 - \dots - x_d)^{\mu_{d+1}} x_1^{\mu_1} \cdots x_d^{\mu_d}$, for every $x = (x_1, \dots, x_d) \in K_d$.

Fleming-Viot type operators have been object of investigation in several other frameworks, such as the d-dimensional hypercube $[0, 1]^d$ (see, for example, [72], [147] and the references quoted therein).

From now on, we shall set

$$Q_d := [0, 1]^d \tag{2.3.26}$$

and we shall denote by $\mathscr{C}^\delta(Q_d)$, $0 < \delta < 1$, the space of all Hölder-continuous functions of exponent δ on Q_d (with respect to $\|\cdot\|_2$) endowed with the norm $\|\cdot\|_\delta$ defined by

$$\|u\|_\delta := \|u\|_\infty + \sup_{\substack{x,y \in Q_d \\ x \neq y}} \frac{|u(x) - u(y)|^\delta}{\|x - y\|_2}. \tag{2.3.27}$$

Moreover, we define on such Q_d the Fleming-Viot type operators

$$L_{Q_d}(u)(x) := \sum_{i=1}^{d} \lambda_i(x) x_i (1 - x_i) \frac{\partial^2 u}{\partial x_i^2}(x) + \sum_{i=1}^{d} c_i(x) \beta_i(x) \frac{\partial u}{\partial x_i}(x) \tag{2.3.28}$$

($u \in \mathscr{C}^2(Q_d), x = (x_1, \dots, x_d) \in Q_d$), where, for every $i = 1, \dots, d$, the functions λ_i, c_i and β_i satisfy the following assumptions: there exists $\delta \in (0, 1)$ such that, for every $i = 1, \dots, d$,

(i) λ_i, $c_i \in \mathscr{C}^\delta(Q_d)$, $\lambda_i(x) > 0$ and $c_i(x) > 0$ for every $x \in Q_d$.

(ii) $\beta_i \in \mathscr{C}^\delta(Q_d)$, $\beta_i(0, \dots, 0) > 0$ and $\beta_i(1, \dots, 1) < 0$.

Then the following result holds (for a proof, see [72, Theorem 2.1]).

Theorem 2.3.24. *If (i) and (ii) hold true, the operator* $(L_{Q_d}, \mathscr{C}^2(Q_d))$ *defined by* (2.3.28) *is closable and its closure* $(B_{Q_d}, D(B_{Q_d}))$ *generates a Markov semigroup* $(T(t))_{t\geq 0}$ *on* $\mathscr{C}(Q_d)$. *Furthermore,* $\mathscr{C}^2(Q_d)$ *is a core for* $(B_{Q_d}, D(B_{Q_d}))$.

On the subset Q_d several authors studied the case when the semigroup $(T(t))_{t\geq 0}$ may be extended to a C_0-semigroup on some L^p-spaces. In particular, Mugnolo and Rhandi (see [147, Theorem 2]) proved that the extension of the operator

$$\widetilde{L}_{Q_d}(u)(x) := \sum_{i=1}^{d} x_i(1-x_i)\frac{\partial^2 u}{\partial x_i^2}(x) + \sum_{i=1}^{d}(\alpha_i(1-x_i) - \alpha_{i+1}x_i)\frac{\partial u}{\partial x_i}(x), \quad (2.3.29)$$

($u \in \mathscr{C}^2(Q_d), x = (x_1, \dots, x_d) \in Q_d$), where $\alpha_i > 0$ for every $i = 1, \dots, d+1$, to $L^p(Q_d, \mu)$ (μ being a suitable invariant measure) is closable and its closure generates a positive contraction semigroup (see [147, Theorem 2]).

In particular, consider the differential operator

$$\widetilde{V}_1(u)(x) := \sum_{i=1}^{d} \frac{x_i(1-x_i)}{2}\frac{\partial^2 u}{\partial x_i^2}(x) + \sum_{i=1}^{d}\left(\frac{1}{2} - x_i\right)\frac{\partial u}{\partial x_i}(x), \quad (2.3.30)$$

for every $u \in \mathscr{C}^2([0,1]^d)$ and $x = (x_i)_{1\leq i\leq d} \in [0,1]^d$; then the following result holds (see [20, Theorem 3.3]).

Theorem 2.3.25. *For every* $p \geq 1$, *the operator* $(\widetilde{V}_1, \mathscr{C}^2(Q_d))$ *(defined in* (2.3.30)*) is closable in* $L^p(Q_d)$ *and its closure* $(\widetilde{A}, D(\widetilde{A}))$ *generates a positive contraction* C_0*-semigroup* $(\widetilde{T}(t))_{t\geq 0}$ *in* $L^p(Q_d)$. *Further,* $\mathscr{C}^2(Q_d)$ *is a core for* $(\widetilde{A}, D(\widetilde{A}))$ *and*

$$\widetilde{A} = \widetilde{V}_1 \quad \text{on } \mathscr{C}^2(Q_d). \quad (2.3.31)$$

We point out that the semigroup $(\widetilde{T}(t))_{t\geq 0}$ is the extension of the corresponding semigroup on $\mathscr{C}(Q_d)$; moreover, in [20, Theorem 3.3] it was also provided an approximation of the semigroup $(\widetilde{T}(t))_{t\geq 0}$ in L^p-norm by means of suitable iterates of (a modification of) Kantorovich operators on Q_d (for a definition see [210]).

By a completely different method such a generation and approximation result was established for $d = 1$ by J. Nagel (see [149]).

2.4 Notes and comments

The theory of C_0-semigroups of operators represents one of the most elegant examples of how a mathematical theory may be fruitfully employed in order to solve concrete problems arising from different areas such as partial differential equations, ergodic theory, probability theory, mathematical finance or population dynamics.

The systematic study of C_0-semigroups goes back to the 1930s; in particular, in 1948 Hille published the first monograph on this subject (see [115]); that book, together with its second edition coauthored by Phillips (see [116]), is one of the most important references on the subject. For more details on semigroups theory and its wide applications we refer the reader also to [83], [106], [93], [159] and the references quoted therein.

Among the generation theorems, the theorems by Trotter (see [202]) and Schnabl (see [187]) play an important role in this monograph, since they allow not only to determine sufficient conditions in order that a linear operator is the generator of a C_0-semigroup, but they also provide an approximation formula for that semigroup by means of suitable iterates of positive linear operators. Other results on this subject may be found in [185], [153], [155]. More recent generalizations (also to locally convex spaces) may be found in [51], [128], [188]. Moreover, the monograph [61] deals with the approximation of semigroups.

Corollary 2.2.3 and Corollary 2.2.11 are new, whereas a direct proof of Remark 2.2.4 may be found in [27, Theorem 2.1].

Theorem 2.2.6 seems also to be new and generalizes similar results contained in [61, Section 1.2.2].

Feller semigroups, as we have seen, represent the link between stochastic analysis and functional analysis. In particular, Theorem 2.3.4 was established by Feller ([98]) for semigroups acting on $\mathscr{C}(X)$, X being a compact Hausdorff space; later on, it was generalized to Feller semigroups acting on $\mathscr{C}_0(X)$, X being a locally compact Hausdorff space, by Bony, Courrége and Priouret (see [58]). Other results on the subject may be found in [179], [118], [205].

We also refer to [16], [30] [31] where the authors study positive C_0-semigroups acting on weighted spaces of continuous functions defined on a locally compact Hausdorff space X. Among other things, they study those positive C_0-semigroups satisfying the so-called *Feller property*, i.e., they leave invariant the space $\mathscr{C}_0(X)$ and the restrictions to $\mathscr{C}_0(X)$ are contractive and strongly continuous.

Such positive C_0-semigroups are the only possible C_0-semigroups on the bigger weighted continuous function spaces which are associated with probability transition functions, and hence with right-continuous Markov processes.

The theory of Markov processes arises from the need to build a mathematical theory which better describes non-deterministic natural phenomena, such as the variation of the price of an option or the noise in an electric device. In particular, in 1828 the English botanist R. Brown described the first example of non-deterministic phenomenon, the chaotic motion of the pollen grains in the water. After him such phenomenon was called *Brownian motion* and it often occurs in different models arising from physics, mathematical finance and other fields. A first mathematical approach to the Brownian motion was introduced by A. Einstein in 1905 and it allowed also to estimate in an accurate way the Avogadro's number by observing particles undergoing Brownian motion. Between 1906 and 1909, J. Perrin tested Einstein's theory experimentally. Anyway, a rigorous mathematical theory for the existence of a Brownian motion was established by N. Wiener in 1923 and this formalization gave birth to the definition of Markov processes.

Many monographs are devoted to the theory of Markov processes. Without any claim of completeness, we refer the interested reader for example to [48, Chapter 12], [56], [91], [96], [118], [129], [195], [199] and the references quoted therein.

Given a non-degenerate second-order differential operator of the form

$$A(u) = \alpha u'' + \beta u' + \gamma u,$$

on some real interval J and for some continuous coefficients α, β and γ, Feller (see [98], [93, Chapter VI, Section 4.c]) studied under what conditions the operator A is the generator of a Feller semigroup. In particular, he recognized that the properties of generation of A depend on the behaviour of the coefficients α and β on the boundary of the interval J and, in this spirit, he obtained a characterization of the end-points of J, later on called after him *Feller analysis*. The theory was inspired in particular by the application of probability theory in the study of one-dimensional diffusion processes. More details on such subject and on its connection with probability theory may be found in [100] or in [129]. In particular, Theorem 2.3.19 was explicitly stated in the paper [78].

A generalization of Feller analysis for boundary conditions in the higher dimensional case may be found in [195], [196], [197], [198], [199].

As discussed in Subsection 2.3.4, Fleming-Viot operators are second-order elliptic differential operators which occur in the theory of Fleming-Viot processes on the canonical simplex K_d. Fleming-Viot processes, in particular, are measure-valued diffusion approximations of processes associated with discrete Markov chains in population genetics. Ethier proved that, under suitable conditions, a Fleming-Viot operator (2.3.21) is closable and its closure generates a Feller semigroup (see [95], [96]). For additional details on these operators we refer back to Subsection 2.3.4.

3 Bernstein-Schnabl operators associated with Markov operators

> As constructive function theory we want to call the direction of function theory which follows the aim to give the simplest and most pleasant basis for the quantitative investigation and calculation both of empirical and of all other functions occurring as solutions of naturally posed problems of mathematical analysis (for instance, as solutions of differential or functional equations).
>
> *Sergey Natanovich Bernstein*

This chapter plays a central role in the whole of the book. We study the fundamental sequence of operators, called *Bernstein-Schnabl* operators, which are one of the main objects of interest in our investigations and which can be constructed by means of an arbitrary Markov operator T on $\mathscr{C}(K)$, K being a convex compact subset of some (not necessarily finite dimensional) locally convex Hausdorff space.

These operators generalize the classical Bernstein operators defined on the unit interval, on multidimensional simplices and on hypercubes, and they share with them several similar preservation properties which we investigate in detail.

The main interest in their study lies not only in the fact that they are an approximation process for continuous functions on K, but also because, by means of them, it is possible to constructively approximate suitable Markov semigroups of operators. Such Markov semigroups, in finite dimensional settings, are generated by second-order elliptic degenerate differential operators, so that a similar approximation holds for solutions to the initial-boundary value problems associated with such differential operators.

We stress that no special assumptions on the boundary of K nor on the Markov operator T are imposed. Furthermore, our general framework allows to apply the theory to a variety of situations which seem to have interest both to theoretical and applied problems.

The connections with Markov semigroups and evolution equations will be discussed in the next chapter. In the present one, we first show the approximation properties of Bernstein-Schnabl operators by also pointing out some estimates of the rate of convergence. Then we study some preservation properties such as the preservation of Hölder continuity and of convexity. Motivated by similar properties of Bernstein operators on $\mathscr{C}([0,1])$, we also investigate the monotonicity and the majorizing properties for these operators, which lead us to consider other classes of convex functions such as T-axially convex and T-convex ones.

Several examples scattered throughout the chapter illustrate the general results in more concrete cases.

3.1 Generalities, definitions and examples

Throughout this chapter we shall fix a locally convex Hausdorff space X and a metrizable convex compact subset K of X; moreover, we shall consider a Markov operator $T : \mathscr{C}(K) \longrightarrow \mathscr{C}(K)$. Then, as (1.3.1) shows, for every $x \in K$ there exists a (unique) probability Borel measure $\tilde{\mu}_x^T$ such that

$$T(f)(x) = \int_K f \, d\tilde{\mu}_x^T \qquad \text{for every } f \in \mathscr{C}(K) \tag{3.1.1}$$

and the family $(\tilde{\mu}_x^T)_{x \in K}$ defines a continuous selection of probability Borel measures associated with the operator T.

Moreover, we assume that

$$T(h) = h \qquad \text{for every } h \in A(K), \tag{3.1.2}$$

or, equivalently, that

$$\int_K h \, d\tilde{\mu}_x^T = h(x) \qquad \text{for every } h \in A(K) \text{ and } x \in K; \tag{3.1.3}$$

here $A(K)$ designates the space of all affine continuous functions on K.

Accordingly, $A(K) \subset M_T := \{h \in \mathscr{C}(K) \mid T(h) = h\}$ and hence, from Theorem 1.3.1 and from (1.1.36), it follows that

$$\partial_e K \subset \partial_{M_T} K \subset \partial_T K \subset \partial_H K, \tag{3.1.4}$$

where $H := T(\mathscr{C}(K))$ and $\partial_T K$ is the set of interpolation points for T (see (1.3.4)).

We proceed now to introduce the sequence of positive linear operators which will be one of the main objects of interest in this book.

For any $n \geq 1$, consider the mapping $\pi_n : K^n \longrightarrow K$ defined by

$$\pi_n(x_1, \dots, x_n) := \frac{x_1 + \dots + x_n}{n} \qquad ((x_1, \dots, x_n) \in K^n), \tag{3.1.5}$$

where K^n denotes the product space of K with itself n times. Moreover, for each $x \in K$ and $f \in \mathscr{C}(K)$, denoting by $\tilde{\mu}_{x,n}^T$ the tensor product of $\tilde{\mu}_x^T$ with itself n times (see (1.1.26)), set

$$B_n(f)(x) := \int_{K^n} f \circ \pi_n \, d\tilde{\mu}_{x,n}^T. \tag{3.1.6}$$

More explicitly (see (1.1.28)),

$$B_n(f)(x) = \int_K \cdots \int_K f\left(\frac{x_1 + \dots + x_n}{n}\right) d\tilde{\mu}_x^T(x_1) \cdots d\tilde{\mu}_x^T(x_n). \tag{3.1.7}$$

Because of Proposition 1.1.6, part (3), $B_n(f) \in \mathscr{C}(K)$, so that B_n actually turns out to be a positive linear operator from $\mathscr{C}(K)$ into $\mathscr{C}(K)$. It will be referred

to as the *n-th Bernstein-Schnabl operator associated with* T or, for short, the *n-th B-S operator associated with* T. If necessary, we shall also denote it by $B_{n,T}$ in order to show more explicitly the dependence on T. Moreover, if $f \in \mathscr{C}(K)$, then $B_n(f)$ will be often referred to be the *n-th Bernstein-Schnabl function of* f.

The B-S operators B_n, $n \geq 1$, have a nice probabilistic interpretation. Actually, by a theorem of Kolmogorov (see, e.g., [49, Corollary 9.5]), for every $x \in K$ there exists a sequence $(X_{n,x})_{n\geq 1}$ of independent random variables, on a suitable probability space $(\Omega, \mathscr{F}, P)$ with values in K, such that $\tilde{\mu}_x^T$ is the distribution of each $X_{n,x}$, $n \geq 1$. Formula (3.1.3) says that the "mean value" of each $X_{n,x}$ is x. Intuitively, $(X_{n,x})_{n\geq 1}$ is associated with the experiment of choosing randomly a point in K and $\tilde{\mu}_x^T(B)$ is the probability that a point is chosen in a Borel subset B of K. This experiment is repeated countably often and independently. Then $X_{n,x}$ denotes the random point reached at the n-th trial.

Moreover, $B_n(f)(x)$ can be interpreted as

$$B_n(f)(x) = \int_\Omega f\left(\frac{1}{n}\sum_{h=1}^{n} X_{h,x}\right) dP, \tag{3.1.8}$$

i.e., $B_n(f)(x)$ represents the mean value of the real random variable $f(\frac{1}{n}\sum_{h=1}^{n} X_{h,x})$.

We also point out that an equivalent way to construct the sequence $(B_n)_{n\geq 1}$ is to start with a continuous selection $\mathscr{S} = (\tilde{\mu}_x)_{x\in K}$ of probability Borel measures on K, i.e., for every $f \in \mathscr{C}(K)$ the function $x \mapsto \int_K f\, d\tilde{\mu}_x$ is continuous on K, satisfying (3.1.3). In such a case, every B_n will be referred to as the *n-th Bernstein-Schnabl operator associated with* $\mathscr{S}$.

Coming back to the general definition (3.1.7), note that $B_n(\mathbf{1}) = \mathbf{1}$ and hence $||B_n|| = 1$. Moreover, $B_1 = T$.

Another expression of the B_n's can be obtained in terms of tensorial product of operators.

Actually, let $T^{(n)} : \mathscr{C}(K^n) \longrightarrow \mathscr{C}(K^n)$ be the Markov operator defined as the tensor product of T with itself n times, i.e., $T^{(n)} := \underbrace{T \otimes \cdots \otimes T}_{n \text{ times}}$ (see (1.1.30)). Moreover, consider the function $i_n : K \longrightarrow K^n$ defined as

$$i_n(x) := (x, \dots, x) \qquad (x \in K). \tag{3.1.9}$$

Then

$$B_n(f) = T^{(n)}(f \circ \pi_n) \circ i_n \qquad \text{for every } f \in \mathscr{C}(K). \tag{3.1.10}$$

We finally point out that, in defining Bernstein-Schnabl operators, assumption (3.1.2) is not essential. In fact, it will be needed in order to show the approximation properties of them (see Section 3.2).

In what follows we show some examples of Bernstein-Schnabl operators which include, in particular, the classical Bernstein operators on $[0, 1]$, on the d-dimensional cube and on the d-dimensional simplex.

Furthermore, we also discuss some simple constructions which generate new sequences starting from given ones.

3.1.1 Bernstein-Schnabl operators on $[0,1]$

1. **Bernstein operators on** $[0,1]$**.** Consider the Markov projection $T_1 : \mathscr{C}([0,1]) \longrightarrow \mathscr{C}([0,1])$ defined by (1.1.49).

Then, by induction, it is easy to show that the corresponding Bernstein-Schnabl operators (3.1.7) turn into the classical *Bernstein operators on* $\mathscr{C}([0,1])$ defined, for every $f \in \mathscr{C}([0,1]), x \in [0,1]$ and $n \geq 1$, as

$$B_n(f)(x) = \sum_{k=0}^{n} \binom{n}{k} x^k (1-x)^{n-k} f\left(\frac{k}{n}\right). \tag{3.1.11}$$

Note that, in this case, $\partial_{T_1}[0,1] = \{0,1\} = \partial_e[0,1]$.

Other examples are shown below.

2. Let $0 = a_0 < a_1 < \dots < a_p = 1$ be a partition of $[0,1]$ and consider the positive linear operator $T : \mathscr{C}([0,1]) \longrightarrow \mathscr{C}([0,1])$ defined by setting, for every $f \in \mathscr{C}([0,1])$ and for every $x \in [0,1]$,

$$T(f)(x) := \frac{x - a_k}{a_{k+1} - a_k} f(a_{k+1}) + \frac{a_{k+1} - x}{a_{k+1} - a_k} f(a_k), \tag{3.1.12}$$

whenever $x \in [a_k, a_{k+1}]$ $(0 \leq k \leq p-1)$.

Then T is a Markov operator which satisfies assumption (3.1.2). The corresponding Bernstein-Schnabl operators are defined, for every $n \geq 1$, $f \in \mathscr{C}([0,1])$ and $x \in [a_k, a_{k+1}]$, $0 \leq k \leq p-1$, as

$$\begin{aligned} B_n(f)(x) = & \frac{1}{(a_{k+1} - a_k)^n} \sum_{h=0}^{n} \binom{n}{h} (x - a_k)^h (a_{k+1} - x)^{n-h} \\ & \times f\left(\frac{h}{n} a_{k+1} + \frac{n-h}{n} a_k\right). \end{aligned} \tag{3.1.13}$$

Moreover, $\partial_T[0,1] = \{0, a_1, \dots, a_{p-1}, 1\}$ and $\partial_e[0,1] = \{0,1\}$.

3. Given a function $b \in \mathscr{C}([0,1])$ such that $0 \leq b(x) \leq \min\{2x, 2(1-x)\}$ for each $x \in [0,1]$, consider the positive linear operator $T : \mathscr{C}([0,1]) \longrightarrow \mathscr{C}([0,1])$ defined by setting, for every $f \in \mathscr{C}([0,1])$ and $x \in [0,1]$,

$$T(f)(x) = \left(1 - x - \frac{b(x)}{2}\right) f(0) + b(x) f\left(\frac{1}{2}\right) + \left(x - \frac{b(x)}{2}\right) f(1). \tag{3.1.14}$$

Then T is a Markov operator which satisfies (3.1.2). Moreover,

$$\partial_T[0,1] = \begin{cases} \{0,1\} & \text{if } b\left(\frac{1}{2}\right) \neq 1; \\ \\ \left\{0, \frac{1}{2}, 1\right\} & \text{if } b\left(\frac{1}{2}\right) = 1. \end{cases} \tag{3.1.15}$$

The Bernstein-Schnabl operators associated with T are given by

$$\begin{aligned} B_n(f)(x) &= \sum_{h=0}^{n}\sum_{k=0}^{n-h} \binom{n}{h}\binom{n-h}{k}\left(1-x-\frac{b(x)}{2}\right)^{n-h-k} \\ &\times b(x)^h\left(x-\frac{b(x)}{2}\right)^k f\left(\frac{h+2k}{2n}\right) \end{aligned} \tag{3.1.16}$$

($f \in \mathscr{C}([0,1])$, $x \in [0,1]$, $n \geq 1$).

Below we show some examples of Bernstein-Schnabl operators defined on not necessarily finite dimensional settings.

3.1.2 Bernstein-Schnabl operators on Bauer simplices

Given a Bauer simplex K (see Subsection 1.1.3), then there exists a natural Markov projection $T : \mathscr{C}(K) \longrightarrow \mathscr{C}(K)$ which is defined by

$$T(f)(x) := \int_K f\, d\tilde{\mu}_x \qquad (f \in \mathscr{C}(K), x \in K), \tag{3.1.17}$$

where $\tilde{\mu}_x$ is the unique probability Borel measure on K such that $\tilde{\mu}_x\left(K \setminus \overline{\partial_e K}\right) = 0$ and

$$\int_K h\, d\tilde{\mu}_x = h(x) \qquad \text{for every } h \in A(K).$$

Therefore, T satisfies (3.1.2) and $\partial_T K = \partial_e K$. Accordingly, we can consider the corresponding Bernstein-Schnabl operators which will be referred to as the *canonical Bernstein-Schnabl operators associated with the Bauer simplex K*.

As shown in Subsection 3.1.1, if $K = [0,1]$, they turn into the classical Bernstein operators. A similar result holds true for d-dimensional simplices.

More precisely, consider the canonical simplex K_d in $\mathbf{R}^d$ (see (1.1.45)) and the canonical projection T_d associated with K_d and defined by (1.1.48).

Then, again by induction, one can check that the Bernstein-Schnabl operators associated with T_d coincide with the so called *Bernstein operators on* $\mathscr{C}(K_d)$, first studied by Dinghas (see [87]) and Lorentz (see [132]) and defined, for every $f \in \mathscr{C}(K_d)$, $x = (x_1, \dots, x_d) \in K_d$ and $n \geq 1$, as

$$\begin{aligned} B_n(f)(x) &:= \sum_{\substack{h_1,\dots,h_d=0,\dots,n \\ h_1+\dots+h_d \leq n}} f\left(\frac{h_1}{n},\dots,\frac{h_d}{n}\right) \frac{n!}{h_1!\cdots h_d!\,(n-h_1-\dots-h_d)!} \\ &\times\ x_1^{h_1}\cdots x_d^{h_d}\left(1-\sum_{i=1}^{d} x_i\right)^{n-\sum\limits_{i=1}^{d} h_i}. \end{aligned} \tag{3.1.18}$$

3.1.3 Bernstein operators on polytopes

Given a polytope K of $\mathbf{R}^d$, $d \geq 2$, with vertices $x_1, \dots, x_N$, $N \geq 2$, then there are several continuous selections of probability Borel measures $\mathscr{S} = (\tilde{\mu}_x)_{x \in K}$ verifying (3.1.3) such that $\mathrm{Supp}(\tilde{\mu}_x) \subset \{x_1, \dots, x_N\}$ (or $\mathrm{Supp}(\tilde{\mu}_x) \subset \partial K$) for every $x \in K$ (see, e.g., [113] and [208] and the references therein).

Each of such continuous selections generates a corresponding sequence of Bernstein-Schnabl operators.

3.1.4 Bernstein-Schnabl operators associated with strictly elliptic differential operators

Let K be a convex compact subset of $\mathbf{R}^d$ having non-empty interior $\mathrm{int}(K)$ and consider a symmetric matrix $(a_{ij})_{\substack{1 \leq i \leq d \\ 1 \leq j \leq d}}$ of Hölder continuous functions on $\mathrm{int}(K)$ with exponent $\beta \in]0, 1[$.

Let L be the differential operator

$$L(u)(x) := \sum_{i,j=1}^{d} a_{ij}(x) \frac{\partial^2 u(x)}{\partial x_i \partial x_j} \tag{3.1.19}$$

$(u \in \mathscr{C}^2(\mathrm{int}(K)), x \in \mathrm{int}(K))$ and assume that it is *strictly elliptic*, i.e., for every $x \in \mathrm{int}(K)$ the matrix $(a_{ij}(x))_{\substack{1 \leq i \leq d \\ 1 \leq j \leq d}}$ is positive-definite and, denoted by $\sigma(x)$ its smallest eigenvalue, we have $\sigma(x) \geq \sigma_0 > 0$, for some $\sigma_0 \in \mathbf{R}$.

Denote by $T_L : \mathscr{C}(K) \longrightarrow \mathscr{C}(K)$ the *Poisson operator* associated with L. Thus, for every $f \in \mathscr{C}(K)$, $T_L(f)$ denotes the unique solution to the Dirichlet problem

$$\begin{cases} Lu = 0 & \text{on } \mathrm{int}(K), \quad u \in \mathscr{C}(K) \cap \mathscr{C}^2(\mathrm{int}(K)); \\ u = f & \text{on } \partial K. \end{cases} \tag{3.1.20}$$

T_L is a Markov projection satisfying (3.1.2) and $\partial_T K = \partial K$. The corresponding Bernstein-Schnabl operators will be referred to as the *Bernstein-Schnabl operators associated with L and the domain K*.

For instance, if K is the closed ball of center the origin of $\mathbf{R}^d$ and radius 1 and L is the *Laplace operator* Δ, i.e.,

$$\Delta u = \sum_{i=1}^{d} \frac{\partial^2 u}{\partial x_i^2} \tag{3.1.21}$$

$(u \in \mathscr{C}^2(\mathrm{int}(K)))$, then

$$T_\Delta(f)(x) = \begin{cases} \frac{1 - \|x\|_2^2}{\sigma_d} \int_{\partial K} \frac{f(z)}{\|z - x\|_2^d} \, d\sigma(z) & \text{if } \|x\|_2 < 1; \\ f(x) & \text{if } \|x\|_2 = 1 \end{cases} \tag{3.1.22}$$

$(f \in \mathscr{C}(K), x \in K)$, where σ_d and σ denote the surface area of the unit sphere in $\mathbf{R}^d$ and the surface measure on ∂K, respectively.

Therefore, for every $f \in \mathscr{C}(K)$, $x \in K$ and $n \geq 1$, we get

$$B_n(f)(x) = f(x) \qquad \text{if } x \in \partial K \tag{3.1.23}$$

and, if $x \in \operatorname{int}(K)$,

$$\begin{aligned} B_n(f)(x) &= \left(\frac{1 - \|x\|_2^2}{\sigma_d}\right)^n \\ &\times \int_{\partial K} \cdots \int_{\partial K} \frac{f\left(\frac{x_1 + \ldots + x_n}{n}\right)}{\|x_1 - x\|_2^d \cdots \|x_n - x\|_2^d} \, d\sigma(x_1) \cdots d\sigma(x_n). \end{aligned} \tag{3.1.24}$$

3.1.5 Bernstein-Schnabl operators associated with tensor products of Markov operators

Consider a finite family $(K_i)_{1 \leq i \leq d}$ of convex compact subsets and, for every $i = 1, \ldots, d$, let $T_i : \mathscr{C}(K_i) \longrightarrow \mathscr{C}(K_i)$ be a Markov operator satisfying (3.1.2). Set $K := \prod_{i=1}^{d} K_i$ and denote by T the tensor product of $(T_i)_{1 \leq i \leq d}$ (see (1.1.30)). Then T is a Markov operator. Moreover, if $h \in A(K)$ and $x = (x_i)_{1 \leq i \leq d} \in K$, on account of (1.1.31) and of the fact that h is affine with respect to each variable, we obtain

$$T(h)(x) = \int_{K_1} \cdots \int_{K_d} h(y_1, \ldots, y_d) \, d\tilde{\mu}_{x_1}^{T_1}(y_1) \cdots d\tilde{\mu}_{x_d}^{T_d}(y_d) = h(x),$$

i.e., T satisfies (3.1.2) as well. Moreover, by using (1.1.30) and (1.1.31) it is easy to show that

$$\partial_T K = \prod_{i=1}^{d} \partial_{T_i} K_i. \tag{3.1.25}$$

Finally, note that T is a Markov projection if each T_i is so.

Now consider the Bernstein-Schnabl operators B_n, $n \geq 1$, associated with T and, for every $i = 1, \ldots, d$, let $B_{n,i}$ be the Bernstein-Schnabl operators associated with each T_i.

From the commutativity as well as the associativity properties of tensor products of measures (see Proposition 1.1.6), it follows that

$$B_n\left(\bigotimes_{i=1}^{d} f_i\right) = \bigotimes_{i=1}^{d} B_{n,i}(f_i) \tag{3.1.26}$$

for every $(f_i)_{1\le i\le d} \in \prod_{i=1}^{d} \mathscr{C}(K_i)$ (see (1.1.22)) and hence, since $\bigotimes_{i=1}^{d} \mathscr{C}(K_i)$ is dense in $\prod_{i=1}^{d} \mathscr{C}(K_i)$,

$$B_n = \bigotimes_{i=1}^{d} B_{n,i}. \tag{3.1.27}$$

In particular, if $K_i := [0,1]$ for every $i = 1, \dots, d$ and all the operators T_i coincide with the operator T_1 defined by (1.1.49), setting again $Q_d := [0,1]^d$ (see (2.3.26)), then the tensor product $\bigotimes_{i=1}^{d} T_i$ is the Markov operator $S_d : \mathscr{C}(Q_d) \longrightarrow \mathscr{C}(Q_d)$ defined by

$$S_d(f)(x) := \sum_{h_1,\dots,h_d=0}^{1} f(\delta_{h_1 1}, \dots, \delta_{h_d 1}) x_1^{h_1}(1-x_1)^{1-h_1} \cdots x_d^{h_d}(1-x_d)^{1-h_d} \tag{3.1.28}$$

$(f \in \mathscr{C}(Q_d),\ x = (x_1, \dots, x_d) \in Q_d)$, where δ_{ij} stands for the Kronecker symbol. From (3.1.25) it also follows that

$$\partial_{S_d} Q_d = \{(\delta_{h_1 1}, \dots, \delta_{h_d 1}) \mid h_1, \dots, h_d \in \{0,1\}\}. \tag{3.1.29}$$

Moreover, the corresponding Bernstein-Schnabl operators turn into the Bernstein operators on $\mathscr{C}(Q_d)$ defined by

$$B_n(f)(x) := \sum_{h_1,\dots,h_d=0}^{n} \binom{n}{h_1} \cdots \binom{n}{h_d} f\left(\frac{h_1}{n}, \dots, \frac{h_d}{n}\right) \times x_1^{h_1}(1-x_1)^{n-h_1} \cdots x_d^{h_d}(1-x_d)^{n-h_d} \tag{3.1.30}$$

$(f \in \mathscr{C}(Q_d),\ x = (x_1, \dots, x_d) \in Q_d,\ n \ge 1)$.

3.1.6 Bernstein-Schnabl operators associated with convex combinations of Markov operators

Consider two Markov operators T and S on $\mathscr{C}(K)$ satisfying (3.1.2) and let $\lambda : K \setminus (\partial_T K \cap \partial_S K) \longrightarrow \mathbf{R}$ be a continuous function such that $0 \le \lambda \le 1$.

For every $f \in \mathscr{C}(K)$ and $x \in K$ set

$$U_\lambda(f)(x) := \begin{cases} \lambda(x)T(f)(x) + (1-\lambda(x))S(f)(x) & \text{if } x \notin \partial_T K \cap \partial_S K; \\ f(x) & \text{if } x \in \partial_T K \cap \partial_S K. \end{cases} \tag{3.1.31}$$

It is easily seen that $U_\lambda(f) \in \mathscr{C}(K)$ for every $f \in \mathscr{C}(K)$ and that the operator $U_\lambda : \mathscr{C}(K) \longrightarrow \mathscr{C}(K)$ is a Markov operator satisfying (3.1.2). Moreover, $\partial_T K \cap \partial_S K \subset \partial_{U_\lambda} K$.

In order to describe the Bernstein-Schnabl operators $(B_{n,U_\lambda})_{n\geq 1}$ in terms of $(B_{n,T})_{n\geq 1}$ and $(B_{n,S})_{n\geq 1}$, for every $f \in \mathscr{C}(K), \alpha \in [0,1]$ and $z \in K$, denote by $f_{z,\alpha} : K \longrightarrow \mathbf{R}$ the function defined by

$$f_{z,\alpha}(x) = f(\alpha x + (1-\alpha)z) \qquad (x \in K). \tag{3.1.32}$$

Then, for every $n \geq 1$, $f \in \mathscr{C}(K)$ and $x \in K$,

$$B_{n,U_\lambda}(f)(x) = \begin{cases} \sum\limits_{k=0}^{n} \binom{n}{k} \lambda(x)^k (1-\lambda(x))^{n-k} & \\ \times B_{k,T}(B_{n-k,S}(f_{x,k/n}))(x) & \text{if } x \notin \partial_T K \cap \partial_S K; \\ & \\ f(x) & \text{if } x \in \partial_T K \cap \partial_S K. \end{cases} \tag{3.1.33}$$

The particular case where S is the identity operator I on $\mathscr{C}(K)$ is of special interest. In this case $\partial_S K = K$ and $B_{n,S} = I$ for every $n \geq 1$. Moreover, if $\lambda(x) > 0$ for every $x \in K \setminus \partial_T K$, then

$$\partial_{U_\lambda} K = \partial_T K. \tag{3.1.34}$$

The Bernstein-Schnabl operators B_{n,U_λ}, $n \geq 1$, will be referred to as the *Lototsky-Schnabl operators associated with* T *and* $\lambda \in \mathscr{C}(K \setminus \partial_T K)$ and they will be simply denoted by $B_{n,\lambda}$, $n \geq 1$.

Actually, if $f \in \mathscr{C}(K)$, $x \in K$ and $n \geq 1$, then

$$B_{n,\lambda}(f)(x) = \begin{cases} \sum\limits_{k=0}^{n} \binom{n}{k} \lambda(x)^k (1-\lambda(x))^{n-k} B_{k,T}(f_{x,k/n})(x) & \text{if } x \notin \partial_T K; \\ & \\ f(x) & \text{if } x \in \partial_T K. \end{cases} \tag{3.1.35}$$

3.1.7 Bernstein-Schnabl operators associated with convex convolution products of Markov operators

Once again consider two Markov operators S and T on $\mathscr{C}(K)$ satisfying (3.1.2) and the relevant selections of probability Borel measures $(\tilde{\mu}_x^S)_{x\in K}$ and $(\tilde{\mu}_x^T)_{x\in K}$ associated with them according to (1.3.1).

Considering the mapping $\pi_2 : K \times K \longrightarrow K$ defined by

$$\pi_2(x_1, x_2) := \frac{x_1 + x_2}{2} \qquad ((x_1, x_2) \in K \times K), \tag{3.1.36}$$

for each $f \in \mathscr{C}(K)$ we define the following function on K by setting

$$U(f)(x) := \int_{K\times K} f \circ \pi_2 \, d\tilde{\mu}_x^S \otimes d\tilde{\mu}_x^T = \int_K \int_K f\left(\frac{x_1 + x_2}{2}\right) d\tilde{\mu}_x^S(x_1)\, d\tilde{\mu}_x^T(x_2) \tag{3.1.37}$$

$(x \in K)$. Then $U(f) \in \mathscr{C}(K)$ and the operator U is a Markov operator satisfying (3.1.2). Moreover, $\partial_S K \cap \partial_T K \subset \partial_U K$.

The operator U will be called the *convex convolution product of* S *and* T. The Bernstein-Schnabl operators associated with U are given by

$$B_{n,U}(f)(x) = B_{n,S}(B_{n,T,f,x})(x) \qquad (f \in \mathscr{C}(K),\ x \in K), \tag{3.1.38}$$

where

$$B_{n,T,f,x}(x_1) := B_{n,T}(f(\pi_2(x_1,\cdot)))(x) \qquad (x_1 \in K). \tag{3.1.39}$$

3.2 Approximation properties and rate of convergence

In this section we study some approximation properties of the sequence $(B_n)_{n\geq 1}$ of Bernstein-Schnabl operators associated with a Markov operator T satisfying (3.1.2). In particular, we prove that the sequence $(B_n)_{n\geq 1}$ is an approximation process in $\mathscr{C}(K)$ and we give some estimates of the rate of convergence by means of different moduli of smoothness.

We recall once again that K denotes a metrizable convex compact subset. We also point out that

$$B_n(f) = f \quad \text{on } \partial_T K \quad (\text{and hence on } \partial_e K) \tag{3.2.1}$$

for every $f \in \mathscr{C}(K)$.

Theorem 3.2.1. *Let $(B_n)_{n\geq 1}$ be the sequence of Bernstein-Schnabl operators associated with a Markov operator T on $\mathscr{C}(K)$ satisfying (3.1.2). Then, for every $h, k \in A(K)$ and $n \geq 1$,*

$$B_n(h) = h \tag{3.2.2}$$

and

$$B_n(hk) = \frac{1}{n}T(hk) + \frac{n-1}{n}hk. \tag{3.2.3}$$

In particular,

$$B_n(h^2) = \frac{1}{n}T(h^2) + \frac{n-1}{n}h^2. \tag{3.2.4}$$

Moreover, for every $f \in \mathscr{C}(K)$,

$$\lim_{n\to\infty} B_n(f) = f \tag{3.2.5}$$

uniformly on K.

Proof. Fix $n \geq 1$ and $h \in A(K)$. Then (3.2.2) easily follows taking (3.1.2) and (3.1.7) into account.

Moreover, for every $h, k \in A(K)$, $n \geq 1$ and $x_1, \dots, x_n \in K$,

$$h\left(\frac{x_1 + \dots + x_n}{n}\right) k\left(\frac{x_1 + \dots + x_n}{n}\right) = \frac{1}{n^2}\left(\sum_{i=1}^{n} h(x_i) \sum_{j=1}^{n} k(x_j)\right)$$
$$= \frac{1}{n^2}\left(\sum_{i=1}^{n} h(x_i)k(x_i) + \sum_{\substack{i,j\in\{1,\dots,n\}\\ i\neq j}} h(x_i)k(x_j)\right);$$

hence, for every $x \in K$, we get

$$B_n(hk)(x) = \int_K \cdots \int_K (hk)\left(\frac{x_1 + \dots + x_n}{n}\right) d\tilde{\mu}_x^T(x_1) \cdots d\tilde{\mu}_x^T(x_n)$$
$$= \frac{1}{n^2}\int_K \sum_{i=1}^{n} (hk)(x_i)\, d\tilde{\mu}_x^T + \frac{n-1}{n}(hk)(x),$$

and this completes the proof of (3.2.3) and (3.2.4).

Finally, (3.2.5) immediately follows from (3.2.2) and (3.2.4) because of Theorem 1.2.8 applied to the sequence $(B_n)_{n\geq 1}$ as well as to the subspace $M := A(K)$, which separates the points of K by the Hahn-Banach theorem. □

We proceed to give some estimates of the rate of convergence in (3.2.5) by means of the total modulus of smoothness $\Omega(f,\delta)$ defined by (1.6.10).

Proposition 3.2.2. *For every $n \geq 1$ and $f \in \mathscr{C}(K)$,*

$$\|B_n(f) - f\|_\infty \leq 2\Omega\left(f, \frac{1}{\sqrt{n}}\right). \tag{3.2.6}$$

Proof. From (1.6.18) and (1.6.19) it follows that for every $n \geq 1$, $x \in K$ and $\delta > 0$,

$$|B_n(f)(x) - f(x)| \leq (1 + \sigma_n(\delta, x))\Omega(f, \delta),$$

where

$$\sigma_n(\delta, x) := \sup\left\{\sum_{j=1}^{m}(B_n(h_j^2)(x) - h_j^2(x)) \mid m \geq 1, h_1, \dots, h_m \in L(K)\right.$$
$$\left.\text{and } \left\|\sum_{j=1}^{m} h_j^2\right\|_\infty = \frac{1}{\delta^2}\right\}.$$

On the other hand, given $\delta > 0$, $m \geq 1$ and $h_1, \dots, h_m \in L(K)$ such that $\left\|\sum_{j=1}^{m} h_j^2\right\|_\infty = \frac{1}{\delta^2}$, from (3.2.4) we get

$$\sum_{j=1}^{m}(B_n(h_j^2)(x) - h_j^2(x)) \leq \frac{1}{n}T\left(\sum_{j=1}^{m} h_j^2\right)(x) \leq \frac{1}{n}\left\|\sum_{j=1}^{m} h_j^2\right\|_\infty = \frac{1}{n\delta^2}.$$

From this latter formula, for $\delta = \sqrt{1/n}$, (3.2.6) easily follows. □

Let now K be a convex compact subset of $\mathbf{R}^d$, $d \geq 1$. Then it is possible to furnish further estimates of the rate of convergence in (3.2.5), by means of the first (see (1.4.8)) and the second modulus of smoothness (see (1.6.2)).

Actually, we preliminarily notice that, since (1.6.12) holds, from the previous result it follows that

$$\|B_n(f) - f\|_\infty \leq 2\omega\left(f, \frac{r(K)}{\sqrt{n}}\right) \tag{3.2.7}$$

$(f \in \mathscr{C}(K), n \geq 1)$, where $r(K) := \max\{\|x\|_2 \mid x \in K\}$.

Now, we show an estimate of the rate of convergence in (3.2.5) by means of $\omega_2(f, \delta)$.

Proposition 3.2.3. *For every $f \in \mathscr{C}(K)$ and $n \geq 1$,*

$$\|B_n(f) - f\|_\infty \leq \max\{2, C\} \left(\frac{r(K)^2}{n} \|f\|_\infty + \omega_2 \left(f, \frac{r(K)}{\sqrt{n}} \right) \right),$$

where C is an absolute constant, which depends on d and K, only.

Proof. We shall apply Theorem 1.6.3. Since, for every $n \geq 1$, $B_n(\mathbf{1}) = \mathbf{1}$ (and, hence, $\|B_n\| = 1$) and $B_n(pr_i) = pr_i$ for every $i = 1, \dots, d$ (see Theorem 3.2.1), then the quantity defined by (1.6.5) becomes

$$\lambda_{n,\infty} = \|B_n(e) - e\|_\infty$$

(see (1.6.4)).

Therefore, from (1.6.6) we infer that

$$\|B_n(f) - f\|_\infty \leq C\{\lambda_{n,\infty} \|f\|_\infty + \omega_2(f, \lambda_{n,\infty}^{1/2})\}.$$

On the other hand, for every $n \geq 1$, $i = 1, \dots d$ and $x \in K$, we get

$$B_n(pr_i^2)(x) - pr_i^2(x) = \frac{1}{n}(T(pr_i^2)(x) - pr_i^2(x)) \leq \frac{1}{n} T(pr_i^2)(x)$$

(see (3.2.4)), so that

$$|B_n(e)(x) - e(x)| \leq \frac{1}{n} T \left(\sum_{i=1}^{d} pr_i^2 \right)(x) \leq \frac{r(K)^2}{n},$$

and this completes the proof. □

Below we give another simple (pointwise) estimate, again in finite dimensional settings but for Lipschitz continuous functions.

Proposition 3.2.4. *Let K be a convex compact subset of $\mathbf{R}^d$, $d \geq 1$. Consider $f \in \mathscr{C}(K)$ such that $|f(x) - f(y)| \leq M\|x - y\|_2$ $(x, y \in K)$ for some $M \geq 0$.*

Then, for each $n \geq 1$ and $x \in K$,

$$|B_n(f)(x) - f(x)| \leq M \sqrt{\frac{1}{n}(T(e)(x) - e(x))}, \tag{3.2.8}$$

where $e(x)$ is defined by (1.6.4).

Proof. For a given $x \in K$, introducing the function $\Phi_x(y) := \|x - y\|_2$ $(y \in K)$, we obtain $|f - f(x)| \leq M\Phi_x$.

Therefore, taking the Cauchy-Schwarz inequality (1.1.6) into account, we infer that, for every $n \geq 1$,

$$|B_n(f) - f(x)| \leq M B_n(\Phi_x) \leq M \sqrt{B_n(\Phi_x^2)}.$$

On the other hand, $\Phi_x^2 = \sum_{i=1}^{d} \left(pr_i^2 - 2pr_i(x)pr_i + pr_i^2(x)\right)$ and hence, by applying (3.2.2) and (3.2.4), we finally obtain

$$|B_n(f)(x) - f(x)| \leq M\sqrt{B_n(\Phi_x^2)(x)} = M\sqrt{\frac{T(e)(x) - e(x)}{n}}. \qquad \square$$

In some particular cases it is also possible to describe the asymptotic behaviour of the iterates of Bernstein-Schnabl operators.

Consider a Markov operator $T : \mathscr{C}(K) \longrightarrow \mathscr{C}(K)$ satisfying (3.1.2). Hence

$$A(K) \subset H := T(\mathscr{C}(K)). \tag{3.2.9}$$

Furthermore, assume that

$$h_{z,\alpha} \in H \text{ for every } z \in K, \alpha \in [0,1], h \in H, \tag{3.2.10}$$

where $h_{z,\alpha}$ is defined by (3.1.32).

For instance, all the Markov operators considered in Subsections 3.1.1, 1, 3.1.2, 3.1.4, 3.1.5, (3.1.28) satisfy (3.2.10).

Theorem 3.2.5. *Given a Markov operator T on $\mathscr{C}(K)$ satisfying* (3.2.10), *consider a further Markov operator $S : \mathscr{C}(K) \longrightarrow \mathscr{C}(K)$ such that*
(i) $S \circ T = T$, i.e., $S(h) = h$ for every $h \in H$;
(ii) $\partial_S K \subset \partial_T K$, i.e., for every $x \in K \setminus \partial_T K$ there exists $f \in \mathscr{C}(K)$ such that $S(f)(x) \neq f(x)$.
If $(B_{n,S})_{n\geq 1}$ denotes the sequence of Bernstein-Schnabl operators associated with S, then, for every $n \geq 1$ and $f \in \mathscr{C}(K)$,

$$\lim_{m\to\infty} B_{n,S}^m(f) = T(f) \qquad \text{uniformly on } K.$$

Proof. Given $n \geq 1$, by (3.2.10) and assumption (i) it is easy to check that $B_{n,S}(h) = h$ for every $h \in H$ and hence $B_{n,S} \circ T = T$.

Let $(h_n)_{n\geq 1}$ be a sequence in $A(K)$ separating the points of K and such that the series $\Phi := \sum_{i=1}^{\infty} h_n^2$ is uniformly convergent on K (see Remark 1.2.9).

By Theorem 1.3.1, $\partial_S K = \{x \in K \mid S(\Phi)(x) = \Phi(x)\}$. Furthermore, $\Phi \leq B_{n,S}(\Phi)$ and, on account of (3.2.4), $B_{n,S}(\Phi) = \frac{1}{n}S(\Phi) + \frac{n-1}{n}\Phi$. Therefore, if for some $x \in K$ we assume that $B_{n,S}(\Phi)(x) = \Phi(x)$, then $S(\Phi)(x) = \Phi(x)$ and hence $x \in \partial_S K \subset \partial_T K$. The result now follows by applying Theorem 1.3.9 to the Markov operator $B_{n,S}$. $\square$

Below we mention some particular consequences of the above Theorem 3.2.5.

Corollary 3.2.6. *Let K be a metrizable Bauer simplex and consider its canonical Markov projection T on $\mathscr{C}(K)$ (see Subsection 3.1.3).*

Consider a Markov operator S on $\mathscr{C}(K)$ satisfying (3.1.2) *and such that $\partial_S K \subset \partial_e K$, and denote by $B_{n,S}$, $n \geq 1$, the corresponding Bernstein-Schnabl operators. Then, for every $n \geq 1$ and $f \in \mathscr{C}(K)$,*

$$\lim_{m\to\infty} B^m_{n,S}(f) = T(f) \quad \textit{uniformly on } K$$

and, in particular,

$$\lim_{m\to\infty} S^m(f) = T(f) \quad \textit{uniformly on } K.$$

Example 3.2.7. *Consider the Bernstein-Schnabl operators $(B_n)_{n\geq 1}$ on $\mathscr{C}([0,1])$ associated with a function $b \in \mathscr{C}([0,1])$ satisfying $b\left(\frac{1}{2}\right) \neq 1$ and defined by* (3.1.16). *Then, for every $f \in \mathscr{C}([0,1])$ and $n \geq 1$,*

$$\lim_{m\to\infty} B^m_n(f)(x) = T_1(f)(x) = xf(1) + (1-x)f(0)$$

uniformly with respect to $x \in [0,1]$.

Corollary 3.2.8. *Consider the Poisson operator $T_L : \mathscr{C}(K) \longrightarrow \mathscr{C}(K)$ as in Subsection 3.1.4 and let S be a Markov operator on $\mathscr{C}(K)$ satisfying* (3.1.2) *such that $\partial_S K \subset \partial K$ and $S(h) = h$ for every $h \in T_L(\mathscr{C}(K)) := \{u \in \mathscr{C}(K) \mid u \in \mathscr{C}^2(\mathrm{int}(K))$ and $L(u) = 0\}$ Then, for every $n \geq 1$ and $f \in \mathscr{C}(K)$,*

$$\lim_{m\to\infty} B^m_{n,S}(f) = T_L(f) \quad \textit{uniformly on } K$$

and, in particular,

$$\lim_{m\to\infty} S^m(f) = T_L(f) \quad \textit{uniformly on } K.$$

Corollary 3.2.9. *Let T be a Markov projection on K satisfying* (3.1.2) *and* (3.2.10). *Consider the sequence $(B_{n,\lambda})_{n\geq 1}$ of Lototsky-Schnabl operators associated with T and a given function $\lambda \in \mathscr{C}(K \setminus \partial_T K)$ as in* (3.1.35) *and assume that $\lambda(x) > 0$ for every $x \in K \setminus \partial_T K$. Then, for every $n \geq 1$ and $f \in \mathscr{C}(K)$,*

$$\lim_{m\to\infty} B^m_{n,\lambda}(f) = T(f) \quad \textit{uniformly on } K.$$

In particular, for $\lambda = \mathbf{1}$,

$$\lim_{m\to\infty} B^m_{n,T}(f) = T(f) \quad \textit{uniformly on } K.$$

Proof. The operators $B_{n,\lambda}$ are generated by the Markov operator U_λ defined by

$$U_\lambda(f)(x) := \begin{cases} \lambda(x)T(f)(x) + (1-\lambda(x))f(x) & \text{if } x \notin \partial_T K; \\ \\ f(x) & \text{if } x \in \partial_T K \end{cases}$$

($f \in \mathscr{C}(K), x \in K$). Since T is a projection, then $U_\lambda \circ T = T$ and, by (3.1.34), $\partial_{U_\lambda} K = \partial_T K$. Therefore, the result follows from Theorem 3.2.5. □

When the function λ is constant and, as above, T is a Markov projection satisfying (3.1.2) and (3.2.10), then the asymptotic behaviour of other kinds of operators $B_{n,\lambda}$ and, in particular, of Bernstein-Schnabl operators $B_{n,T}$ can be easily described.

Actually, for every $n, m \geq 1$ and $h \in H$,

$$B_{n,\lambda}^{m}(h) = h \tag{3.2.11}$$

and from (3.2.4) it follows that for every $h \in A(K)$,

$$B_{n,\lambda}^{m}(h^2) = \left(1 - \left(1 - \frac{\lambda}{n}\right)^m\right) T(h^2) + \left(1 - \frac{\lambda}{n}\right)^m h^2. \tag{3.2.12}$$

Therefore, applying Theorem 1.2.8 and Corollary 1.3.6, we get the following result.

Theorem 3.2.10. *Under the above hypotheses on T, given $\lambda \in \mathbf{R}$, $0 < \lambda \leq 1$ and $f \in \mathscr{C}(K)$, then*
(1) $\lim_{n\to\infty} B_{n,\lambda}^{m}(f) = f$ uniformly on K for every $m \geq 1$.
(2) $\lim_{\lambda\to 0^+} B_{n,\lambda}^{m}(f) = f$ uniformly on K for every $n, m \geq 1$.
(3) For every sequence $(k(n))_{n\geq 1}$ of positive integers,

$$\lim_{n\to\infty} B_{n,\lambda}^{k(n)}(f) = \begin{cases} f & \text{uniformly on } K \text{ if } \lim_{n\to\infty} \frac{k(n)}{n} = 0; \\[2ex] T(f) & \text{uniformly on } K \text{ if } \lim_{n\to\infty} \frac{k(n)}{n} = +\infty. \end{cases}$$

Similar results hold true for the Bernstein-Schnabl operators associated with T as well (i.e., for $\lambda = 1$).

3.3 Preservation of Hölder continuity

Both from a theoretical and a constructive point of view, a significant problem for approximation processes is to ascertain whether they preserve some regularity properties of the functions they approximate.

In this section, as well as in the next ones, we investigate such preservation properties for Bernstein-Schnabl operators associated with Markov operators.

We begin with discussing their behaviour on Hölder continuous functions.

Let us denote again by K a convex compact subset of a metrizable locally convex space X. Denoting by ρ_0 the metric on X which induces its topology, we shall assume that (1.6.23) holds true.

Note that from (1.6.23) it also follows that

$$\rho_0(kx, ky) \le k\rho_0(x, y) \tag{3.3.1}$$

for every $k \in [0, 1]$ and $x, y \in X$.

Denoting by ρ the restriction of ρ_0 to K, consider the modulus of continuity associated with it and defined by (1.6.20). Then

$$\omega_\rho(f, t\delta) \le (1+t)\omega_\rho(f, \delta)$$

for every $f \in \mathscr{C}(K)$, $\delta, t > 0$ (see [155]).

Given $M \ge 0$ and $0 < \alpha \le 1$, according to (1.6.21), we set

$$\mathrm{Lip}(M, \alpha) := \{f \in \mathscr{C}(K) \mid |f(x) - f(y)| \le M\rho(x, y)^\alpha \text{ for every } x, y \in K\}.$$

We are interested in investigating the behaviour of Bernstein-Schnabl operators on the functions lying in $\mathrm{Lip}(M, \alpha)$.

We consider a Markov operator T on $\mathscr{C}(K)$ (not necessarily satisfying (3.1.2)) and we assume that there exists $c \ge 1$ such that

$$T(\mathrm{Lip}(1, 1)) \subset \mathrm{Lip}(c, 1), \tag{3.3.2}$$

or, equivalently,

$$T(\mathrm{Lip}(M, 1)) \subset \mathrm{Lip}(cM, 1) \tag{3.3.3}$$

for every $M \ge 0$.

At the end of this section we shall discuss more closely some examples where (3.3.2) is satisfied. We also mention that the case $c = 1$ in (3.3.2) is particularly important because of its consequences in the study of Hölder continuity of the solutions to some initial-boundary value evolution problems which will be treated in the subsequent chapters.

We proceed further by fixing $n \ge 2$. For each $f \in \mathscr{C}(K)$ and $x_1, \dots, x_{n-1} \in K$, consider the function $f_{x_1,\dots,x_{n-1}} : K \longrightarrow \mathbf{R}$ defined by

$$f_{x_1,\dots,x_{n-1}}(t) := f\left(\frac{x_1 + \dots + x_{n-1} + t}{n}\right) \qquad (t \in K). \tag{3.3.4}$$

Furthermore, for every $x \in K$ consider the function

$$f^x_{x_1,\dots,x_{n-2}}(t) := T\left(f_{x_1,\dots,x_{n-2},t}\right)(x) \qquad (t \in K) \tag{3.3.5}$$

and, for every $k = 3, \dots, n-1$, define recursively the functions $f^x_{x_1,\dots,x_{n-k}} : K \longrightarrow \mathbf{R}$ by setting

$$f^x_{x_1,\dots,x_{n-k}}(t) := T\left(f^x_{x_1,\dots,x_{n-k},t}\right)(x) \qquad (t \in K). \tag{3.3.6}$$

Finally, set

$$f^x(t) = T\left(f^x_t\right)(x) \qquad (t \in K). \tag{3.3.7}$$

The next result stresses the behaviour of the sequence $(B_n)_{n\geq 1}$ of Bernstein-Schnabl operators on Lipschitz continuous functions.

Theorem 3.3.1. *Consider the sequence $(B_n)_{n\geq 1}$ of Bernstein-Schnabl operators associated with a Markov operator T. Under assumption (3.3.2), for any $n \geq 1$ and $f \in \mathrm{Lip}(M, 1)$,*

$$B_n(f) \in \mathrm{Lip}(cM, 1). \tag{3.3.8}$$

Proof. For $n = 1$, we have that $B_1 = T$, so that, by means of (3.3.3), we get (3.3.8).

Fix $n \geq 2$, $f \in \mathrm{Lip}(M, 1)$, $x \in K$ and consider the functions defined by (3.3.4)-(3.3.7). By finite induction it is easy to prove that, by means of (3.3.1), $f_{x_1,\dots,x_{n-1}} \in \mathrm{Lip}(M/n, 1)$ and that

$$\|f^x_{x_1,\dots,x_{k-1},u} - f^x_{x_1,\dots,x_{k-1},v}\|_\infty \leq \frac{M}{n}\rho(u, v),$$

for any $u, v \in K$ and $k = 2, \dots, n-2$; hence, $f^x_{x_1,\dots,x_k} \in \mathrm{Lip}(M/n, 1)$ and $f^x \in \mathrm{Lip}(M/n, 1)$.

Moreover, for every $y \in K$ and $k = 1, \dots, n-1$, from (3.3.3) it follows that

$$T\left(f^y_{x_1,\dots,x_k}\right)(x) \leq T\left(f^y_{x_1,\dots,x_k}\right)(y) + \frac{cM}{n}\rho(x, y),$$

as well as

$$T(f^y)(x) \leq T(f^y)(y) + \frac{cM}{n}\rho(x, y)$$

and

$$T(f_{x_1,\dots,x_{n-1}})(x) \leq T(f_{x_1,\dots,x_{n-1}})(y) + \frac{cM}{n}\rho(x, y).$$

Therefore,

$$\begin{aligned} B_n(f)(x) &= \int_K \cdots \int_K f\left(\frac{x_1 + \dots + x_n}{n}\right) d\tilde{\mu}^T_x(x_1) \cdots d\tilde{\mu}^T_x(x_n) \\ &= \int_K \cdots \int_K T\left(f_{x_1,\dots,x_{n-1}}\right)(x)\, d\tilde{\mu}^T_x(x_1) \cdots d\tilde{\mu}^T_x(x_{n-1}) \end{aligned}$$

$$
\begin{aligned}
&\le \int_K \cdots \int_K T\left(f_{x_1,\dots,x_{n-1}}\right)(y)\, d\tilde{\mu}_x^T(x_1)\cdots d\tilde{\mu}_x^T(x_{n-1}) + \frac{cM}{n}\rho(x,y)\\
&= \int_K \cdots \int_K f^y_{x_1,\dots,x_{n-2}}(x_{n-1})\, d\tilde{\mu}_x^T(x_1)\cdots d\tilde{\mu}_x^T(x_{n-1}) + \frac{cM}{n}\rho(x,y)\\
&= \int_K \cdots \int_K T\left(f^y_{x_1,\dots,x_{n-2}}\right)(x)\, d\tilde{\mu}_x^T(x_1)\cdots d\tilde{\mu}_x^T(x_{n-2}) + \frac{cM}{n}\rho(x,y)\\
&\le \int_K \cdots \int_K T\left(f^y_{x_1,\dots,x_{n-2}}\right)(y)\, d\tilde{\mu}_x^T(x_1)\cdots d\tilde{\mu}_x^T(x_{n-2}) + \frac{2cM}{n}\rho(x,y)\\
&= \int_K \cdots \int_K f^y_{x_1,\dots,x_{n-3}}(x_{n-2})\, d\tilde{\mu}_x^T(x_1)\cdots d\tilde{\mu}_x^T(x_{n-2}) + \frac{2cM}{n}\rho(x,y)\\
&= \int_K \cdots \int_K T\left(f^y_{x_1,\dots,x_{n-3}}\right)(x)\, d\tilde{\mu}_x^T(x_1)\cdots d\tilde{\mu}_x^T(x_{n-3}) + \frac{2cM}{n}\rho(x,y)\\
&\le \dots \le \int_K T\left(f^y_{x_1}\right)(y)\, d\tilde{\mu}_x^T(x_1) + \frac{(n-1)cM}{n}\rho(x,y)\\
&= \int_K f^y(x_1)\, d\tilde{\mu}_x^T(x_1) + \frac{(n-1)cM}{n}\rho(x,y)\\
&= T(f^y)(x) + \frac{(n-1)cM}{n}\rho(x,y) \le T(f^y)(y) + cM\rho(x,y).
\end{aligned}
$$

Moreover,

$$
\begin{aligned}
B_n(f)(y) &= \int_K \cdots \int_K f\left(\frac{x_1+\dots+x_n}{n}\right) d\tilde{\mu}_y^T(x_1)\cdots d\tilde{\mu}_y^T(x_n)\\
&= \int_K \cdots \int_K T\left(f_{x_1,\dots,x_{n-1}}\right)(y)\, d\tilde{\mu}_y^T(x_1)\cdots d\tilde{\mu}_y^T(x_{n-1})\\
&= \dots = \int_K T(f^y_{x_1})(y)\, d\tilde{\mu}_y^T(x_1) = T(f^y)(y).
\end{aligned}
$$

Accordingly,

$$
|B_n(f)(x) - B_n(f)(y)| \le cM\rho(x,y)
$$

and this completes the proof. □

The previous result, together with Theorem 1.6.6, leads to the following corollary.

Corollary 3.3.2. *If $T(\mathrm{Lip}(1,1)) \subset \mathrm{Lip}(c,1)$ for some $c \ge 1$, then, for every $f \in \mathscr{C}(K)$, $\delta > 0$ and $n \ge 1$,*

$$
\omega_\rho(B_n(f),\delta) \le (1+c)\omega_\rho(f,\delta) \tag{3.3.9}
$$

and

$$
B_n(\mathrm{Lip}(M,\alpha)) \subset \mathrm{Lip}(c^\alpha M,\alpha) \tag{3.3.10}
$$

for every $M > 0$ and $\alpha \in]0,1]$.

In particular, if $c = 1$, then

$$
\omega_\rho(B_n(f),\delta) \le 2\omega_\rho(f,\delta) \tag{3.3.11}
$$

and

$$B_n(\mathrm{Lip}(M,\alpha)) \subset \mathrm{Lip}(M,\alpha) \tag{3.3.12}$$

for every $M > 0$ *and* $\alpha \in]0,1]$.

Before discussing some examples concerning assumption (3.3.2), it is not superfluous to point out that, in the case where K is a convex compact subset of a normed space X, if T satisfies (3.1.2) and if $T(\mathrm{Lip}(1,1)) \subset \mathrm{Lip}(c,1)$ for some $c > 0$, then necessarily $c \geq 1$.

Indeed, assuming on the contrary that $c < 1$, consider a non-constant (on K) linear continuous functional ϕ on X and denote by h its restriction to K. Then $h \in \mathrm{Lip}(\|\varphi\|_\infty, 1)$ and hence $h = T(h) \in \mathrm{Lip}(c\|\varphi\|_\infty, 1)$.

By induction, we infer that $h \in \mathrm{Lip}(c^n\|\varphi\|_\infty, 1)$ for every $n \geq 1$. Since $c^n \to 0$ as $n \to \infty$, this clearly forces h to be constant on K, a contradiction.

As regards (3.3.2), actually if we consider the Poisson operator T_Δ associated with the Laplace operator Δ (see (3.1.22)), then it is false for every $c > 0$ (see, for instance [117, Example (5.11), Section 5], [177, p. 34], [82, p. 519]).

On the contrary, considering the canonical simplex K_d, $d \geq 1$, (see (1.1.45)), endowed with the l_p-metric, $1 \leq p \leq +\infty$, i.e., the metric generated by the l_p-norms, and the natural projection T_d on $\mathscr{C}(K_d)$ (see (1.1.48)), then

$$T_d(\mathrm{Lip}(1,1)) \subset \mathrm{Lip}(d^{1-1/p},1) \tag{3.3.13}$$

(see [18, p. 453]). Thus, if $p = 1$, then

$$T_d(\mathrm{Lip}(1,1)) \subset \mathrm{Lip}(1,1). \tag{3.3.14}$$

The last inclusion holds true in particular for $d = 1$, i.e., for $K_d = [0,1]$ and for the Markov projection T_1 defined by (1.1.49). Therefore Theorem 3.3.1 applies to Bernstein operators on $\mathscr{C}([0,1])$ as well.

It is not difficult to show that the operator T on $\mathscr{C}([0,1])$ defined by (3.1.12) satisfies (3.3.2) with $c = 1$ as well as the operator (3.1.14) satisfies

$$T(\mathrm{Lip}(1,1)) \subset \mathrm{Lip}(c,1) \tag{3.3.15}$$

with $c = \max\{1, M\}$, provided $b \in \mathrm{Lip}(2M,1)$, $M \geq 0$.

A simple calculation shows that the operator S_d defined by (3.1.28) satisfies (3.3.2) with $c = 1$, provided that the hypercube Q_d is endowed with the l_1-metric.

More generally, for each $i = 1, \dots, d$, $d \geq 2$, let K_i be a convex compact subset of a normed space $(E_i, \|\cdot\|_i)$ and consider a Markov operator T_i on $\mathscr{C}(K_i)$ such that

$$T_i(\mathrm{Lip}(1,1)) \subset \mathrm{Lip}(c_i,1) \tag{3.3.16}$$

for some $c_i \geq 1$.

Considering the product space $K := \prod_{i=1}^{d} K_i$, endowed with the l_1-metric, along with the tensor product T on $\mathscr{C}(K)$ of $(T_i)_{1\le i\le d}$ (see (1.1.30)), then

$$T(\mathrm{Lip}(1,1)) \subset \mathrm{Lip}(c,1), \tag{3.3.17}$$

where $c = c_1 \cdot c_2 \cdots c_d$.

In the setting of Subsection 3.1.6, if λ is a constant function, then U_λ verifies (3.3.2) with

$$c = \max\{c_S, c_T\}, \tag{3.3.18}$$

provided that S and T satisfy the analogous inclusion with constants c_S and c_T, respectively.

Under the same assumption on S and T, considering the convex convolution product U of S and T (see (3.1.37)), then

$$U(\mathrm{Lip}(1,1)) \subset \mathrm{Lip}\left(\frac{c_S + c_T}{2}, 1\right). \tag{3.3.19}$$

We sketch below a short proof. To simplify notation we shall write

$$U(f)(x) := S\left(T\left(f\left(\frac{s+t}{2}\right)\right)(x)\right)(x)$$

($f \in \mathscr{C}(K), x \in K$), with the convention that S acts on functions of variable $s \in K$ and T on functions of variable $t \in K$.

Let $f \in \mathrm{Lip}(1,1)$. For $a, b \in K$, we have

$$\begin{aligned}&\left\|f\left(\frac{a+\cdot}{2}\right) - f\left(\frac{b+\cdot}{2}\right)\right\|_\infty = \max_{u\in K}\left|f\left(\frac{a+u}{2}\right) - f\left(\frac{b+u}{2}\right)\right|\\ &\le \max_{u\in K}\rho\left(\frac{a+u}{2}, \frac{b+u}{2}\right) \le \frac{1}{2}\max_{u\in K}\rho(a+u, b+u) = \frac{1}{2}\rho(a,b),\end{aligned}$$

where ρ is the metric on K (see (1.6.23) and (3.3.1)).

Thus

$$\left\|f\left(\frac{a+\cdot}{2}\right) - f\left(\frac{b+\cdot}{2}\right)\right\|_\infty \le \frac{1}{2}\rho(a,b). \tag{3.3.20}$$

Let $y \in K$. Then, for every $a, b \in K$, from (3.3.20) we get

$$\begin{aligned}&\left|T\left(f\left(\frac{a+t}{2}\right)\right)(y) - T\left(f\left(\frac{b+t}{2}\right)\right)(y)\right|\\ &\le \left\|T\left(f\left(\frac{a+t}{2}\right)\right) - T\left(f\left(\frac{b+t}{2}\right)\right)\right\|_\infty\\ &\le \left\|f\left(\frac{a+\cdot}{2}\right) - f\left(\frac{b+\cdot}{2}\right)\right\|_\infty \le \frac{1}{2}\rho(a,b).\end{aligned}$$

This yields

$$T\left(f\left(\frac{\cdot+t}{2}\right)\right)(y) \in \mathrm{Lip}\left(\frac{1}{2}, 1\right) \qquad (y \in K),$$

which implies

$$S\left(T\left(f\left(\frac{s+t}{2}\right)\right)(y)\right) \in \mathrm{Lip}\left(\frac{1}{2}c_S, 1\right),$$

and so,

$$\left|S\left(T\left(f\left(\frac{s+t}{2}\right)\right)(y)\right)(y) - S\left(T\left(f\left(\frac{s+t}{2}\right)\right)(y)\right)(x)\right| \le \frac{1}{2}c_S\rho(x,y). \tag{3.3.21}$$

On the other hand, for all $u \in K$, $f\left(\frac{u+\cdot}{2}\right) \in \mathrm{Lip}\left(\frac{1}{2}, 1\right)$ and hence $T\left(f\left(\frac{u+t}{2}\right)\right) \in \mathrm{Lip}\left(\frac{1}{2}c_T, 1\right)$ $(u \in K)$.

Therefore, for every $x, y, u \in K$,

$$\left|T\left(f\left(\frac{u+t}{2}\right)\right)(y) - T\left(f\left(\frac{u+t}{2}\right)\right)(x)\right| \le \frac{1}{2}c_T\rho(x,y). \tag{3.3.22}$$

Now we have, by using (3.3.22),

$$\begin{aligned}
&\left|S\left(T\left(f\left(\frac{s+t}{2}\right)\right)(y)\right)(x) - S\left(T\left(f\left(\frac{s+t}{2}\right)\right)(x)\right)(x)\right| \\
&\le \left\|S\left(T\left(f\left(\frac{s+t}{2}\right)\right)(y)\right) - S\left(T\left(f\left(\frac{s+t}{2}\right)\right)(x)\right)\right\|_\infty \\
&\le \left\|T\left(f\left(\frac{\cdot+t}{2}\right)\right)(y) - T\left(f\left(\frac{\cdot+t}{2}\right)\right)(x)\right\|_\infty \\
&= \max_{u\in K}\left|T\left(f\left(\frac{u+t}{2}\right)\right)(y) - T\left(f\left(\frac{u+t}{2}\right)\right)(x)\right| \le \frac{1}{2}c_T\rho(x,y),
\end{aligned}$$

i.e., for every $x, y \in K$,

$$\begin{aligned}
&\left|S\left(T\left(f\left(\frac{s+t}{2}\right)\right)(y)\right)(x) - S\left(T\left(f\left(\frac{s+t}{2}\right)\right)(x)\right)(x)\right| \\
&\le \frac{1}{2}c_T\rho(x,y).
\end{aligned} \tag{3.3.23}$$

Combining (3.3.21) and (3.3.23), we deduce

$$|U(f)(x) - U(f)(y)| \le \frac{1}{2}(c_S + c_T)\rho(x,y) \qquad (x, y \in K),$$

and this completes the proof.

We end this section by pointing out another problem related to inclusion (3.3.2) which seems to be not devoid of interest.

Assume that K is a compact convex subset of a real normed space $(E, \|\cdot\|)$ and consider a Markov operator T on $\mathscr{C}(K)$ satisfying (3.3.2). The problem consists in determining the quantity

$$c(\|\cdot\|, T) := \inf\{c \in \mathbf{R}^+ \mid T(\mathrm{Lip}(1,1)) \subset \mathrm{Lip}(c,1)\}. \tag{3.3.24}$$

$c(\|\cdot\|, T)$ is called the *smallest Lipschitz constant* and, clearly, it is the smallest real constant satisfying (3.3.2); as observed before, $c(\|\cdot\|, T) \ge 1$.

This problem was first presented in [174], where the reader may find some solutions in particular settings. Further results in this respect would be desirable. Here we state some contributions in the simple setting of $\mathbf{R}^2$ endowed with an arbitrary norm $\|\cdot\|$.

If $u, v \in \mathbf{R}^2$ are linearly independent vectors, we set

$$K(\|\cdot\|, u, v) := \max\left\{\frac{(1-t)\|u\| + t\|v\|}{\|(1-t)u + tv\|} \mid 0 \le t \le 1\right\} \tag{3.3.25}$$

and

$$Q(\|\cdot\|, u, v) := \max\left\{\left\|s\frac{u}{\|u\|} + t\frac{v}{\|v\|}\right\|^{-1} \mid s, t \in \mathbf{R}, |s| + |t| = 1\right\}. \tag{3.3.26}$$

3.3.1 Smallest Lipschitz constants and triangles

In this subsection K denotes a triangle with vertices $u, v, w \in \mathbf{R}^2$ and angles α, β, γ, $\alpha + \beta + \gamma = \pi$. Let T be the canonical projection associated with K as in Subsection 1.1.3.

Theorem 3.3.3. *The following statements hold true:*
(i) For an arbitrary norm $\|\cdot\|$ on $\mathbf{R}^2$,

$$\begin{aligned} c(\|\cdot\|, T) = \max\{&K(\|\cdot\|, v-u, w-u),\\ &K(\|\cdot\|, u-v, w-v), K(\|\cdot\|, u-w, v-w)\}. \end{aligned} \tag{3.3.27}$$

(ii) With respect to the Euclidean norm $\|\cdot\|_2$ on $\mathbf{R}^2$,

$$c(\|\cdot\|_2, T) = \max\left\{\left(\cos\frac{\alpha}{2}\right)^{-1}, \left(\cos\frac{\beta}{2}\right)^{-1}, \left(\cos\frac{\gamma}{2}\right)^{-1}\right\}. \tag{3.3.28}$$

Proof. (i). Set

$$M(u, v, w) := \max\{K(\|\cdot\|, v-u, w-u), K(\|\cdot\|, u-v, w-v), K(\|\cdot\|, u-w, v-w)\}.$$

Let $c \ge 1$ such that $T(\mathrm{Lip}(1,1)) \subset \mathrm{Lip}(c,1)$. Take $f \in \mathscr{C}(K)$, $f(x) := \|x - u\|$, $x \in K$. Obviously, $f \in \mathrm{Lip}(1,1)$, so that $T(f) \in \mathrm{Lip}(c,1)$. In particular,

$$|T(f)(x) - T(f)(u)| \le c\|x - u\| \quad (x \in K). \tag{1}$$

Fix $0 \le t \le 1$ and set $x := (1-t)v + tw$. Since $T(f)$ is affine, we have $T(f)(x) = (1-t)T(f)(v) + tT(f)(w)$. Moreover, $T(f)(u) = f(u) = 0$, $T(f)(v) = f(v) = \|u - v\|$ and $T(f)(w) = f(w) = \|w - u\|$, so that (1) becomes

$$(1-t)\|v-u\| + t\|w-u\| \le c\|(1-t)(v-u) + t(w-u)\|.$$

It follows that $K(\|\cdot\|, v-u, w-u) \le c$ and hence $K(\|\cdot\|, v-u, w-u) \le c(\|\cdot\|, T)$. Reasoning in the same way, we infer that

$$c(\|\cdot\|, T) \ge M(u, v, w). \tag{2}$$

To prove the reverse inequality, fix $x, y \in K$, $x \neq y$ and set $x = x_1 u + x_2 v + x_3 w$ and $y = y_1 u + y_2 v + y_3 w$, where for every $i = 1, 2, 3$, $x_i, y_i \ge 0$ and $\sum_{i=1}^{3} x_i = \sum_{i=1}^{3} y_i = 1$; suppose that $x_2 - y_2 \ge 0$ and $x_3 - y_3 \ge 0$, and hence $y_1 - x_1 > 0$ (the other cases may be treated similarly).

Fix $f \in \mathrm{Lip}(1, 1)$. Then

$$\begin{aligned}|T(f)(x) - T(f)(y)| &= |(x_2 - y_2)(f(v) - f(u)) + (x_3 - y_3)(f(w) - f(u))| \\ \le (x_2 - y_2)\|v - u\| + (x_3 - y_3)\|w - u\| &= \frac{(1-t)\|v-u\| + t\|w-u\|}{\|(1-t)(v-u) + t(w-u)\|}\|x - y\|,\end{aligned}$$

where $t := \dfrac{x_3 - y_3}{y_1 - x_1}$.

It follows that $c(\|\cdot\|, T) \le K(\|\cdot\|, v-u, w-u)$, and hence the reverse inequality of (2) is established. This completes the proof of statement (i).

(ii). Set $a := \|v-u\|_2$, $b := \|w-u\|_2$ and let α be the angle of the vectors $v-u$ and $w-u$.

Fix $0 \le t \le 1$; then

$$\begin{aligned}\left(\frac{\|(1-t)(v-u) + t(w-u)\|_2}{(1-t)\|v-u\|_2 + t\|w-u\|_2}\right)^2 &= \frac{a^2(1-t)^2 + b^2t^2 + 2abt(1-t)\cos\alpha}{a^2(1-t)^2 + b^2t^2 + 2abt(1-t)} \\ &= 1 - \frac{4abt(1-t)}{a^2(1-t)^2 + b^2t^2 + 2abt(1-t)}\sin^2\frac{\alpha}{2}.\end{aligned}$$

The minimal value of the last expression, with respect to $t \in [0, 1]$, is equal to 1 (and it is achieved for $t = a/(a+b)$). It follows that $K(\|\cdot\|, v-u, w-u)^{-2} = 1 - \sin^2\dfrac{\alpha}{2} = \cos^2\dfrac{\alpha}{2}$; consequently, $K(\|\cdot\|, v-u, w-u) = \left(\cos\frac{\alpha}{2}\right)^{-1}$ and (3.3.28) is a consequence of (3.3.27). □

3.3.2 Smallest Lipschitz constants and parallelograms

Let $u, v \in \mathbf{R}^2$ be linearly independent. In this subsection P will be the parallelogram

$$P := \{au + bv \mid a, b \in [0, 1]\}.$$

For $x = au + bv \in P$, where $a, b \in [0, 1]$, and $f \in \mathscr{C}(P)$, set

$$T(f)(x) := (1-a)(1-b)f(\mathbf{0}) + a(1-b)f(u) + b(1-a)f(v) + abf(u+v), \tag{3.3.29}$$

where $\mathbf{0}$ denotes the null vector.

Then T is a Markov operator on $\mathscr{C}(P)$ and it satisfies assumptions (3.1.2) and (3.3.2).

Let $\theta \in (0, \pi)$ be the angle of the vectors u and v. Then the following result holds.

Theorem 3.3.4. *The following statements hold true:*
(i) For an arbitrary norm $\|\cdot\|$ on $\mathbf{R}^2$,

$$c(\|\cdot\|, T) \le Q(\|\cdot\|, u, v). \tag{3.3.30}$$

(ii) With respect to the Euclidean norm $\|\cdot\|_2$ on $\mathbf{R}^2$,

$$c(\|\cdot\|_2, T) = \left(\frac{2}{1 - |\cos\theta|}\right)^{1/2}. \tag{3.3.31}$$

Proof. (i). Fix $f \in \mathrm{Lip}(1, 1)$ and $x = au + bv$, $y = cu + dv$, with $a, b, c, d \in [0, 1]$. Then

$$\begin{aligned}|T(f)(x) - T(f)(y)| &= |(d - b)((1 - a)(f(\mathbf{0}) - f(v)) + a(f(u) - f(u + v)))\\ &+ (c - a)(((1 - d)(f(\mathbf{0}) - f(u)) + d(f(u) - f(u + v)))|\\ &\le |d - b|((1 - a)\|v\| + a\|v\|) + |c - a|((1 - d)\|u\| + d\|u\|)\\ &= |d - b|\|v\| + |c - a|\|u\|.\end{aligned}$$

Therefore, for $x \ne y$,

$$\begin{aligned}|T(f)(x) - T(f)(y)| &\le \frac{|a - c|\|u\| + |b - d|\|v\|}{\|(a - c)u + (b - d)v\|}\|x - y\|\\ &= \left\|\frac{(a - c)\|u\|\frac{u}{\|u\|} + (b - d)\|v\|\frac{v}{\|v\|}}{|a - c|\|u\| + |b - d|\|v\|}\right\|^{-1}\|x - y\| \le Q(\|\cdot\|, u, v)\|x - y\|,\end{aligned}$$

which proves (3.3.30).

(ii). Fix $c \ge 1$ such that $T(\mathrm{Lip}(1, 1)) \subset \mathrm{Lip}(c, 1)$. First, suppose that $\cos\theta \ge 0$. Then take $f(x) := \|x - u\|_2$ $(x \in P)$ and set $w(t) := (1 - t)u + t\frac{\|u\|_2}{\|v\|_2}v$, $t > 0$.

Since $f \in \mathrm{Lip}(1, 1)$, then $T(f) \in \mathrm{Lip}(c, 1)$, so that, for sufficiently small $t > 0$, we get

$$c \ge \frac{|T(f)(w(t)) - T(f)(u)|}{\|w(t) - u\|_2} \to \left(\frac{2}{1 - \cos\theta}\right)^{1/2} \quad (t \to 0^+).$$

If $\cos\theta < 0$, take $f := \|\cdot\|_2$ and $w(t) := t\left(u + \frac{\|u\|_2}{\|v\|_2}v\right)$, $t > 0$. Then

$$c \ge \frac{|T(f)(w(t)) - T(f)(0)|}{\|w(t)\|_2} \to \left(\frac{2}{1 + \cos\theta}\right)^{1/2} \quad (t \to 0^+).$$

Summing up, we have proved that

$$c(\|\cdot\|_2, T) \ge \left(\frac{2}{1 - |\cos\theta|}\right)^{1/2}. \tag{1}$$

Now we prove that

$$Q(\|\cdot\|_2, u, v) = \left(\frac{2}{1-|\cos\theta|}\right)^{1/2}. \tag{2}$$

Indeed,

$$\begin{aligned}
Q(\|\cdot\|_2, u, v)^2 &= \max\left\{\left\|s\frac{u}{\|u\|_2}+t\frac{v}{\|v\|_2}\right\|^{-2} \;\middle|\; |s|+|t|=1\right\}\\
&= \max\left\{(s^2+t^2+2st\cos\theta)^{-1} \;\middle|\; |s|+|t|=1\right\}\\
&= (\min\{1-2|st|+2st\cos\theta \mid |s|+|t|=1\})^{-1}.
\end{aligned}$$

If $|s|+|t|=1$, then $-\frac{1}{4}\le st\le\frac{1}{4}$; thus, if $\cos\theta\ge 0$, we get

$$(Q(\|\cdot\|_2, u, v))^2 = \left(1-2\frac{1}{4}-2\frac{1}{4}\cos\theta\right)^{-1} = \frac{2}{1-\cos\theta}.$$

If $\cos\theta<0$, then

$$Q(\|\cdot\|_2, u, v)^2 = \left(1-2\frac{1}{4}+2\frac{1}{4}\cos\theta\right)^{-1} = \frac{2}{1+\cos\theta}.$$

This proves (2); now (3.3.31) is a consequence of (3.3.30), (1) and (2). □

3.4 Bernstein-Schnabl operators and convexity

Bernstein operators on $\mathscr{C}([0,1])$ map convex functions into convex functions. This property was first recognized by Popoviciu ([163]) (see also Example 3.4.5). However, a similar property does not hold any more in two or more variables, so the problem of investigating when it holds seems to be not devoid of interest.

In this section we shall discuss such aspects for the more general sequences of Bernstein-Schnabl operators.

As in the preceding sections, we keep fixed a metrizable convex compact subset K and a Markov operator T on $\mathscr{C}(K)$ satisfying (3.1.2). Consider also the relevant continuous selection $(\tilde{\mu}_x^T)_{x\in K}$ of probability Borel measures on K and the associated Bernstein-Schnabl operators B_n, $n \geq 1$.

Let $f \in \mathscr{C}(K)$. For $n \geq 2$ and $x_1, \dots, x_{n-1} \in K$, consider the function $f_{x_1,\dots,x_{n-1}}$ defined by (3.3.4). If $n \geq 3$ and $x_1, \dots, x_{n-2} \in K$, define

$$\tilde{f}_{x_1,\dots,x_{n-2}}(s,t) := f\left(\frac{x_1 + \dots + x_{n-2} + s + t}{n}\right) \qquad (s,t \in K). \tag{3.4.1}$$

Consider also the function

$$\tilde{f}(s,t) := f\left(\frac{s+t}{2}\right) \qquad (s,t \in K). \tag{3.4.2}$$

Let $N \geq 2$ be given. Suppose that

(i) $T(g)$ is convex for all $g \in \{f, f_{x_1}, f_{x_1,x_2}, \dots, f_{x_1,\dots,x_{N-1}}\}$ and for all $x_1, \dots, x_{N-1} \in K$;

(ii) For every $x, y \in K$, the inequality

$$\begin{aligned}\Delta(\varphi; x, y) :=& \iint\limits_{K^2} \varphi(s,t)\, d\tilde{\mu}_x^T(s) d\tilde{\mu}_x^T(t) + \\ &+ \iint\limits_{K^2} \varphi(s,t)\, d\tilde{\mu}_y^T(s) d\tilde{\mu}_y^T(t) - 2 \iint\limits_{K^2} \varphi(s,t)\, d\tilde{\mu}_x^T(s) d\tilde{\mu}_y^T(t) \geq 0\end{aligned} \tag{3.4.3}$$

holds for $\varphi = \tilde{f}$ if $N = 2$ and for every $\varphi \in \{\tilde{f}, \tilde{f}_{x_1}, \dots, \tilde{f}_{x_1,\dots,x_{N-2}}\}$ and $x_1, \dots, x_{N-2} \in K$ if $N \geq 3$.

For a given $n \geq 1$ and $x_1, \dots, x_n \in K$, set

$$F_n(f; x_1, \dots, x_n) := \int_K \cdots \int_K f\left(\frac{t_1 + \dots + t_n}{n}\right) d\tilde{\mu}_{x_1}^T(t_1) \cdots d\tilde{\mu}_{x_n}^T(t_n). \tag{3.4.4}$$

Observe that, if $n \geq 2$, the function $F_n(f; \dots)$ is invariant with respect to any permutation of the indices $1, \dots, n$. Combined with the above condition (ii), this fact leads to

$$\begin{aligned}&2F_n(f; x_1, \dots, x_i, \dots, x_j, \dots, x_n) \\ &\leq F_n(f; x_1, \dots, x_i, \dots, x_i, \dots, x_n) + F_n(f; x_1, \dots, x_j, \dots, x_j, \dots, x_n),\end{aligned} \tag{3.4.5}$$

for all $1 \le i \le j \le n$; in the first summand only x_j is replaced by x_i, and in the second summand only x_i is replaced by x_j.

Moreover, for every $p, q \ge 0$, $p + q = n$, and for every $x, y \in K$ set

$$S_{n,p,q}(f; x, y) := F_n(f; \underbrace{x, \dots, x}_{p}, \underbrace{y, \dots, y}_{q}) + F_n(f; \underbrace{x, \dots, x}_{q}, \underbrace{y, \dots, y}_{p}). \tag{3.4.6}$$

Note that

$$S_{n,p,q}(f; x, y) = S_{n,q,p}(f; x, y). \tag{3.4.7}$$

Lemma 3.4.1. *Under hypotheses (i) and (ii), for every $k \in \{1, \dots, N\}$ and $x, y \in K$, we have*

$$S_{k,k-1,1}(f; x, y) \le S_{k,k,0}(f; x, y). \tag{3.4.8}$$

Proof. First note that, if $k = 1$, (3.4.8) is satisfied by virtue of (3.4.7). Assume that $k = 2m + 1$ with $1 \le m \le (N - 1)/2$. By (3.4.5) and (3.4.6) we obtain

$$\begin{aligned}
&2S_{2m+1,2m,1}(f; x, y) = 2F_{2m+1}(f; x, \dots, x, y) + 2F_{2m+1}(f; x, y, \dots, y)\\
&\le F_{2m+1}(f; x, \dots, x, x, x) + F_{2m+1}(f; x, \dots, x, y, y) + F_{2m+1}(f; x, x, y, \dots, y)\\
&+ F_{2m+1}(f; y, y, y, \dots, y) = S_{2m+1,2m+1,0}(f; x, y) + S_{2m+1,2m-1,2}(f; x, y).
\end{aligned}$$

Hence $2S_{2m+1,2m,1}(f; x, y) \le S_{2m+1,2m+1,0}(f; x, y) + S_{2m+1,2m-1,2}(f; x, y)$. With an analogue reasoning it is possible to prove the following inequalities:

$$\begin{aligned}
2S_{2m+1,2m-1,2}(f; x, y) &\le S_{2m+1,2m,1}(f; x, y) + S_{2m+1,2m-2,3}(f; x, y),\\
2S_{2m+1,2m-2,3}(f; x, y) &\le S_{2m+1,2m-1,2}(f; x, y) + S_{2m+1,2m-3,4}(f; x, y),\\
&\vdots\\
2S_{2m+1,m+1,m}(f; x, y) &\le S_{2m+1,m+2,m-1}(f; x, y) + S_{2m+1,m,m+1}(f; x, y).
\end{aligned}$$

From (3.4.7) it follows that $S_{2m+1,m+1,m}(f; x, y) = S_{2m+1,m,m+1}(f; x, y)$ and so

$$S_{2m+1,m+1,m}(f; x, y) \le S_{2m+1,m+2,m-1}(f; x, y).$$

Subsequently, we have

$$\begin{aligned}
&S_{2m+1,m+1,m}(f; x, y) \le S_{2m+1,m+2,m-1}(f; x, y)\\
&\le S_{2m+1,m+3,m-2}(f; x, y) \le \cdots \le S_{2m+1,2m,1}(f; x, y) \le S_{2m+1,2m+1,0}(f; x, y)
\end{aligned}$$

and the last inequality actually gives (3.4.8).

Assume now that $k = 2m$, $1 \le m \le N/2$. By an analogue reasoning we get

$$\begin{aligned}
2S_{2m,2m-1,1}(f; x, y) &\le S_{2m,2m,0}(f; x, y) + S_{k,2m-2,2}(f; x, y),\\
2S_{2m,2m-2,2}(f; x, y) &\le S_{2m,2m-1,1}(f; x, y) + S_{2m,2m-3,3}(f; x, y),\\
&\vdots\\
2S_{2m,m,m}(f; x, y) &\le S_{2m,m+1,m-1}(f; x, y) + S_{2m,m-1,m+1}(f; x, y).
\end{aligned}$$

Since $S_{2m,m+1,m-1}(f;x,y) = S_{2m,m-1,m+1}(f;x,y)$, we have

$$S_{2m,m,m}(f;x,y) \le S_{2m,m+1,m-1}(f;x,y)$$

and hence

$$\begin{aligned}&S_{2m,m,m}(f;x,y)\\&\le S_{2m,m+1,m-1}(f;x,y) \le \dots \le S_{2m,2m-1,1}(f;x,y) \le S_{2m,2m,0}(f;x,y).\end{aligned}$$

Again, the last inequality gives (3.4.8). □

The convexity of (some) $B_n(f)$ is established by the next result.

Theorem 3.4.2. *Let $f \in \mathscr{C}(K)$ and $N \ge 2$ be given. Suppose that the above conditions (i) and (ii) are satisfied. Then $B_1(f), \dots, B_N(f)$ are convex functions.*

Proof. Since $B_1(f) = T(f)$, the convexity of $B_1(f)$ follows from (i).

Let's prove it for $N = 2$; due to the continuity of $B_2(f)$, we have to verify that

$$B_2(f)\left(\frac{x+y}{2}\right) \le \frac{1}{2}(B_2(f)(x) + B_2(f)(y)) \qquad (x, y \in K). \tag{1}$$

Indeed,

$$\begin{aligned}B_2(f)\left(\frac{x+y}{2}\right) &= \iint\limits_{K^2} f\left(\frac{t_1+t_2}{2}\right) d\tilde{\mu}^T_{\frac{x+y}{2}}(t_1) d\tilde{\mu}^T_{\frac{x+y}{2}}(t_2)\\
&= \int_K \left(\int_K f\left(\frac{t_1+t_2}{2}\right) d\tilde{\mu}^T_{\frac{x+y}{2}}(t_1)\right) d\tilde{\mu}^T_{\frac{x+y}{2}}(t_2)\\
&= \int_K T(f_{t_2})\left(\frac{x+y}{2}\right) d\tilde{\mu}^T_{\frac{x+y}{2}}(t_2)\\
&\le \frac{1}{2}\int_K T(f_{t_2})(x) d\tilde{\mu}^T_{\frac{x+y}{2}}(t_2) + \frac{1}{2}\int_K T(f_{t_2})(y) d\tilde{\mu}^T_{\frac{x+y}{2}}(t_2)\\
&= \frac{1}{2}\iint\limits_{K^2} f\left(\frac{t_1+t_2}{2}\right) d\tilde{\mu}^T_x(t_1) d\tilde{\mu}^T_{\frac{x+y}{2}}(t_2) + \frac{1}{2}\iint\limits_{K^2} f\left(\frac{t_1+t_2}{2}\right) d\tilde{\mu}^T_y(t_1) d\tilde{\mu}^T_{\frac{x+y}{2}}(t_2)\\
&\le \frac{1}{2}\iint\limits_{K^2} f\left(\frac{t_1+t_2}{2}\right) d\tilde{\mu}^T_{\frac{x+y}{2}}(t_1) d\tilde{\mu}^T_{\frac{x+y}{2}}(t_2) + \frac{1}{4}\iint\limits_{K^2} f\left(\frac{t_1+t_2}{2}\right) d\tilde{\mu}^T_x(t_1) d\tilde{\mu}^T_x(t_2)\\
&\quad + \frac{1}{4}\iint\limits_{K^2} f\left(\frac{t_1+t_2}{2}\right) d\tilde{\mu}^T_y(t_1) d\tilde{\mu}^T_y(t_2),\end{aligned}$$

where we have used (i) with $g = f_{t_2}$ and (ii) with $\varphi = \tilde{f}$. It follows that

$$\frac{1}{2}B_2(f)\left(\frac{x+y}{2}\right) \le \frac{1}{4}B_2(f)(x) + \frac{1}{4}B_2(f)(y),$$

and this leads to (1).

Suppose now that $N \ge 3$ and, for $n \in \{1, \dots, N\}$, consider the function F_n defined by (3.4.4).

Let's prove by induction on n the inequality

$$B_n(f)\left(\frac{x+y}{2}\right) \le \frac{1}{2}(B_n(f)(x) + B_n(f)(y)) \qquad (x, y \in K), \tag{2}$$

where $n \in \{1, \dots, N\}$. We already know that it holds for $n \in \{1, 2\}$. Suppose that (2) is true for a certain $n-1 < N$ and, for every $z \in K$, consider the function $\delta_n(f, z)$ defined by

$$\delta_n(f, z) := f\left(\frac{n-1}{n}t + \frac{1}{n}z\right) \qquad (t \in K).$$

Then

$$B_n(f)\left(\frac{x+y}{2}\right) = \int_K B_{n-1}(\delta_n(f, x_n))\left(\frac{x+y}{2}\right) d\tilde{\mu}^T_{\frac{x+y}{2}}(x_n).$$

Since (2) is true for $n-1$, we get

$$\begin{aligned}
B_n(f)\left(\frac{x+y}{2}\right) &\le \frac{1}{2}\int_K B_{n-1}(\delta_n(f, x_n))(x)\, d\tilde{\mu}^T_{\frac{x+y}{2}}(x_n) \\
&+ \frac{1}{2}\int_K B_{n-1}(\delta_n(f, x_n))(y)\, d\tilde{\mu}^T_{\frac{x+y}{2}}(x_n) \\
&= \frac{1}{2}\int_K \cdots \int_K T(f_{x_1,\dots,x_{n-1}})\left(\frac{x+y}{2}\right) d\tilde{\mu}^T_x(x_1)\cdots d\tilde{\mu}^T_x(x_{n-1}) \\
&+ \frac{1}{2}\int_K \cdots \int_K T(f_{x_1,\dots,x_{n-1}})\left(\frac{x+y}{2}\right) d\tilde{\mu}^T_y(x_1)\cdots d\tilde{\mu}^T_y(x_{n-1})
\end{aligned}$$

and, by assumption (i) and (3.4.6),

$$\begin{aligned}
B_n(f)\left(\frac{x+y}{2}\right) &\le \frac{1}{4}\int_K \cdots \int_K T(f_{x_1,\dots,x_{n-1}})(x)\, d\tilde{\mu}^T_x(x_1)\cdots d\tilde{\mu}^T_x(x_{n-1}) \\
&+ \frac{1}{4}\int_K \cdots \int_K T(f_{x_1,\dots,x_{n-1}})(y)\, d\tilde{\mu}^T_x(x_1)\cdots d\tilde{\mu}^T_x(x_{n-1}) \\
&+ \frac{1}{4}\int_K \cdots \int_K T(f_{x_1,\dots,x_{n-1}})(x)\, d\tilde{\mu}^T_y(x_1)\cdots d\tilde{\mu}^T_y(x_{n-1}) \\
&+ \frac{1}{4}\int_K \cdots \int_K T(f_{x_1,\dots,x_{n-1}})(y)\, d\tilde{\mu}^T_y(x_1)\cdots d\tilde{\mu}^T_y(x_{n-1}) \\
&= \frac{1}{4}B_n(f)(x) + \frac{1}{4}B_n(f)(y) + \frac{1}{4}F_n(f; x, \dots, x, y) + \frac{1}{4}F_n(f; y, \dots, y, x) \\
&= \frac{1}{4}B_n(f)(x) + \frac{1}{4}B_n(f)(y) + \frac{1}{4}S_{n,n-1,1}(f; x, y).
\end{aligned}$$

By using (3.4.8) for $k = n$ we get

$$\begin{aligned}
B_n(f)\left(\frac{x+y}{2}\right) &\le \frac{1}{4}B_n(f)(x) + \frac{1}{4}B_n(f)(y) + \frac{1}{4}S_{n,n,0}(f; x, y) \\
&= \frac{1}{4}B_n(f)(x) + \frac{1}{4}B_n(f)(y) + \frac{1}{4}\left(F_n(f; x, \dots, x) + F_n(f; y, \dots, y)\right) \\
&= \frac{1}{2}B_n(f)(x) + \frac{1}{2}B_n(f)(y),
\end{aligned}$$

and this concludes the proof. □

Let us mention an important consequence of Theorem 3.4.2.

Theorem 3.4.3. *Assume that*
(c_1) T maps continuous convex functions into (continuous) convex functions.
(c_2) For every convex function $f \in \mathscr{C}(K)$ and for every $x, y \in K$

$$\Delta(\tilde{f}; x, y) \geq 0$$

(see (3.4.2) *and* (3.4.3)*).*

Then $B_n(f)$ is convex for every $n \geq 1$ and for every convex $f \in \mathscr{C}(K)$.

Proof. If $f \in \mathscr{C}(K)$ is convex, then the functions $f_{x_1,\dots,x_{N-1}}$, defined by (3.3.4), are convex as well for each $N \geq 2$ and $x_1, \dots, x_{N-1} \in K$, so that $T(f_{x_1,\dots,x_{N-1}})$ is convex by (c_1).

Moreover, for $N \geq 3$ and $x_1, \dots, x_{N-2} \in K$, considering the function

$$g(t) = f\left(\frac{x_1 + \dots + x_{N-2} + 2t}{N}\right) \qquad (t \in K),$$

it yields $\tilde{f}_{x_1,\dots,x_{N-2}} = \tilde{g}$ and g is convex as well.

Therefore, by (c_2), inequality (3.4.3) is fulfilled for $\tilde{f}_{x_1,\dots,x_{N-2}}$ and hence the result follows from Theorem 3.4.2. □

Remarks 3.4.4.

1. We point out that, for any given $f \in \mathscr{C}(K)$,

$$\Delta(\tilde{f}; x, y) = B_2(f)(x) + B_2(f)(y) - 2 \iint\limits_{K^2} f\left(\frac{s+t}{2}\right) d\tilde{\mu}_x^T(s)\, d\tilde{\mu}_y^T(t)$$

($x, y \in K$).

Therefore, condition (c_2) is satisfied provided that we require that
(c_2') For every convex function $f \in \mathscr{C}(K)$, $B_2(f)$ is convex and, for every $x, y \in K$,

$$B_2(f)\left(\frac{x+y}{2}\right) \geq \iint\limits_{K^2} f\left(\frac{s+t}{2}\right) d\tilde{\mu}_x^T(s)\, d\tilde{\mu}_y^T(t)$$

2. Another useful way to look at condition (c_2) is to consider, for $x, y \in K$ fixed, the linear functional on $\mathscr{C}(K)$ defined by

$$\Lambda^*(f) := \Delta(\tilde{f}; x, y) \qquad (f \in \mathscr{C}(K)). \tag{3.4.9}$$

Thus, (c_2) means that

$$\Lambda^*(f) \geq 0 \qquad \text{for every convex } f \in \mathscr{C}(K). \tag{3.4.10}$$

When $K = [0, 1]$, according to a result of T. Popoviciu (see, e.g., [122, Chapter XI, Theorem 5.1 and Example 1.2, (1.5) and (2.5)]), then (3.4.10) holds true if and only if

$$\Lambda^*(\mathbf{1}) = \Lambda^*(e_1) = 0 \tag{3.4.11}$$

and

$$\Lambda^*(\psi_t^+) \geq 0 \qquad \text{for every } t \in [0,1], \tag{3.4.12}$$

where the function e_1 is defined by (1.2.6) and $\psi_t^+(s) := \sup\{\psi_t(s), 0\} = \sup\{s - t, 0\}$ $(0 \leq s \leq 1)$.

Below we show some cases where Theorems 3.4.2 and 3.4.3 can be applied.

Example 3.4.5. *Consider the Bernstein operators $(B_n)_{n\geq1}$ on $\mathscr{C}([0,1])$ defined by (3.1.11) and associated with the Markov operator T_1 defined by (1.1.49). Clearly T_1 satisfies condition (c_1).*

Moreover, since

$$B_2(f)(x) = (1-x)^2 f(0) + 2x(1-x) f\left(\frac{1}{2}\right) + x^2 f(1)$$

$(f \in \mathscr{C}([0,1]), 0 \leq x \leq 1)$, then, for every convex function $f \in \mathscr{C}([0,1])$, we get

$$B_2(f)''(x) = 2\left(f(0) - 2f\left(\frac{1}{2}\right) + f(1)\right) \geq 0$$

and, for $x, y \in [0,1]$,

$$\begin{aligned} B_2(f)\left(\frac{x+y}{2}\right) &- \iint\limits_{[0,1]^2} f\left(\frac{s+t}{2}\right) d\tilde{\mu}_x^{T_1}(s)\, d\tilde{\mu}_y^{T_1}(t) \\ &= \frac{(x-y)^2}{4}\left(f(0) - 2f\left(\frac{1}{2}\right) + f(1)\right) \geq 0. \end{aligned}$$

Therefore, condition (c_2') is satisfied and hence, according to Theorem 3.4.3 and Remark 3.4.4, 1, each B_n maps continuous convex functions on $[0,1]$ into continuous convex ones.

However, this important property of Bernstein operators on $\mathscr{C}([0,1])$ can be also directly obtained by evaluating the second derivative of each $B_n(f)$ $(f \in \mathscr{C}([0,1])$, f convex) is given by

$$\begin{aligned} B_n(f)''(x) = n(n-1) \sum_{h=0}^{n-2} &\left[f\left(\frac{h+2}{n}\right) - 2f\left(\frac{h+1}{n}\right) + f\left(\frac{h}{n}\right)\right] \\ \times \binom{n-2}{h} x^h (1-x)^{n-1-h} & \end{aligned}$$

$(0 \leq x \leq 1, n \geq 2)$.

Example 3.4.6. Consider the Bernstein-Schnabl operators $(B_n)_{n\geq1}$ defined by (3.1.16) and associated with the Markov operator T defined by (3.1.14). Furthermore, assume that the involved function b is concave. Then, for $f \in \mathscr{C}([0,1])$

and $x, y \in [0,1]$, we get

$$\frac{T(f)(x)+T(f)(y)}{2} - T(f)\left(\frac{x+y}{2}\right)$$
$$= \frac{1}{2}\left[b\left(\frac{x+y}{2}\right) - \frac{b(x)+b(y)}{2}\right]\left[f(0) - 2f\left(\frac{1}{2}\right) + f(1)\right]$$

which shows that T maps continuous convex functions into convex ones.

In order to verify condition (c_2), given $x, y \in [0,1]$, we shall consider the linear functional Λ^* on $\mathscr{C}([0,1])$ defined by (3.4.9) and we shall show conditions (3.4.11) and (3.4.12).

By elementary calculations, for every $f \in \mathscr{C}([0,1])$ we have

$$\Lambda^*(f) = \alpha(x,y)f(0) + \beta(x,y)f\left(\frac{1}{4}\right)$$
$$+ \gamma(x,y)f\left(\frac{1}{2}\right) + \delta(x,y)f\left(\frac{3}{4}\right) + \eta(x,y)f(1),$$

where

$$\alpha(x,y) := \left(y - x + \frac{b(y)-b(x)}{2}\right)^2,$$
$$\beta(x,y) := -2(b(y)-b(x))(y-x) - (b(y)-b(x))^2,$$
$$\gamma(x,y) := \frac{3}{2}(b(y)-b(x))^2 - 2(y-x)^2,$$
$$\delta(x,y) := 2(b(y)-b(x))(y-x) - (b(y)-b(x))^2$$

and

$$\eta(x,y) := \left(x - y + \frac{b(y)-b(x)}{2}\right)^2.$$

Therefore, $\Lambda^*(\mathbf{1}) = \Lambda^*(e_1) = 0$.

On the other hand, for $t \in [0,1]$, we get

$$\Lambda^*(\psi_t^+) = \begin{cases} \beta(x,y)\,(1/4-t) + \gamma(x,y)\,(1/2-t) \\ +\delta(x,y)\,(3/4-t) + \eta(x,y)\,(1-t) & \text{if } 0 \le t \le 1/4; \\[1ex] \gamma(x,y)\,(1/2-t) + \delta(x,y)\,(3/4-t) + \eta(x,y)\,(1-t) & \text{if } 1/4 \le x \le 1/2; \\[1ex] \delta(x,y)\,(3/4-t) + \eta(x,y)\,(1-t) & \text{if } 1/2 \le x \le 3/4; \\[1ex] \eta(x,y)\,(1-t) & \text{if } 3/4 \le x \le 1. \end{cases}$$

In order to show (3.4.12), it suffices to prove it for $t = 0, 1/4, 1/2, 3/4, 1$. Actually,

$$\Lambda^*(\psi_0^+) = \Lambda^*(e_1) = 0,$$

$$\Lambda^*(\psi^+_{1/4}) = \frac{1}{4}\alpha(x,y) \geq 0,$$
$$\Lambda^*(\psi^+_{3/4}) = \Lambda^*(e_1) = \frac{1}{4}\eta(x,y) \geq 0,$$
$$\Lambda^*(\psi^+_1) = \Lambda^*(0) = 0$$

and, finally,

$$\Lambda^*(\psi^+_{1/2}) = \frac{1}{2}\left((y-x)^2 - \left(\frac{b(y)-b(x)}{2}\right)^2\right),$$

which is positive because, b being concave and $0 \leq \frac{b}{2} \leq \min_{0\leq s\leq 1}\{s, 1-s\}$, for $x \neq y$

$$\left|\frac{b(y)-b(x)}{2(y-x)}\right| \leq 1.$$

Summing up, according to Theorem 3.4.3, all the operators B_n, $n \geq 1$, map continuous convex functions on $[0,1]$ into convex ones.

Example 3.4.7. Consider again the operator T defined by (3.1.14) with $b(x) := 2x(1-x)^2, x \in [0,1]$. Explicitly, we have

$$T(f)(x) = (1-x)(1-x+x^2)f(0) + 2x(1-x)^2 f\left(\frac{1}{2}\right) + x^2(2-x)f(1) \quad (3.4.13)$$

for all $f \in \mathscr{C}([0,1])$ and $x \in [0,1]$.

Using Example 3.4.6, we find that

$$\frac{1}{2}\left(T(f)\left(\frac{1}{2}\right) + T(f)(1)\right) - T(f)\left(\frac{3}{4}\right) = -\frac{1}{64}\left(f(0) - 2f\left(\frac{1}{2}\right) + f(1)\right).$$

So, there are convex functions f for which $T(f)$ is not convex.

On the other hand, $|b'(x)| \leq 2$ for all $x \in [0,1]$, so that $b \in \mathrm{Lip}(2,1)$. Therefore, $T(\mathrm{Lip}(1,1)) \subset \mathrm{Lip}(1,1)$ (see (3.3.15)).

Moreover,

$$\frac{d}{dx}T(f)(x) = (3x^2 - 4x + 2)\left(f\left(\frac{1}{2}\right) - f(0)\right) + (4x - 3x^2)\left(f(1) - f\left(\frac{1}{2}\right)\right),$$

which implies that $T(f)$ is increasing whenever $f \in \mathscr{C}([0,1])$ is increasing.

Example 3.4.8. *Consider a Markov operator T on $\mathscr{C}([0,1])$ satisfying conditions (c_1) and (c_2) and $\lambda \in [0,1]$. Then the operator*

$$U_\lambda := \lambda T + (1-\lambda)I$$

verifies (c_1) and (c_2) as well. Therefore, all the Lototsky-Schnabl operators $B_{n,\lambda}$, $n \geq 1$, associated with T and λ and defined by (3.1.35), map continuous convex functions on $[0,1]$ into convex ones.

Clearly, U_λ verifies (c_1). As regards condition (c_2), denoting by $(\tilde{\mu}_x^T)_{0\le x\le 1}$ the continuous selection of probability Borel measures on $[0,1]$ associated with T, then the corresponding one associated with U_λ is given by

$$\tilde{\nu}_x := \lambda\tilde{\mu}_x^T + (1-\lambda)\varepsilon_x.$$

Therefore, for any $f \in \mathscr{C}([0,1])$ and $x, y \in [0,1]$,

$$\iint\limits_{[0,1]^2} f\left(\frac{s+t}{2}\right) d\tilde{\nu}_x(s)\, d\tilde{\nu}_y(t) = \lambda^2 \iint\limits_{[0,1]^2} f\left(\frac{s+t}{2}\right) d\tilde{\mu}_x(s)\, d\tilde{\mu}_y(t)$$
$$+ \lambda(1-\lambda)\left[xf\left(\frac{1+y}{2}\right) + (1-x)f\left(\frac{y}{2}\right) + yf\left(\frac{1+x}{2}\right) + (1-y)f\left(\frac{x}{2}\right)\right]$$
$$+ (1-\lambda)^2 f\left(\frac{x+y}{2}\right).$$

Thus, denoting by $\Delta_T(\tilde{f}; x, y)$ and $\Delta_{U_\lambda}(\tilde{f}; x, y)$ the relevant quantities (3.4.3) associated with T and U_λ, respectively, and assuming, in addition, that f is convex, we get

$$\Delta_{U_\lambda}(\tilde{f}; x, y) = \lambda^2 \Delta_T(\tilde{f}; x, y) + (1-\lambda)^2\left(f(x) + f(y) - 2f\left(\frac{x+y}{2}\right)\right)$$
$$+ 2\lambda(1-\lambda)(y-x)\left(f\left(\frac{1+y}{2}\right) - f\left(\frac{y}{2}\right) - \left(f\left(\frac{1+x}{2}\right) - f\left(\frac{x}{2}\right)\right)\right)$$

and hence (c_2) follows because

$$\frac{f\left(\frac{1+u}{2}\right) - f\left(\frac{u}{2}\right)}{\frac{1+u}{2} - \frac{u}{2}} \le \frac{f\left(\frac{1+v}{2}\right) - f\left(\frac{v}{2}\right)}{\frac{1+v}{2} - \frac{v}{2}}$$

for every $0 \le u < v \le 1$.

Remark 3.4.9. We point out that the result of Example 3.4.8 fails if λ is not constant. Consider, indeed, $T = T_1$ (see (1.1.49)), $\lambda(x) = \sin^2(4\pi x)$ and $f(x) = x^2$ $(0 \le x \le 1)$. Then the function

$$B_{1,\lambda}(f)(x) = x^2 + x(1-x)\sin^2(4\pi x) \qquad (0 \le x \le 1)$$

is not convex.

Example 3.4.10. Let K_2 be the canonical simplex in $\mathbf{R}^2$ and consider the canonical projection T_2 on $\mathscr{C}(K_2)$ and the corresponding Bernstein operators defined by (1.1.48) and (3.1.18), respectively. In the sequel it will be more convenient to describe K_2, T_2 and the Bernstein operators in terms of *barycentric coordinates.*

Actually, $K_2 := \{(u_1, u_2) \in \mathbf{R}^2 \mid u_1 \ge 0,\, u_2 \ge 0,\, u_1 + u_2 \le 1\}$ can be identified with

$$K := \{(x_1, x_2, x_3) \in \mathbf{R}^3 \mid x_1 \ge 0,\, x_2 \ge 0,\, x_3 \ge 0,\, x_1 + x_2 + x_3 = 1\}$$

by means of the mapping $\Phi : K_2 \longrightarrow K$ defined by

$$\Phi(u_1, u_2) := (1 - u_1 - u_2, u_1, u_2) \qquad ((u_1, u_2) \in K_2).$$

Moreover, every $f \in \mathscr{C}(K)$ can be identified with the function $F \in \mathscr{C}(K_2)$ defined by

$$F(u_1, u_2) := f(\Phi(u_1, u_2)) = f(1 - u_1 - u_2, u_1, u_2)$$

and the Markov operator T_2 on $\mathscr{C}(K_2)$ turns into the Markov operator T on $\mathscr{C}(K)$ defined by

$$T(f)(x) = x_1 f(1,0,0) + x_2 f(0,1,0) + x_3 f(0,0,1)$$

$(f \in \mathscr{C}(K), x = (x_1, x_2, x_3) \in K)$.

Finally, the Bernstein operators reduce to

$$B_n(f)(x) = \sum_{h_1+h_2+h_3=n} \frac{n!}{h_1!\,h_2!\,h_3!} x_1^{h_1} x_2^{h_2} x_3^{h_3} f\left(\frac{h_1}{n}, \frac{h_2}{n}, \frac{h_3}{n}\right)$$

$(f \in \mathscr{C}(K), x = (x_1, x_2, x_3) \in K, n \geq 1)$.

We now proceed to discuss conditions (i) and (ii) in this framework.

Condition (i) is obviously satisfied since $T(\mathscr{C}(K)) \subset A(K)$.

Note also that $\partial_T K = \partial_e K = \{(1,0,0), (0,1,0), (0,0,1)\}$ and hence, if $f \in \mathscr{C}(K)$ vanishes on $\partial_e K$, then $T(f)(x) = 0$ for every $x \in K$.

Due to this fact, it suffices to check condition (ii) only for points $x_1, \dots, x_{N-1} \in \partial_e K$, $N \geq 2$.

Now let's study condition (ii) for $x, y \in K$, $f \in \mathscr{C}(K)$ and $\varphi := \tilde{f}_{z_1,\dots,z_{n-2}}$ $(2 \leq n \leq N$, $z_1, \dots, z_{n-2} \in \partial_e K)$ (we recall that if $N = 2$, then $\varphi := \tilde{f}$).

Fix $m = 1, \dots, n-2$ and consider $z_m = (z_{m1}, z_{m2}, z_{m3}) \in \partial_e K$ and $i := \sum_{m=1}^{n-2} z_{m1}$, $j := \sum_{m=1}^{n-2} z_{m2}$, $k := \sum_{m=1}^{n-2} z_{m3}$. Then i, j, k are nonnegative integers and $i + j + k = n - 2$. Moreover,

$$\varphi(s,t) = f\left(\frac{i + s_1 + t_1}{n}, \frac{j + s_2 + t_2}{n}, \frac{k + s_3 + t_3}{n}\right)$$

for all $s = (s_1, s_2, s_3) \in K$, $t = (t_1, t_2, t_3) \in K$.

Keeping n fixed, we shall use the notation $f_{u,v,w} := f(u/n, v/n, w/n)$ for all nonnegative integers u, v, w with $u + v + w = n$.

It is not difficult to prove that, with notation from (3.4.3),

$$\begin{aligned}
\Delta(\varphi; x, y) &:= (x_1 - y_1)^2 \left(f_{i+2,j,k} + f_{i,j,k+2} - 2f_{i+1,j,k+1}\right) \\
&+ (x_2 - y_2)^2 \left(f_{i,j+2,k} + f_{i,j,k+2} - 2f_{i,j+1,k+1}\right) \\
&+ 2(x_1 - y_1)(x_2 - y_2) \left(f_{i,j,k+2} + f_{i+1,j+1,k} - f_{i,j+1,k+1} - f_{i+1,j,k+1}\right).
\end{aligned}$$

Now let's assume that the inequalities

$$\begin{cases} f_{i+2,j,k} + f_{i,j+1,k+1} \geq f_{i+1,j+1,k} + f_{i+1,j,k+1} \\ f_{i,j+2,k} + f_{i+1,j,k+1} \geq f_{i,j+1,k+1} + f_{i+1,j+1,k} \\ f_{i,j,k+2} + f_{i+1,j+1,k} \geq f_{i,j+1,k+1} + f_{i+1,j,k+1} \end{cases} \tag{C_n}$$

are satisfied for all $n \in \{2, \dots, N\}$ and all nonnegative integers i, j, k with $i+j+k = n-2$. Then it is easy to check that

$$\Delta(\varphi; x, y) \geq$$
$$\geq (x_1 + x_2 - y_1 - y_2)^2 \left(f_{i,j,k+2} + f_{i+1,j+1,k} - f_{i,j+1,k+1} - f_{i+1,j,k+1}\right) \geq 0.$$

In conclusion, if the inequalities (C_n) are satisfied for all $n \in \{2, \dots, N\}$, then condition (ii) in Theorem 3.4.2 is also satisfied, and so $B_1(f), \dots, B_N(f)$ are convex functions.

Example 3.4.11. Let $Q_2 := [0,1]^2$. Consider the positive linear operator $S_2 : \mathscr{C}(Q_2) \longrightarrow \mathscr{C}(Q_2)$ defined by

$$S_2(f)(x) = (1-x_1)(1-x_2)f(0,0) + (1-x_1)x_2 f(0,1)$$
$$+ x_1(1-x_2)f(1,0) + x_1 x_2 f(1,1)$$

for all $f \in \mathscr{C}(Q_2)$ and all $x = (x_1, x_2) \in Q_2$ (see (2.3.26)-(3.1.28)).

In this case $\partial_{S_2} Q_2 = \partial_e Q_2 = \{(0,0), (0,1), (1,0), (1,1)\}$ so that, if $f \in \mathscr{C}(Q_2)$ vanishes on $\partial_e Q_2$, then $S_2(f) = 0$.

Consequently, it suffices to check the hypotheses of Theorem 3.4.2 only for points $x_1, \dots, x_{N-1} \in \partial_e Q_2$, $N \geq 2$.

In order to check (i), it suffices to impose to $f \in \mathscr{C}(Q_2)$ the condition

$$f_{i,j} - f_{i+1,j} - f_{i,j+1} + f_{i+1,j+1} = 0 \tag{H_n}$$

for all $n \in \{1, \dots, N\}$ and all nonnegative integers $0 \leq i \leq n-1$, $0 \leq j \leq n-1$, where $f_{i,j} := f\,(i/n, j/n)$.

Indeed, let $n \in \{1, \dots, N\}$ be given, and let $g = f_{z_1, \dots, z_{n-1}}$, $z_1, \dots, z_{n-1} \in \partial_e Q_2$; of course, if $n = 1$, then $g = f$. It is easy to verify that

$$T(g)(x) = f_{i,j} + (f_{i+1,j} - f_{i,j})x_1 + (f_{i,j+1} - f_{i,j})x_2$$
$$+ (f_{i,j} - f_{i,j+1} - f_{i+1,j} + f_{i+1,j+1})x_1 x_2 \qquad (x = (x_1, x_2) \in Q_2)$$

for suitable integers $0 \leq i \leq n-1$, $0 \leq j \leq n-1$, depending on the points $z_1, \dots, z_{n-1}$.

According to (H_n), $T(g)$ is affine, and so (i) is satisfied.

To check (ii), we need, besides the hypotheses (H_n),

$$\begin{cases} f_{i+1,j} + f_{i+1,j+2} - 2f_{i+1,j+1} \geq 0 \\ f_{i,j+1} + f_{i+2,j+1} - 2f_{i+1,j+1} \geq 0 \end{cases} \tag{G_n}$$

for all $n \in \{2, \dots, N\}$ and all nonnegative integers $0 \le i \le n-2$, $0 \le j \le n-2$.

Indeed, fix $n \in \{2, \dots, N\}$ and let $\varphi := \tilde{f}_{z_1, \dots, z_{n-2}}$, with $\varphi := \tilde{f}$ if $n = 2$. Then

$$\Delta(\varphi; x, y) = (f_{i,j+1} + f_{i+2,j+1} - 2f_{i+1,j+1})(x_1 - y_1)^2 + (f_{i+1,j} + f_{i+1,j+2} - 2f_{i+1,j+1})(x_2 - y_2)^2$$

for certain integers $0 \le i \le n-2$, $0 \le j \le n-2$, depending on the points $z_1, \dots, z_{n-2}$.

According to (G_n), $\Delta(\varphi; x, y) \ge 0$ for all $x, y \in Q_2$, and so (ii) is satisfied.

In conclusion, if the inequalities (H_n) $(1 \le n \le N)$ *and* (G_n) $(2 \le n \le N)$ *are satisfied, then* $B_1(f), \dots, B_N(f)$ *are convex functions.*

3.5 Monotonicity properties

In this section we shall discuss some additional properties of the sequence $(B_n(f))_{n\geq 1}$ for special functions $f \in \mathscr{C}(K)$.

The first one is concerned with its monotonicity. As in the preceding sections, we shall fix a metrizable convex compact subset K and a Markov operator T on $\mathscr{C}(K)$. Once again let us denote by $(\tilde{\mu}_x^T)_{x\in K}$ the continuous selection of Borel measures associated with T (see (1.3.1)).

Definition 3.5.1. *For $x \in K$ we set*

$$D_x := \{(u, v) \in K^2 \mid \text{ there exist } \beta \geq 0 \text{ and } p, q \in \operatorname{Supp}(\tilde{\mu}_x^T) \text{ such that } u - v = \beta(p - q)\}. \tag{3.5.1}$$

Then a function $f \in \mathscr{C}(K)$ is called T-axially convex *if*

$$f(\alpha u + (1-\alpha)v) \leq \alpha f(u) + (1-\alpha) f(v) \tag{3.5.2}$$

for every $x \in K$, $(u, v) \in D_x$ and $0 \leq \alpha \leq 1$.

Thus, a T-axially convex function is convex on each segment parallel to a segment joining two arbitrary points of $\operatorname{Supp}(\tilde{\mu}_x^T)$, for every $x \in K$.

If K is a simplex and T is the associated canonical projection, then a T-axially convex function is convex on each segment parallel to a segment joining two extreme points of K.

Every convex function is T-axially convex (for every Markov operator T), but the converse is not true, in general.

However, the two notions coincide if $\partial K \subset \operatorname{Supp}(\tilde{\mu}_x^T)$ for some $x \in K$. This is the case, for instance, of Example 3.1.1, 1) and 3), and Example 3.1.4.

The T-axially convexity guarantees the monotonicity of the sequence of the Bernstein-Schnabl functions.

Theorem 3.5.2. *If $f \in \mathscr{C}(K)$ is T-axially convex, then*

$$B_{n+1}(f) \leq B_n(f) \leq T(f) \tag{3.5.3}$$

for any $n \geq 1$. If, in addition, T satisfies (3.1.2) (or (3.1.3)), then

$$f \leq B_n(f) \qquad (n \geq 1). \tag{3.5.4}$$

Proof. For every $i = 1, \dots, n+1$ consider the mapping $\sigma_i : K^{n+1} \longrightarrow K$ defined by

$$\sigma_i(x_1, \dots, x_{n+1}) := \frac{1}{n} \sum_{\substack{j=1 \\ j\neq i}}^{n+1} x_j \qquad ((x_1, \dots, x_{n+1}) \in K^{n+1}).$$

Then

$$\sum_{i=1}^{n+1} \sigma_i(x_1, \dots, x_{n+1}) = \sum_{i=1}^{n+1} x_i \qquad ((x_1, \dots, x_{n+1}) \in K^{n+1}). \tag{1}$$

Moreover, for each $i \in \{1, \dots, n+1\}$ and $x \in K$,

$$\int_K \cdots \int_K f(\sigma_i(x_1, \dots, x_{n+1}))\, d\tilde{\mu}_x^T(x_1) \cdots d\tilde{\mu}_x^T(x_{n+1}) =$$
$$\int_K \cdots \int_K f\left(\frac{y_1 + \dots + y_n}{n}\right) d\tilde{\mu}_x^T(y_1) \cdots d\tilde{\mu}_x^T(y_n).$$

Therefore, for $x \in K$,

$$\begin{aligned} &B_n(f)(x) - B_{n+1}(f)(x) \\ &= \int_K \cdots \int_K \left(f\left(\frac{x_1 + \dots + x_n}{n}\right) - f\left(\frac{x_1 + \dots + x_{n+1}}{n+1}\right)\right) d\tilde{\mu}_x^T(x_1) \cdots d\tilde{\mu}_x^T(x_{n+1}) \qquad (2) \\ &= \int_{K^{n+1}} \frac{1}{n+1} \sum_{i=1}^{n+1} f \circ \sigma_i - f\left(\frac{1}{n+1}\sum_{i=1}^{n+1} \sigma_i\right) d\tilde{\mu}^T_{x,n+1}, \end{aligned}$$

where by $\tilde{\mu}^T_{x,n+1}$ we denote the tensor product of the measure $\tilde{\mu}_x^T$ with itself $n+1$ times.

Now, for every $i = 0, \dots, n$ and $j = i+1, \dots, n+1$, we set

$$\sigma_{i,j} := \begin{cases} \sigma_j & \text{if } i = 0; \\ \dfrac{\sigma_1 + n\sigma_j}{n+1} & \text{if } i = 1; \\ \dfrac{\sigma_{i-1,j} + (n-i+1)\sigma_{i-1,j}}{n-i+2} & \text{if } i > 1. \end{cases}$$

Then

$$\sigma_{n,n+1} = \frac{1}{2}(\sigma_{n-1,n} + \sigma_{n-1,n+1}) = \frac{1}{3}(\sigma_{n-2,n-1} + \sigma_{n-2,n} + \sigma_{n-2,n+1})$$
$$= \dots = \frac{1}{n+1}(\sigma_{0,1} + \sigma_{0,2} + \dots + \sigma_{0,n+1}) = \frac{1}{n+1}\sum_{j=1}^{n+1} \sigma_j.$$

On the other hand, if $(x_1, \dots, x_{n+1}) \in \operatorname{Supp}(\tilde{\mu}_x^T)^{n+1}$, then

$$(\sigma_{i,j}(x_1, \dots, x_{n+1}), \sigma_{i,k}(x_1, \dots, x_{n+1})) \in D_x$$

for each $i = 0, \dots, n$ and $j, k = i+1, \dots, n+1$, so that

$$\begin{aligned} &f\left(\frac{1}{n+1}\sum_{i=1}^{n+1} \sigma_i(x_1, \dots, x_{n+1})\right) = f(\sigma_{n,n+1}(x_1, \dots, x_{n+1})) \\ &\le \frac{1}{2}\left(f(\sigma_{n-1,n}(x_1, \dots, x_{n+1})) + f(\sigma_{n-1,n+1}(x_1, \dots, x_{n+1}))\right) \\ &\le \frac{1}{3}\left(f(\sigma_{n-2,n-1}(x_1, \dots, x_{n+1})) + f(\sigma_{n-2,n}(x_1, \dots, x_{n+1}))\right) \end{aligned}$$

$$+f(\sigma_{n-2,n+1}(x_1,\dots,x_{n+1}))) \le \dots \le$$
$$\le \frac{1}{n+1}\left(f(\sigma_{0,1}(x_1,\dots,x_{n+1})) + \dots + f(\sigma_{0,n+1}(x_1,\dots,x_{n+1}))\right)$$
$$= \frac{1}{n+1}\sum_{i=1}^{n+1} f(\sigma_i(x_1,\dots,x_{n+1})).$$

By using (2) together with the above inequality, we get (3.5.3).

To get (3.5.4), it suffices to apply (3.5.3) recursively to obtain

$$B_{n+p}(f) \le B_{n+1}(f) \le B_n(f) \le B_1(f) = T(f)$$

$(n \ge 1, p \ge 1)$. Letting $p \to \infty$, we get $f \le B_n(f)$ on account of (3.2.5). □

When $f \in \mathscr{C}(K)$ is convex, then it is possible to characterize those points $x \in K$ satisfying $B_{n+1}(f)(x) = B_n(f)(x)$ for every $n \ge 1$.

Before doing it, we recall that we denote by $\mathrm{co}(B)$ the convex hull of a subset B of a (real or complex) vector space E, i.e., the smallest convex subset of E containing B (see (1.1.44)).

If E is a topological vector space, the closure of $\mathrm{co}(B)$ will be denoted by $\overline{\mathrm{co}}(B)$. Actually, it can be proved that

$$\overline{\mathrm{co}}(B) = \overline{m}(B) \tag{3.5.5}$$

where

$$m(B) := \left\{\frac{x_1 + \dots + x_n}{n} \mid n \ge 1,\, x_1,\dots,x_n \in B\right\} \tag{3.5.6}$$

(see [18, Lemma 6.1.6]).

Note also that, if T is a Markov operator on $\mathscr{C}(K)$ satisfying (3.1.2) (or (3.1.3)), then every $x \in K$ is the barycenter of the measure $\tilde{\mu}_x^T$ (see, e.g., [18, Section 1.5]) and hence

$$x \in \overline{\mathrm{co}}\left(\mathrm{Supp}(\tilde{\mu}_x^T)\right). \tag{3.5.7}$$

Moreover, if $f \in \mathscr{C}(K)$ is convex, then

$$f(x) \le \int_K f\, d\tilde{\mu}_x^T, \quad \text{i.e.,} \quad f(x) \le T(f)(x). \tag{3.5.8}$$

If, in particular, $T(f)(x) = f(x)$, then

$$f \text{ is affine on } \overline{\mathrm{co}}\left(\mathrm{Supp}(\tilde{\mu}_x^T)\right). \tag{3.5.9}$$

This last result is a consequence of a more general one obtained by I. Raşa (see [164]).

The next result yields information on those points of K at which the Bernstein-Schnabl functions can coincide.

Corollary 3.5.3. *Let $f \in \mathscr{C}(K)$ be a convex function and $x \in K$. If T satisfies (3.1.2), then the following statements are equivalent:*

(i) $B_{n+1}(f)(x) = B_n(f)(x)$ *for every* $n \geq 1$.
(ii) $f(x) = T(f)(x)$.
(iii) f *is affine on* $\overline{\text{co}}\left(\text{Supp}(\tilde{\mu}_x^T)\right)$.

Proof. By using the inequalities

$$f \leq B_{n+p}(f) \leq B_n(f) \leq B_1(f) = T(f)$$

and the fact that $\lim_{p\to\infty} B_{n+p}(f) = f$, it is easy to prove that (i) is equivalent to (ii).

The implication (ii) $\Rightarrow$(iii) follows from (3.5.9). Finally, assuming that condition (iii) holds, then, with the same notation as in the proof of Theorem 3.5.2, for every $n \geq 1$ we get

$$\frac{1}{n+1}\sum_{i=1}^{n+1} f \circ \sigma_i = f\left(\frac{1}{n+1}\sum_{i=1}^{n+1}\sigma_i\right)$$

on $\left(\text{Supp}(\tilde{\mu}_x^T)\right)^{n+1} = \text{Supp}(\tilde{\mu}_{x,n+1}^T)$ and hence formula (2) in the proof of Theorem 3.5.2 yields condition (i). Here, $\tilde{\mu}_{x,n+1}^T$ denote the tensor product of the measure $\tilde{\mu}_x^T$ with itself $n+1$ times. □

The majorizing property (3.5.4) is worth being examined more closely not only for the geometric information about the behaviour of the functions $B_n(f)$, $n \geq 1$, but also for its connection with a kind of maximum principle.

Theorem 3.5.4. *Let* T *be a Markov operator on* $\mathscr{C}(K)$ *and denote by* $(B_n)_{n\geq 1}$ *the corresponding sequence of Bernstein-Schnabl operators. Consider* $f \in \mathscr{C}(K)$ *such that*

$$f \leq B_n(f) \qquad \text{for every } n \geq 1.$$

If $x_0 \in K$ *and if* $f(x_0) = M_f := \sup_{x\in K} f(x)$, *then*

$$f(x) = M_f \quad \text{for every } x \in \text{Supp}(\tilde{\mu}_{x_0}^T).$$

In particular, if T *is a Markov projection and the subspace* $\{h \in \mathscr{C}(K) \mid T(h) = h\}$ *separates the points of* K, *then* f *attains its maximum on* $\partial_T K = \partial_H K$, *where* $H := T(\mathscr{C}(K))$.

Proof. By assumptions,

$$0 \leq B_n(f)(x_0) - f(x_0) = \int_{K^n} (f \circ \pi_n - f(x_0))\, d\tilde{\mu}_{x_0,n}^T \leq 0$$

(see (3.1.5)).

Therefore, $f \circ \pi_n = f(x_0) = M_f$ on $\text{Supp}(\tilde{\mu}_{x_0,n}^T) = \left(\text{Supp}(\tilde{\mu}_{x_0}^T)\right)^n$ (see Proposition 1.1.6, (4)). From (3.5.5) it follows that

$$f = M_f \qquad \text{on } \overline{\text{co}}\left(\text{Supp}(\tilde{\mu}_{x_0}^T)\right)$$

and the result follows. The last part of the statement is a consequence of (1.3.9).

□

By surveying Examples 3.1.1-3.1.7, the reader can easily obtain more explicit maximum principles in the various settings described there, especially when the involved operator is a Markov projection. Below we discuss some of them.

Example 3.5.5. *Consider two Markov projections S and T on $\mathscr{C}(K)$ such that the subspaces $\{h \in \mathscr{C}(K) \mid S(h) = h\}$ and $\{h \in \mathscr{C}(K) \mid T(h) = h\}$ separate the points of K. Let $\lambda \in \mathscr{C}(K)$, $0 \le \lambda \le \mathbf{1}$ and consider the Markov operator $U_\lambda = \lambda T + (\mathbf{1} - \lambda)S$ together with its corresponding Bernstein-Schnabl operators $(B_{n,U_\lambda})_{n\ge 1}$ given by (3.1.33).*

If $f \in \mathscr{C}(K)$ and $f \le B_{n,U_\lambda}(f)$ for every $n \ge 1$, then f attains its maximum on the smallest subset of K containing all the supports $\mathrm{Supp}(\tilde{\mu}_x^T)$ and $\mathrm{Supp}(\tilde{\mu}_x^S)$ $(x \in K)$.

This is a consequence of Theorem 3.5.4 because, for every $x \in K$,

$$\mathrm{Supp}(\tilde{\mu}_x^{U_\lambda}) = \begin{cases} \mathrm{Supp}(\tilde{\mu}_x^S) & \text{if } \lambda(x) = 0; \\ \mathrm{Supp}(\tilde{\mu}_x^T) & \text{if } \lambda(x) = 1; \\ \mathrm{Supp}(\tilde{\mu}_x^S) \cup \mathrm{Supp}(\tilde{\mu}_x^T) & \text{if } 0 < \lambda(x) < 1. \end{cases}$$

Example 3.5.6. *Consider the two-dimensional canonical simplex $K_2 := \{(x, y) \in \mathbf{R}^2 \mid x \ge 0,\ y \ge 0,\ x + y \le 1\}$ and the Markov projection T on $\mathscr{C}(K_2)$ defined by*

$$T(f)(x,y) := \begin{cases} (1-x-y)f(0,0) + (x+y)f\left(\frac{x}{x+y}, \frac{y}{x+y}\right) & \text{if } x+y > 0; \\ f(0,0) & \text{if } x = y = 0 \end{cases}$$

$(f \in \mathscr{C}(K_2), (x, y) \in K_2)$.

The operator T verifies (3.1.2) and

$$\partial_T K_2 = \{(0,0)\} \cup \{(x,y) \in K_2 \mid x + y = 1\}. \tag{3.5.10}$$

The Bernstein-Schnabl operators associated with T are given by

$$B_n(f)(x,y) = \begin{cases} \sum\limits_{h=0}^{n} \binom{n}{h}(1-x-y)^{n-h}(x+y)^h f\left(\frac{hx}{n(x+y)}, \frac{hy}{n(x+y)}\right) & \text{if } x+y > 0; \\ f(0,0) & \text{if } x = y = 0 \end{cases}$$

$(f \in \mathscr{C}(K_2), (x, y) \in K_2, n \ge 1)$.

Therefore, each $f \in \mathscr{C}(K_2)$ satisfying $f \le B_n(f)$ for every $n \ge 1$ attains its maximum on $\partial_T K_2$.

Note that, in general, property (3.5.4) fails to characterize T-axially convex functions.

For instance, if $K = K_2$ denotes the two-dimensional canonical simplex, considering the canonical projection T_2 on $\mathscr{C}(K_2)$ (see (1.1.48)) and the corresponding Bernstein-Schnabl operators (3.1.18), then the function

$$f(x,y) = x^2y^2 - \frac{1}{2}x^2y + \frac{3}{64}x^2 - 2xy \qquad ((x,y) \in K_2) \tag{3.5.11}$$

satisfies (3.5.4) but it is not T_2-axially convex.

Below we shall discuss another notion of convexity, more general than the T-axially one, which again implies (3.5.4).

For every $f \in \mathscr{C}(K)$, $z \in K$ and $\alpha \in [0,1]$, consider the function $f_{z,\alpha} \in \mathscr{C}(K)$ defined by (3.1.32).

Definition 3.5.7. *Given a Markov operator T on $\mathscr{C}(K)$, then a function $f \in \mathscr{C}(K)$ is said to be* T-convex *if*

$$f_{z,\alpha} \le T(f_{z,\alpha}) \qquad \text{for every } z \in K \text{ and } \alpha \in [0,1]. \tag{3.5.12}$$

Note that from (3.5.12), setting $\alpha = 1$, it follows that $f \le T(f)$.

We recall that, if $f \in \mathscr{C}(K)$ is convex, then $f \le T(f)$ (see (3.5.8)). Moreover, every $f_{z,\alpha}$ is convex too and hence $f_{z,\alpha} \le T(f_{z,\alpha})$ $(z \in K, \alpha \in [0,1])$.

Therefore, every convex function $f \in \mathscr{C}(K)$ is also T-convex. On the other hand, if f is T-axially convex, then each $f_{z,\alpha}$ is T-axially convex, so that, according to Theorem 3.5.2, $f_{z,\alpha} \le T(f_{z,\alpha})$ $(z \in K, 0, \le \alpha \le 1)$.

In other words, T-axially convex functions are T-convex as well.

The converse generally does not hold true. Consider, indeed, the closed unit ball K of $\mathbf{R}^2$ centered at the origin together with the Poisson operator T_Δ associated with the Laplacian Δ (see (3.1.21) and (3.1.22)). Since $\partial_{T_\Delta} K = \partial K$, the T_Δ-axially convex functions are the convex ones. The function

$$f(x,y) := x^2 - y^2 \qquad ((x,y) \in K) \tag{3.5.13}$$

is harmonic on K together with all the functions $f_{z,\alpha}$, so that $f_{z,\alpha} = T(f_{z,\alpha})$ $(z \in K, \alpha \in [0,1])$. Therefore, f is T_Δ-convex, but it fails to be convex.

On the other hand, if K is a Bauer simplex and T is its canonical projection, then the two notions coincide (see Theorem 3.5.9).

The next result shows in particular that the T-convexity generates not only property (3.5.4), but also the monotonicity with respect to the parameter $\lambda \in \mathscr{C}(K)$ for the family of operators $B_{n,\lambda}$ defined by (3.1.35).

Theorem 3.5.8. *Let T be a Markov operator on $\mathscr{C}(K)$ and denote by $(B_n)_{n\ge1}$ the corresponding Bernstein-Schnabl operators. Given $f \in \mathscr{C}(K)$, the following statements hold true:*

(a) *If S is a Markov operator on $\mathscr{C}(K)$ and if $S(f_{z,\alpha}) \le T(f_{z,\alpha})$ for every $z \in K$ and $0 \le \alpha \le 1$, then*

$$B_{n,S}(f) \le B_n(f) \quad \text{for every } n \ge 1. \tag{3.5.14}$$

(b) *If S is a Markov operator on $\mathscr{C}(K)$ such that $S(u) \le T(u)$ for every convex $u \in \mathscr{C}(K)$ and if f is convex, then (3.5.14) holds true.*

(c) *If f is T-convex and if $\eta, \lambda \in \mathscr{C}(K)$ satisfy $0 \le \eta \le \lambda \le \mathbf{1}$, then*

$$B_{n,\eta}(f) \le B_{n,\lambda}(f) \quad \text{for every } n \ge 1,$$

and hence

$$f \le B_{n,\lambda}(f) \le B_n(f) \quad \text{for every } n \ge 1.$$

Proof. As regards statement (a), first we observe that

$$\int_K f\left(x_0 + \frac{u}{n}\right) d\tilde{\mu}_x^S(u) \le \int_K f\left(x_0 + \frac{u}{n}\right) d\tilde{\mu}_x^T(u) \tag{1}$$

for every $n \ge 1$, $x_0 \in \frac{n-1}{n}K$ and $x \in K$.

Indeed, (1) is certainly true if $n = 1$ because $x_0 = 0$ and $S(f) \le T(f)$. Assume $n > 1$ and write $x_0 = \frac{n-1}{n} z_0$, for some $z_0 \in K$. Then, for every $u \in K$, $f\left(x_0 + \frac{u}{n}\right) = f_{z_0, \frac{1}{n}}(u)$, and hence

$$\int_K f\left(x_0 + \frac{u}{n}\right) d\tilde{\mu}_x^S(u) = S\left(f_{z_0,\frac{1}{n}}\right)(x) \le T\left(f_{z_0,\frac{1}{n}}\right)(x) = \int_K f\left(x_0 + \frac{u}{n}\right) d\tilde{\mu}_x^T(u).$$

Taking these preliminaries into account, for every $n \ge 1$ and $x \in K$, we get

$$\begin{aligned}
&B_{n,S}(f)(x) \\
&= \int_K \cdots \int_K \left(\int_K f\left(\frac{x_1 + \dots + x_{n-1}}{n} + \frac{x_n}{n}\right) d\tilde{\mu}_x^S(x_n)\right) d\tilde{\mu}_x^S(x_{n-1}) \cdots d\tilde{\mu}_x^S(x_1) \\
&\le \int_K \cdots \int_K \left(\int_K f\left(\frac{x_1 + \dots + x_{n-1}}{n} + \frac{x_n}{n}\right) d\tilde{\mu}_x^T(x_n)\right) d\tilde{\mu}_x^S(x_{n-1}) \cdots d\tilde{\mu}_x^S(x_1) \\
&= \int_K \cdots \int_K \left(\int_K f\left(\frac{x_1 + \dots + x_{n-2} + x_n}{n} + \frac{x_{n-1}}{n}\right) d\tilde{\mu}_x^S(x_{n-1})\right) \\
&\times d\tilde{\mu}_x^T(x_n)\, d\tilde{\mu}_x^S(x_{n-2}) \cdots d\tilde{\mu}_x^S(x_1) \\
&\le \int_K \cdots \int_K \left(\int_K f\left(\frac{x_1 + \dots + x_{n-2} + x_n}{n} + \frac{x_{n-1}}{n}\right) d\tilde{\mu}_x^T(x_{n-1})\right) \\
&\times d\tilde{\mu}_x^T(x_n)\, d\tilde{\mu}_x^S(x_{n-2}) \cdots d\tilde{\mu}_x^S(x_1) \le \dots \le \\
&\le \int_K \cdots \int_K f\left(\frac{x_2 + \dots + x_n}{n} + \frac{x_1}{n}\right) d\tilde{\mu}_x^T(x_1) \cdots d\tilde{\mu}_x^T(x_n) = B_n(f).
\end{aligned}$$

Statement (b) follows from (1), because, if $f \in \mathscr{C}(K)$ is convex, then all $f_{z,\alpha}$ are convex too.

Statement (c) also follows from (1), because if $f \in \mathscr{C}(K)$ is T-convex, then, for every $z \in K$ and $\alpha \in [0,1]$,

$$U_\eta(f_{z,\alpha}) \leq U_\lambda(f_{z,\alpha}),$$

where $U_\eta := \eta T + (\mathbf{1} - \eta)I$ and $U_\lambda := \lambda T + (\mathbf{1} - \lambda)I$ (see Example 3.1.6). □

We proceed now to describe this notion in the setting of Bauer simplices and of the products of convex compact sets. For a proof of the next result we refer to [18, Theorem 6.3.2].

Theorem 3.5.9. *Given a Bauer simplex, consider its canonical projection and the corresponding Bernstein-Schnabl operators B_n, $n \geq 1$ (see Subsection 3.1.2). Then for every $f \in \mathscr{C}(K)$ the following statements are equivalent:*
(a) f is T-convex.
(b) f is T-axially convex.
(c) $B_n(f)$ is T-axially convex for every $n \geq 1$.

As regards product spaces, as in Section 3.1.5, consider a finite family $(K_i)_{1 \leq i \leq d}$ of convex compact subsets and, for every $i = 1, \dots, d$, let $T_i : \mathscr{C}(K_i) \longrightarrow \mathscr{C}(K_i)$ be a Markov operator satisfying (3.1.2). Set $K := \prod_{i=1}^{d} K_i$ and denote by T the tensor product of $(T_i)_{1 \leq i \leq d}$ (see (1.1.30)). Then T is a Markov operator on $\mathscr{C}(K)$ and it satisfies (3.1.2) as well.

For $f \in \mathscr{C}(K)$, $x = (x_1, \dots, x_d) \in K$ and $i = 1, \dots, d$, let us consider the function $f_x^i \in \mathscr{C}(K_i)$ defined by

$$f_x^i(t) := f(x_1, \dots, x_{i-1}, t, x_{i+1}, \dots, x_d) \qquad (t \in K_i).$$

Moreover, consider the operator $T_i^* : \mathscr{C}(K) \longrightarrow \mathscr{C}(K)$ such that, for every $f \in \mathscr{C}(K_i)$ and $x \in K$,

$$T_i^*(f)(x) := T_i(f_x^i)(x_i).$$

Then $T = T_1^* \circ \dots \circ T_d^*$.

Proposition 3.5.10. *Let $f \in \mathscr{C}(K)$. If f_x^i is T_i-convex for every $x \in K$ and $i = 1, \dots, d$, then f is T-convex.*

Proof. Fix $z = (z_1, \dots, z_d) \in K$ and $\alpha \in [0,1]$. Then, for every $x \in K$ and $i = 1, \dots, d$,

$$\begin{aligned} T_i^*(f_{z,\alpha})(x) &= T_i\left(\left(f_{z,\alpha}\right)_x^i\right)(x_i) = T_i\left(\left(f^i_{\alpha x + (1-\alpha z)}\right)_{z_i,\alpha}\right)(x_i) \\ &\geq \left(f^i_{\alpha x+(1-\alpha z)}\right)_{z_i,\alpha}(x_i) = \left(f_{z,\alpha}\right)_x^i(x_i) = f_{z,\alpha}(x). \end{aligned}$$

This means that f is T_i^*-convex for each $i = 1, \dots, d$. Since $T = T_1^* \circ \dots \circ T_d^*$, it easily follows that f is T-convex as well. □

Example 3.5.11. *For every $i = 1, 2$, let $K_i = [0,1]$ and $T_i : \mathscr{C}([0,1]) \longrightarrow \mathscr{C}([0,1])$ be the canonical projection (see (1.1.49)). Then $K = Q_2$ (see (2.3.26)) and $T : \mathscr{C}(Q_2) \longrightarrow \mathscr{C}(Q_2)$ is given by*

$$T(f)(x) = (1-x_1)(1-x_2)f(0,0) + x_1(1-x_2)f(1,0) \\ + (1-x_1)x_2 f(0,1) + x_1 x_2 f(1,1)$$

for all $x = (x_1, x_2) \in Q_2$ and all $f \in \mathscr{C}(Q_2)$.

We shall prove that the converse of Proposition 3.5.10 is also true in this setting, i.e., if $f \in \mathscr{C}(Q_2)$ is T-convex, then f_x^i is T_i-convex for every $x \in Q_2$ and $i = 1, 2$.

Let $x = (x_1, x_2) \in Q_2$ and $\alpha \in [0, 1/2]$. Then there exist $y \in \{0,1\}$ and $v \in [0,1]$ such that

$$x_2 = \alpha y + (1-\alpha)v. \tag{3.5.15}$$

(If $x_2 \le 1/2$, take $y = 0$ and $v = x_2/(1-\alpha)$; otherwise, $y = 1$ and $v = (x_2 - \alpha)/(1-\alpha)$).

Let $f \in \mathscr{C}(Q_2)$ be T-convex and $z \in [0,1]$. Consider the function $g := f_{(z,v),\alpha} \in \mathscr{C}(Q_2)$. By using (3.5.15), we get

$$\left(f_x^1\right)_{z,\alpha} = g(\cdot, y) \tag{3.5.16}$$

On the other hand, let $t \in [0,1]$. Since $y \in \{0,1\}$, we have

$$T(g)(t,y) = (1-t)\left((1-y)g(0,0) + yg(0,1)\right) + t\left((1-y)g(1,0) + yg(1,1)\right) \\ = (1-t)g(0,y) + tg(1,y) = T_1(g(\cdot,y))(t).$$

Combined with (3.5.16), this gives

$$T_1\left(\left(f_x^1\right)_{z,\alpha}\right)(t) = T(g)(t,y).$$

Recalling the definition of g and the fact that f is T-convex, we get $T(g) \ge g$, so that

$$T_1\left(\left(f_x^1\right)_{z,\alpha}\right)(t) \ge g(t,y) = \left(f_x^1\right)_{z,\alpha}(t).$$

This entails

$$(1-t)f((1-\alpha)z, x_2) + tf(\alpha + (1-\alpha)z, x_2) \ge f(\alpha t + (1-\alpha)z, x_2) \tag{3.5.17}$$

for all $t, z, x_2 \in [0,1]$ and $\alpha \in [0, 1/2]$.

Setting $h := f(\cdot, x_2)$, we obtain from (3.5.17)

$$(1-t)h((1-\alpha)z) + th(\alpha + (1-\alpha)z) \ge h(\alpha t + (1-\alpha)z) \tag{3.5.18}$$

for all $t, z \in [0,1]$ and $\alpha \in [0, 1/2]$.

Let now $0 \le p \le q \le 1$, $q - p \le 1/2$. Take $\alpha = q - p$, $z = p/(1-\alpha)$. Then (3.5.18) gives

$$(1-t)h(p) + th(q) \ge h((1-t)p + tq) \tag{3.5.19}$$

for all $t \in [0,1]$ and for every $0 \leq p \leq q \leq 1$ with $q-p \leq 1/2$.

It is easy to infer from (3.5.19) that h is convex on $[0,1]$; since $f_x^1 = h$, we have that f_x^1 is convex, i.e., T_1-convex. Similarly, we can prove that f_x^2 is T_2-convex.

Remarks 3.5.12.

1. The result contained in the above example can be extended to a more general setting.

Indeed, let $K_i \subset \mathbf{R}^{n_i}$ be a simplex and let $T_i : \mathscr{C}(K_i) \longrightarrow \mathscr{C}(K_i)$ be the canonical projection (see (1.1.48)), $i = 1, \dots, d$. Then $K := \prod_{i=1}^{d} K_i$ is said to be a *d-simploid* (see [80]).

Let T be the tensor product of $(T_i)_{1\leq i\leq d}$. Then the following statements are equivalent:

(a) $f \in \mathscr{C}(K)$ is T-convex.
(b) f_x^i is T_i-convex for every $x \in K$ and $i = 1, \dots, d$.
(c) f_x^i is axially convex for every $x \in K$ and $i = 1, \dots, d$.

For a proof, see [168, Theorem 6.1].

2. If K is a closed ball of $\mathbf{R}^d$ or, more generally, a closed convex compact subset of $\mathbf{R}^d$ whose boundary is an ellipsoid (see formula (4.2.19) of this monograph) and if T_L denotes the Poisson operator associated with a strictly elliptic differential operator L as in (3.1.19) and (3.1.20), then a function $f \in \mathscr{C}(K)$ is T_L-convex if and only if f is L-subharmonic (see [176, Corollary 4.2].)

3.6 Notes and comments

Bernstein-Schnabl operators were first introduced by Schnabl ([185]) in the context of the sets of all probability Radon measures on a compact Hausdorff space, i.e., in the context of Bauer simplices of Example 3.1.2 (see, e.g., [135, Proposition 6.38]). They were subsequently studied by Grossman ([112]) who named them as Bernstein-Schnabl operators, Nishishiraho ([150] and [156]) and Raşa ([165]-[166]) mainly in connection with approximation and saturation problems and with preservation properties.

In [8] Altomare started several investigations on Bernstein-Schnabl operators associated with a positive projection together with their connections with Markov semigroups and degenerate evolution equations. A comprehensive survey on the main results obtained up to 1994 is documented in [18, Chapter 6 and the references contained in the relevant notes]. More recent results can be found in [10], [25], [19], [26], [28], [34], [35], [64], [167], [168], [169], [170], [174], [176].

Lototsky-Schnabl operators (3.1.35) were first introduced by Schempp ([183]) and, in a more general form, by Grossman ([113]). For a deep analysis on these operators, provided they are associated with a positive projection, we refer to [18, Chapter 6 and the references contained in the relevant notes] and, more recently, to [2], [35], [167], [168], [170], [176].

All the results of Sections 3.2-3.5 are new and generalize similar ones which have been obtained in the case where the generating operator is a Markov projection (see once again [18, Chapter 6 and the references contained in the relevant notes]).

The property that the space of Hölder (Lipschitz, respectively) continuous functions on $[0,1]$ is invariant under Bernstein operators was first discovered by Lindvall ([130]) (Hájek ([114]), respectively). Afterwards, Brown, Elliot and Paget ([60]) gave a simpler proof.

A similar property was proved by Anastassiou, Cottin and Gonska ([37]) for simplices. Raşa ([167]) proved Theorem 3.5.2 and Corollary 3.5.3 in a general framework for Bernstein-Schnabl and Lototsky-Schnabl operators associated with a Markov projection.

In [174] Raşa and Vladislav carried out a detailed study concerning the constant $c(\|\cdot\|, T)$ defined by (3.3.24).

Theorems 3.4.2 and 3.4.3 are new. The second one generalizes Theorem 2.6 of [28] obtained for $K = [0,1]$. Example 3.4.8 generalizes Theorem 1 of [2].

Conditions (C_n) of Example 3.4.10 were considered in [73, Theorem 5] in order to prove that $B_n(f)$ is convex for a given $n \geq 1$. In Theorem 3' of the same paper

a necessary and sufficient condition in order that $B_n(f)$ is convex is also given (see also [74]).

Conditions (H_n) and (G_n) appear in [70, Theorem 2] in order to discuss similar problems on Q_2.

The notion of T-axially convex function is due to Raşa ([167]) (see also [18, Section 6.1]) and it generalizes the notion of axially convex function for simplices of $\mathbf{R}^d$ which appears in [79] and [180] and which seems to have been first introduced in an unpublished manuscript by Schmid in 1975 in the case of two variables.

A version of Theorem 3.5.2 for Markov projections was obtained in [167] (see also [18, Theorem 6.1.14]). It extends several results obtained in different contexts. Particular cases of it are the results obtained in [73] for two dimensional simplices, in [79] and [180] for simplices in $\mathbf{R}^d$, in [80] for the product of a finite number of simplices and in [200] and [39] for the unit interval and the classical Bernstein operators (see also [162]).

The maximum principle described by Theorem 3.5.4 extends a similar one for Markov projections (see [18, Theorem 6.1.17]). It unifies several maximum principles obtained by different and more complicated methods (see, for example, [75] for the simplex in $\mathbf{R}^2$, [80] for the finite products of simplices and [181] for the simplices of $\mathbf{R}^d$, $d \geq 3$).

Raşa introduced in [167] the notion of T-convex function and established part (3) in Theorem 3.5.8 in the particular case where T is a Markov projection (see also [18, Theorem 6.1.14]).

Theorem 3.5.9 is also due to Raşa [167, Theorem 4.1] (see also [18, Theorem 6.3.2]).

In the finite dimensional case Theorem 3.5.9 is due independently to Dahmen ([79, Theorem 4.9 and (4.18)]) and Sauer ([180, Theorem 3]). In the same paper Sauer also showed that the polyhedral convexity is preserved by Bernstein polynomials.

4 Differential operators and Markov semigroups associated with Markov operators

> In recent years important progress has been made in the study of semigroups of operators from the viewpoint of approximation theory... The applications of the theory not only permit integration of a series of diverse questions from many domains of mathematical analysis but also lead to significant new results on classical approximation theory, on the initial and boundary behaviour of solutions of partial differential equations, and on the theory of singular integrals.
>
> *Paul L. Butzer and Hubert Berens*

In this chapter we are concerned with the second fundamental subject of this monograph which pertains to (abstract) differential operators and Markov semigroups. This subject is strongly interrelated with the one we developed in Chapter 3 and the relationship itself between the two topics, namely, positive approximation processes and Markov semigroups, constitutes one of the distinguishing features of the book.

As in Chapter 3, we consider a Markov operator T on $\mathscr{C}(K)$, K being a convex compact subset of some locally convex Hausdorff space, and we associate with it a natural linear operator which, if K is finite dimensional, is indeed a second-order elliptic differential operator which degenerates on a suitable subset of K containing its extreme points.

Our main aim is to show that, under suitable assumptions on T, this (abstract) differential operator is closable and its closure generates a Markov semigroup on $\mathscr{C}(K)$.

One of the difficulties in studying such a problem with the methods of the theory of partial differential equations lies in the fact that the boundary of K is generally non-smooth due to the presence of possible sides and corners. On the other hand, several particular examples of such differential operators are involved in evolution problems arising in biology, financial mathematics and other fields, so that they seem to be worthy of a comprehensive and thorough study.

Due to the circumstance that the above-mentioned operators are related to the Bernstein-Schnabl operators associated with T via an asymptotic formula, we obtain the generation result by means of a Trotter-Schnabl-type theorem by also showing that the Markov semigroup can be described as a limit of suitable iterates of Bernstein-Schnabl operators.

This representation/approximation formula allows us to transfer several preservation properties of Bernstein-Schnabl operators studied in Chapter 3 to the Markov semigroup by stressing, as a consequence, some spatial regularity properties of the solutions to the relevant initial-boundary value evolution

problem as well as some additional properties of the transition functions of the corresponding Markov process. Moreover, in some special cases the asymptotic behavior of the Markov semigroup is exactly described.

The main results rely on the crucial assumption that the Markov operator T maps polynomials into polynomials of the same degree. This preservation property seems to have an independent own interest and in the final part of Section 4.3 we discuss several situations where it is satisfied.

Finally, in Section 4.5 we consider the special case of the interval $[0, 1]$, where additional more complete results are obtained.

4.1 Asymptotic formulae for Bernstein-Schnabl operators

This section will be devoted to some asymptotic formulae for Bernstein-Schnabl operators which will show a saturation phenomenon for such operators, i.e., the order of approximation cannot be improved beyond the critical order $\mathcal{O}(n^{-1})$, where by $\mathcal{O}$ we mean the usual Landau symbol, by improving the smoothness of the approximated functions. Furthermore, the asymptotic formulae are the key tool to show the existence of a Markov semigroup of operators generated by Bernstein-Schnabl operators.

We keep fixed the same notation of the preceding chapter; thus K denotes a metrizable convex compact subset and $T : \mathscr{C}(K) \longrightarrow \mathscr{C}(K)$ stands for a Markov operator satisfying (3.1.2). Furthermore, let us denote by $(B_n)_{n\geq 1}$ the sequence of Bernstein-Schnabl operators associated with T (see (3.1.7)).

For every $m \geq 1$, we denote by $P_m(K)$ the linear subspace generated by products of m continuous affine functions on K, i.e.,

$$P_m(K) := \operatorname{span}\left(\left\{\prod_{i=1}^{m} h_i \mid h_1, \dots, h_m \in A(K)\right\}\right). \tag{4.1.1}$$

Clearly, $P_m(K) \subset P_{m+1}(K)$ and

$$P_\infty(K) := \bigcup_{m\geq 1} P_m(K) \tag{4.1.2}$$

is a subalgebra of $\mathscr{C}(K)$ which separates the points of K and hence, by the Stone-Weierstrass theorem, it is dense in $\mathscr{C}(K)$.

In order to determine the values of Bernstein-Schnabl operators on $P_\infty(K)$, for every $m \geq 1$ and $n \geq 1$ we introduce the symbol $F(m, n)$ to denote the set of all mappings $\sigma : \{1, \dots, m\} \longrightarrow \{1, \dots, n\}$.

If $\sigma \in F(m, n)$, consider the equivalence relation R_σ on $\{1, \dots, m\}$ defined by

$$R_\sigma := \{(i, j) \mid i, j = 1, \dots, m \text{ and } \sigma(i) = \sigma(j)\} \tag{4.1.3}$$

and the corresponding subdivisions $R_1^\sigma, \dots, R_{s_\sigma}^\sigma$ of $\{1, \dots, m\}$ in equivalence classes.

Clearly, $\sum_{j=1}^{s_\sigma} \operatorname{card}(R_j^\sigma) = m$.

Lemma 4.1.1. *Let $h_1, \dots, h_m \in A(K)$, $m \geq 1$. Then, for every $n \geq 1$,*

$$B_n\left(\prod_{j=1}^{m} h_j\right) = \frac{1}{n^m} \sum_{\sigma \in F(m,n)} \prod_{j=1}^{s_\sigma} T\left(\prod_{i \in R_j^\sigma} h_i\right). \tag{4.1.4}$$

Therefore, if $T(P_m(K)) \subset P_m(K)$ for every $m \geq 1$, then

$$B_n(P_m(K)) \subset P_m(K) \tag{4.1.5}$$

for every $n, m \geq 1$.

Proof. First we recall that, if $(\alpha_{ij})_{\substack{1 \leq i \leq n \\ 1 \leq j \leq m}}$ is a matrix of real numbers, then

$$\prod_{j=1}^{m} \sum_{i=1}^{n} \alpha_{ij} = \sum_{\sigma \in F(m,n)} \prod_{j=1}^{m} \alpha_{\sigma(j)j}.$$

Accordingly, if $x_1, \dots x_n \in K$, we get

$$\left(\prod_{j=1}^{m} h_j\right)\left(\frac{x_1 + \dots + x_n}{n}\right) = \frac{1}{n^m} \prod_{j=1}^{m} \sum_{i=1}^{n} h_j(x_i)$$
$$= \frac{1}{n^m} \sum_{\sigma \in F(m,n)} \prod_{j=1}^{m} h_j(x_{\sigma(j)}) = \frac{1}{n^m} \sum_{\sigma \in F(m,n)} \prod_{j=1}^{s_\sigma} \prod_{i \in R_j^\sigma} h_i(x_{\sigma(j)}).$$

On account of (3.1.7) the result follows. □

In order to state our first general asymptotic formula, for every $m \geq 1$ and $h_1, \dots, h_m \in A(K)$ we set

$$\Theta_T(h_1, \dots, h_m) := \begin{cases} 0 & \text{if } m = 1; \\ T(h_1 h_2) - h_1 h_2 & \text{if } m = 2; \\ \sum\limits_{1 \leq i < j \leq m} \left(T(h_i h_j) - h_i h_j\right) \prod\limits_{\substack{r=1 \\ r \neq i,j}}^{m} h_r & \text{if } m \geq 3. \end{cases} \tag{4.1.6}$$

Note that by (3.1.4) and (1.3.4),

$$\Theta_T(h_1, \dots, h_m)(x) = 0 \text{ for every } x \in \partial_e K \text{ or, more generally, for each } x \in \partial_T K. \tag{4.1.7}$$

The next asymptotic formula holds true:

Theorem 4.1.2. *Under the preceding assumptions,*

$$\lim_{n\to\infty} n\left(B_n\left(\prod_{j=1}^{m} h_j\right) - \prod_{j=1}^{m} h_j\right) = \Theta_T(h_1, \dots, h_m) \qquad \textit{uniformly on } K \quad (4.1.8)$$

for every $m \geq 1$ *and* $h_1, \dots, h_m \in A(K)$.

Proof. Fix $h_1, \dots, h_m \in A(K)$, $m \geq 1$. The cases $m = 1, 2$ easily follow from (3.2.2) and (3.2.3). Assuming $m \geq 3$, $n \geq m-1$ and recalling Lemma 4.1.1 and that $\mathrm{card}(F(m,n)) = n^m$, we get

$$\begin{aligned} n\left(B_n\left(\prod_{j=1}^{m} h_j\right) - \prod_{j=1}^{m} h_j\right) &= \frac{1}{n^{m-1}} \sum_{\sigma\in F(m,n)} \left[\prod_{j=1}^{s_\sigma} T\left(\prod_{i\in R_j^\sigma} h_i\right) - \prod_{j=1}^{m} h_j\right] \\ &= \frac{1}{n^{m-1}} \sum_{\sigma\in F^*(m,n)} \left[\prod_{j=1}^{s_\sigma} T\left(\prod_{i\in R_j^\sigma} h_i\right) - \prod_{j=1}^{m} h_j\right], \end{aligned} \tag{1}$$

where $F^*(m,n) := \{\sigma \in F(m,n) \mid \sigma \text{ is not one-to-one}\}$ (see (3.2.2)).

Consider the subset $F_2^*(m,n)$ of all $\sigma \in F^*(m,n)$ such that only one equivalence class contains two distinct elements and the other ones reduce to singletons. The elements of $F_2^*(m,n)$ can be obtained by choosing two distinct indices $i, j = 1, \dots, m$, by assigning to them a unique value between $\{1, \dots, n\}$ and by arranging the remaining $n-1$ numbers as arrangements of size $m-2$, which is possible because $n \geq m-1$.

Therefore

$$\begin{aligned} &\sum_{\sigma\in F_2^*(m,n)} \left[\prod_{j=1}^{s_\sigma} T\left(\prod_{i\in R_j^\sigma} h_i\right) - \prod_{j=1}^{m} h_j\right] \\ &= n(n-1)\cdots(n-m+2) \sum_{1\leq i<j\leq m} (T(h_i h_j) - h_i h_j) \prod_{\substack{r=1 \\ r\neq i,j}}^{m} h_r \\ &= n^{m-1}\left(1-\frac{1}{n}\right)\cdots\left(1-\frac{m-2}{n}\right)\Theta_T(h_1, \dots, h_m). \end{aligned} \tag{2}$$

On the other hand, each element of $F^*(m,n) \setminus F_2^*(m,n)$ can be obtained by choosing a subdivision $R_1, \dots R_p$ of subsets of $\{1, \dots, m\}$ with $p \leq m-2$ and by assigning to the elements of each R_j an element $i_j \in \{1, \dots, n\}$ such that $i_j \neq i_l$ if $j \neq l$. Accordingly,

$$\begin{aligned} &\sum_{\sigma\in F^*(m,n)\setminus F_2^*(m,n)} \left[\prod_{j=1}^{s_\sigma} T\left(\prod_{i\in R_j^\sigma} h_i\right) - \prod_{j=1}^{m} h_j\right] \\ &= \sum_{p=1}^{m-2} n(n-1)\cdots(n-p+1) \sum_{\substack{R_1,\dots,R_p \\ \text{subdivision of } \{1,\dots,m\}}} \left[\prod_{j=1}^{p} T\left(\prod_{i\in R_j} h_i\right) - \prod_{j=1}^{m} h_j\right]. \end{aligned} \tag{3}$$

Now, from (1), (2) and (3) the result follows. □

From Theorem 4.1.2 it also follows that for every $u \in P_\infty(K)$ there exists $\lim_{n\to\infty} n(B_n(u) - u)$ in $\mathscr{C}(K)$ and hence we can consider the linear operator $L_T : P_\infty(K) \longrightarrow \mathscr{C}(K)$

$$L_T(u) := \lim_{n\to\infty} n(B_n(u) - u) \qquad (u \in P_\infty(K)). \tag{4.1.9}$$

Thus, if $h_1, \dots, h_m \in A(K)$, $m \geq 1$, then

$$L_T\left(\prod_{j=1}^{m} h_j\right) = \Theta_T(h_1, \dots, h_m) \tag{4.1.10}$$

Remark 4.1.3. From (4.1.6) and (4.1.10) it also follows that if $p \in P_2(K)$, then $L_T(p) = T(p) - p$. Therefore, if T is a Markov projection and $T(P_2(K)) \subset P_2(K)$, then

$$L_T(T(p)) = 0 \qquad \text{for every } p \in P_2(K). \tag{4.1.11}$$

When K is a convex compact subset of some Euclidean space $\mathbf{R}^d$, $d \geq 1$, then it is possible to state an asymptotic formula, like in Theorem 4.1.2, for functions $u \in \mathscr{C}^2(K)$.

We begin with the following preliminary result.

Lemma 4.1.4. *Let $x \in K$ and $h_1, h_2, h_3, h_4 \in A(K)$ satisfying $h_i(x) = 0$ for $i = 1, \dots, 4$. Then, for every $n \geq 1$,*

$$\begin{aligned} B_n(h_1h_2h_3h_4)(x) &= \frac{1}{n^3}\left[T(h_1h_2h_3h_4)(x) + (n-1)T(h_1h_2)(x)T(h_3h_4)(x)\right. \\ &\left.+(n-1)T(h_1h_3)(x)T(h_2h_4)(x) + (n-1)T(h_1h_4)(x)T(h_2h_3)(x)\right]. \end{aligned} \tag{4.1.12}$$

In particular, if $h, k \in A(K)$ and $h(x) = k(x) = 0$, then

$$\begin{aligned} &B_n(h^2k^2)(x) \\ &= \frac{1}{n^3}\left[T(h^2k^2)(x) + (n-1)T(h^2)(x)T(k^2)(x) + 2(n-1)T(hk)^2(x)\right]. \end{aligned} \tag{4.1.13}$$

Proof. Formula (4.1.12) follows from Lemma 4.1.1 on account of the special hypotheses made on $h_1, \dots, h_4$. Formula (4.1.13) follows from (4.1.12) with $h_1 = h_2 = h$ and $h_3 = h_4 = k$. □

From now on, we shall assume that K is a convex compact subset of $\mathbf{R}^d$, $d \geq 1$, having non-empty interior $\mathrm{int}(K)$. We recall that the symbol $\mathscr{C}^2(K)$ denotes the space of all real-valued continuous functions $u : K \longrightarrow \mathbf{R}$ that are twice-continuously differentiable in $\mathrm{int}(K)$ and whose partial derivatives of order ≤ 2 can be continuously extended to K.

In this case assumption (3.1.2) on the Markov operator T on $\mathscr{C}(K)$ means that

$$T(h) = h \qquad \text{for each } h \in \{\mathbf{1}, pr_1, \dots, pr_d\} \tag{4.1.14}$$

(see (1.1.21)).

Moreover, $P_m(K)$ is the subspace of all (restrictions to K of) polynomials of degree not greater than m and $P_\infty(K)$ is the subalgebra of all polynomials on K.

Consider the differential operator $W_T : \mathscr{C}^2(K) \longrightarrow \mathscr{C}(K)$ defined by

$$W_T(u) := \frac{1}{2}\sum_{i,j=1}^{d} \alpha_{ij}\frac{\partial^2 u}{\partial x_i \partial x_j} \qquad (u \in \mathscr{C}^2(K)) \tag{4.1.15}$$

where, for every $i, j = 1, \dots, d$,

$$\alpha_{ij} := T(pr_i pr_j) - pr_i pr_j. \tag{4.1.16}$$

It turns out that the differential operator W_T is generated by an asymptotic formula for Bernstein-Schnabl operators.

Theorem 4.1.5. *For every $u \in \mathscr{C}^2(K)$,*

$$\lim_{n\to\infty} n(B_n(u) - u) = W_T \qquad \textit{uniformly on } K. \tag{4.1.17}$$

Therefore,

$$L_T(u) = W_T(u) \qquad \textit{for every } u \in P_\infty(K). \tag{4.1.18}$$

Proof. The proof consists in applying Theorem 1.5.2. More precisely, we shall verify conditions (i), (ii), (iii), (v) and (vi), with $q = 4$, of that theorem. We shall also keep the same notation as there.

For any $n \geq 1$, $i = 1, \dots, d$ and $x = (x_i)_{1\leq i\leq d} \in K$, we have that $B_n(\mathbf{1}) = \mathbf{1}$ and, by (3.2.2),

$$B_n(pr_i \circ \Psi_x)(x) = B_n(pr_i - x_i)(x) = pr_i(x) - x_i = 0,$$

which shows (i) and (ii) with $\gamma = 0$ and $\beta_i = 0$.

On the other hand, by applying (3.2.3) and (4.1.14), for $i, j = 1, \dots, d$,

$$\begin{aligned} &B_n((pr_i \circ \Psi_x)(pr_j \circ \Psi_x))(x) = B_n((pr_i - x_i)(pr_j - x_j))(x) \\ &= \frac{1}{n}T((pr_i - x_i)(pr_j - x_j))(x) + \frac{n-1}{n}(pr_i(x) - x_i)(pr_j(x) - x_j) = \frac{1}{n}\alpha_{ij}(x) \end{aligned}$$

and hence (iii) follows.

A similar reasoning shows that

$$\begin{aligned} B_n(\Phi_x^2)(x) &= B_n\left(\sum_{i=1}^{d}(pr_i - x_i)^2\right)(x) \\ &= \frac{1}{n}T\left(\sum_{i=1}^{d}(pr_i - x_i)^2\right)(x) = \frac{1}{n}\left[T(e)(x) - e(x)\right], \end{aligned}$$

where Φ_x is defined by (1.5.2) and $e(x) = \sum_{i=1}^{d} x_i^2$, which implies (v).

Finally, by (4.1.13), we get

$$
\begin{aligned}
nB_n(\Phi_x^4)(x) &= \sum_{i,j=1}^{d} nB_n((pr_i - x_i)^2(pr_j - x_j)^2)(x) \\
&= \frac{1}{n^2}\sum_{i,j=1}^{d} \big[T((pr_i - x_i)^2(pr_j - x_j)^2)(x) \\
&\quad +(n-1)T((pr_i - x_i)^2)(x)T((pr_j - x_j)^2)(x) \\
&\quad +2(n-1)T(((pr_i - x_i)(pr_j - x_j))^2)(x)\big] .
\end{aligned}
$$

Therefore, there exists a suitable positive constant M, depending on K only, such that

$$
nB_n(\Phi_x^4)(x) \le \frac{3n-2}{n^2}M \qquad \text{for every } x \in K,
$$

which implies (vi). The proof of (4.1.17) is now complete.

Clearly (4.1.18) follows from Theorem 4.1.2. □

4.2 Differential operators associated with Markov operators

The differential operator W_T and the operator L_T, which we have introduced by (4.1.9)-(4.1.10) and (4.1.15), naturally arise when studying asymptotic formulae for Bernstein-Schnabl operators.

These operators play a fundamental role in this book and in this section we begin to deepen their study by also discussing several illustrative examples.

We shall mostly consider the finite dimensional cases. Thus, K is a convex compact subset of $\mathbf{R}^d$, $d \geq 1$, having non-empty interior.

For the convenience of the reader, we recall that given a Markov operator T on $\mathscr{C}(K)$ satisfying (3.1.2) (or, equivalently, (4.1.14)) the differential operator $W_T : \mathscr{C}^2(K) \longrightarrow \mathscr{C}(K)$ is defined by

$$W_T(u) := \frac{1}{2} \sum_{i,j=1}^{d} \alpha_{ij} \frac{\partial^2 u}{\partial x_i \partial x_j}$$

($u \in \mathscr{C}^2(K)$), and, for each $i, j = 1, \dots, d$ and $x = (x_i)_{1 \leq i \leq d} \in K$,

$$\alpha_{ij}(x) := T(pr_i pr_j)(x) - (pr_i pr_j)(x) = T((pr_i - x_i)(pr_j - x_j))(x).$$

Accordingly, if $\xi_1, \dots, \xi_d \in \mathbf{R}$, then

$$\sum_{i,j=1}^{d} \alpha_{ij}(x)\xi_i\xi_j = T\left(\left(\sum_{i=1}^{d} \xi_i(pr_i - x_i)\right)^2\right)(x) \geq 0,$$

which implies that W_T is elliptic. Moreover, it degenerates on $\partial_T K$ (see (1.3.4)) and, in particular, on $\partial_e K$ because $\alpha_{ij} = 0$ on $\partial_T K$ for every $i, j = 1, \dots, d$.

The operator W_T will be referred to as the *elliptic second-order differential operator associated with the Markov operator* T.

On account of formula (4.1.18) of Theorem 4.1.5, if K is a more general convex compact subset, the operator L_T will be also referred to as the *abstract differential operator associated with* T.

Below we discuss some examples.

Examples 4.2.1.

1. *Consider a Markov operator T on $\mathscr{C}([0,1])$ satisfying (3.1.2), i.e.,*

$$T(e_1) = e_1, \tag{4.2.1}$$

where $e_1(x) := x$ $(0 \leq x \leq 1)$.

Then, for every $u \in \mathscr{C}^2([0,1])$ and $x \in [0,1]$,

$$W_T(u)(x) = \frac{\alpha(x)}{2} u''(x), \tag{4.2.2}$$

with

$$\alpha(x) := T(e_2)(x) - x^2 \tag{4.2.3}$$

and $e_2(x) := x^2$ $(0 \le x \le 1)$.

In particular, if T *is the canonical projection defined by* (1.1.49), *then*

$$\alpha(x) = x(1-x) \qquad (0 \le x \le 1) \tag{4.2.4}$$

and, if T *is given by* (3.1.14), *then*

$$\alpha(x) = x(1-x) - \frac{b(x)}{4} \qquad (0 \le x \le 1). \tag{4.2.5}$$

Coming back to the general case, since $e_2 \le e_1$ *on* $[0,1]$, *from* (4.2.1) *and* (1.1.7) *it follows that*

$$0 \le \alpha(x) \le x(1-x) \qquad (0 \le x \le 1). \tag{4.2.6}$$

By Theorem 1.3.1, $\partial_T([0,1]) = \{x \in [0,1] \mid T(e_2)(x) = x^2\}$ *and hence, if* $\partial_T([0,1]) = \{0,1\}$, *then*

$$0 < \alpha(x) \qquad (0 < x < 1). \tag{4.2.7}$$

Moreover, $\alpha \in \mathscr{C}([0,1])$ *and* $\alpha(0) = \alpha(1) = 0$.

Conversely, if $\alpha \in \mathscr{C}([0,1])$ *satisfies* (4.2.6), *then there always exists a Markov operator* T *on* $\mathscr{C}([0,1])$ *satisfying* (3.1.2) *such that* $\alpha = T(e_2) - e_2$.

Indeed, it is enough to consider the function $\lambda :]0,1[\longrightarrow \mathbf{R}$ *defined by*

$$\lambda(x) := \frac{\alpha(x)}{x(1-x)} \qquad (0 < x < 1) \tag{4.2.8}$$

and the Markov operator $T : \mathscr{C}([0,1]) \longrightarrow \mathscr{C}([0,1])$ *defined by*

$$T(f)(x) := \begin{cases} \lambda(x)T_1(f)(x) + (1-\lambda(x))f(x) & \text{if } 0 < x < 1; \\ \\ f(x) & \text{if } x = 0,1 \end{cases} \tag{4.2.9}$$

$(f \in \mathscr{C}([0,1]), 0 \le x \le 1)$ *(see* (1.1.49) *and* (3.1.31)*).*

2. *The differential operator associated with the canonical projection* T_d *on the d-dimensional simplex (see* (1.1.45) *and* (1.1.48)*) is given by*

$$\begin{aligned} W_{T_d}(u)(x) &:= \frac{1}{2}\sum_{i,j=1}^{d} x_i(\delta_{ij} - x_j)\frac{\partial^2 u}{\partial x_i \partial x_j}(x) \\ &= \frac{1}{2}\sum_{i,j=1}^{d} x_i(1-x_i)\frac{\partial^2 u}{\partial x_i^2}(x) - \sum_{1\le i<j\le d} x_i x_j \frac{\partial^2 u}{\partial x_i \partial x_j}(x) \end{aligned} \tag{4.2.10}$$

$(u \in \mathscr{C}^2(K_d), x = (x_i)_{1\le i\le d} \in K_d)$, *where* δ_{ij} *stands for the Kronecker symbol.*

The coefficients of W_{T_d} *vanish on the vertices of the simplex. Below we describe other differential operators with different degeneracies.*

Let $S : \mathscr{C}(K_d) \longrightarrow \mathscr{C}(K_d)$ *be the Markov operator defined by*

$$S(f)(x) := \begin{cases} \left(1 - \dfrac{x_1}{1 - \sum\limits_{i=2}^{d} x_i}\right) f(0, x_2, \dots, x_d) \\ + \dfrac{x_1}{1 - \sum\limits_{i=2}^{d} x_i} f\left(1 - \sum\limits_{i=2}^{d} x_i, x_2, \dots, x_d\right) & \text{if } \sum\limits_{i=2}^{d} x_i \neq 1; \\ f(0, x_2, \dots, x_d) & \text{if } \sum\limits_{i=2}^{d} x_i = 1 \end{cases} \tag{4.2.11}$$

$(f \in \mathscr{C}(K_d), x = (x_1, \dots, x_d) \in K_d)$.

Then,

$$S(pr_1 pr_j) = \begin{cases} \left(1 - \sum\limits_{i=2}^{d} pr_i\right) pr_1 & \text{if } j = 1; \\ pr_1 pr_j & \text{if } 1 < j \leq d \end{cases} \tag{4.2.12}$$

and $S(pr_i pr_j) = pr_i pr_j$ *for every* $1 < i \leq j \leq d$.

Therefore the differential operator associated with S *is given by*

$$W_S(u)(x) = \frac{1}{2}\left(x_1(1 - x_1) - x_1 \sum_{i=2}^{d} x_i\right) \frac{\partial^2 u}{\partial x_1^2}(x) \tag{4.2.13}$$

$(u \in \mathscr{C}^2(K_d), x = (x_1, \dots, x_d) \in K_d)$.

Thus, W_S *degenerates on the faces* $\{x = (x_1, \dots, x_d) \in K_d \mid x_1 = 0\}$ *and* $\left\{x = (x_1, \dots, x_d) \in K_d \mid \sum\limits_{i=1}^{d} x_i = 1\right\}$.

Finally, note that the differential operator associated with the Markov operator $G := \frac{T_d + S}{2}$ *is given by*

$$\begin{aligned} W_G(u)(x) = \frac{1}{4}\Bigg(&\left(2x_1(1 - x_1) - x_1 \sum_{i=2}^{d} x_i\right) \frac{\partial^2 u}{\partial x_1^2}(x) \\ &+ \sum_{i=2}^{d} x_i(1 - x_i) \frac{\partial^2 u}{\partial x_i^2}(x) - 2 \sum_{1 \leq i < j \leq d} x_i x_j \frac{\partial^2 u}{\partial x_i \partial x_j}(x)\Bigg) \end{aligned} \tag{4.2.14}$$

$(u \in \mathscr{C}^2(K_d), x = (x_1, \dots, x_d) \in K_d)$.

Therefore, W_G *degenerates on the vertices of* K_d *as well.*

3. *Let* $Q_d := [0, 1]^d$, $d \geq 1$, *and for every* $i = 1, \dots, d$ *consider a Markov operator* T_i *on* $\mathscr{C}([0, 1])$ *satisfying* (4.2.1).

If $T := \bigotimes\limits_{i=1}^{d} T_i$ *(see* (1.1.30)*), then*

$$W_T(u)(x) = \frac{1}{2} \sum_{i=1}^{d} \alpha_i(x) \frac{\partial^2 u}{\partial x_i^2}(x), \tag{4.2.15}$$

$(u \in \mathscr{C}^2(Q_d), x = (x_i)_{1 \le i \le d} \in Q_d)$, *where*

$$\alpha_i(x) := T_i(e_2)(x_i) - x_i^2 \qquad (1 \le i \le d). \tag{4.2.16}$$

Moreover,

$$0 \le \alpha_i(x) \le x_i(1 - x_i) \qquad (1 \le i \le d, x = (x_i)_{1 \le i \le d} \in Q_d). \tag{4.2.17}$$

Conversely, on account of Example 4.2.1, 1, if $(\alpha_i)_{1 \le i \le d}$ *is a family of continuous functions on* $[0,1]$ *satisfying* (4.2.17), *then there always exists a Markov operator* T *on* $\mathscr{C}(Q_d)$ *satisfying* (3.1.2) *whose differential operator satisfies* (4.2.15).

Finally note that, if $T = S_d$ *(see* (3.1.28)*), then*

$$W_{S_d}(u)(x) = \frac{1}{2} \sum_{i=1}^{d} x_i(1 - x_i) \frac{\partial^2 u}{\partial x_i^2}(x) \tag{4.2.18}$$

$(u \in \mathscr{C}^2(Q_d), x = (x_i)_{1 \le i \le d} \in Q_d)$.

In order to determine (in some particular cases) the differential operator associated with the Poisson operator considered in Subsection 3.1.4, we need to point out some preliminary results.

Consider a convex compact subset K of $\mathbf{R}^d$, $d \ge 2$, such that its boundary ∂K is an *ellipsoid*, i.e., there exists a real symmetric and positive-definite matrix $R = (r_{ij})_{\substack{1 \le i \le d \\ 1 \le j \le d}}$ and $\overline{x} = (\overline{x}_i)_{1 \le i \le d} \in \mathbf{R}^d$ such that

$$K = \left\{ x \in \mathbf{R}^d \mid \mathcal{Q}(x - \overline{x}) := \sum_{i,j=1}^{d} r_{ij}(x_i - \overline{x}_i)(x_j - \overline{x}_j) \le 1 \right\}. \tag{4.2.19}$$

Furthermore, consider a strictly elliptic differential operator

$$L(u)(x) := \sum_{i,j=1}^{d} c_{ij} \frac{\partial^2 u}{\partial x_i \partial x_j}(x) \tag{4.2.20}$$

$(u \in \mathscr{C}^2(\mathrm{int}(K)), x \in \mathrm{int}(K))$ associated with a real symmetric and positive-definite matrix $C = (c_{ij})_{\substack{1 \le i \le d \\ 1 \le j \le d}}$ and denote by T_L the relevant Poisson operator on $\mathscr{C}(K)$ (see (3.1.20)).

Theorem 4.2.2. *Under the above assumptions, there exists an isomorphism* $\varphi : \mathbf{R}^d \longrightarrow \mathbf{R}^d$ *such that, considering the mapping* $\Phi : \mathscr{C}(K) \longrightarrow \mathscr{C}(K')$ *defined by*

$$\Phi(u)(y) := u(\varphi^{-1}(y)) \qquad (u \in \mathscr{C}(K), y \in K') \tag{4.2.21}$$

with $K' := \varphi(K)$, *one gets*

$$\Phi \circ T_L = T_\Delta \circ \Phi, \tag{4.2.22}$$

where T_Δ *denotes the Poisson operator on* $\mathscr{C}(K')$ *associated with the Laplacian* Δ *(see* (3.1.21) *and* (3.1.22)*).*

Moreover, for every $m \ge 1$, T_L *maps polynomials of degree* $\le m$ *into polynomials of degree* $\le m$.

Proof. Considering the eigenvalues $\lambda_1, \dots, \lambda_d$ of the matrix C, let M be an orthogonal matrix such that

$$MCM^t = \begin{pmatrix} \lambda_1 & \cdots & 0 \\ \vdots & \cdots & \vdots \\ \vdots & \vdots & \vdots \\ 0 & \cdots & \lambda_d \end{pmatrix}$$

and set

$$D := \begin{pmatrix} \sqrt{\lambda_1} & \cdots & 0 \\ \vdots & \cdots & \vdots \\ \vdots & \vdots & \vdots \\ 0 & \cdots & \sqrt{\lambda_d} \end{pmatrix}.$$

Finally, let $\varphi : \mathbf{R}^d \longrightarrow \mathbf{R}^d$ be the linear transformation associated with the product matrix $((DM)^t)^{-1}$. Then, since $C = (DM)^t DM$, by a direct calculation we get

$$L(u)(x) = \Delta(\Phi(u))(\varphi(x)) \qquad (x \in \mathrm{int}(K), u \in \mathscr{C}^2(K))$$

and (4.2.22) holds true.

In order to show the final part of the statement, note that

$$K' = \{y \in \mathbf{R}^d \mid \mathcal{Q}'(y - \varphi(\overline{x})) \le 1\},$$

where $\mathcal{Q}'$ is the positive-definite quadratic form associated with the matrix

$$DMR(DM)^t,$$

so that $\partial K'$ is an ellipsoid, too.

Given $m \ge 1$, denote by $P_m(K)$ and $P_m(K')$ the subspaces of all (the restrictions to K and K' of) polynomials of degree $\le m$. From [41, Theorem 4], it follows that $T_\Delta(P_m(K')) \subset P_m(K')$ and hence

$$T_L(P_m(K)) = \Phi^{-1}(T_\Delta(\Phi(P_m(K)))) \subset P_m(K),$$

because $\Phi(P_m(K)) \subset P_m(K')$. □

By means of Theorem 4.2.2 we can now describe the differential operator W_{T_L}, defined according to (4.1.15) and (4.1.16), which we shall briefly denote by W_L.

Corollary 4.2.3. *Let K and L be as in* (4.2.19) *and* (4.2.20). *Under the additional assumptions of Theorem 4.2.2, the differential operator W_L associated with T_L is given by*

$$W_L(u)(x) = \begin{cases} \dfrac{1}{2\sum\limits_{i,j=1}^{d} r_{ij}c_{ij}}(1 - \mathcal{Q}(x - \overline{x}))L(u)(x) & \text{if } x \in \mathrm{int}(K); \\ \\ 0 & \text{if } x \in \partial K \end{cases} \tag{4.2.23}$$

$(u \in \mathscr{C}^2(K), x \in K)$.

In particular, if K is the closed ball (with respect to the Euclidean norm $\|\cdot\|_2$) with center $\overline{x} \in \mathbf{R}^d$ and radius $r > 0$ and if $L = \Delta$, then

$$W_\Delta(u)(x) = \begin{cases} \dfrac{r^2 - \|x - \overline{x}\|_2^2}{2d} \Delta(u)(x) & \text{if } \|x - \overline{x}\|_2 < r; \\ 0 & \text{if } \|x - \overline{x}\|_2 = r \end{cases} \tag{4.2.24}$$

$(u \in \mathscr{C}^2(K), x \in K)$.

Proof. Consider $h, k = 1, \dots, d$. According to (4.1.16), it suffices to show that

$$T_L(pr_h pr_k) - pr_h pr_k = \frac{1 - \mathcal{Q} \circ \tau}{\sum\limits_{i,j=1}^{d} r_{ij} c_{ij}} c_{hk},$$

where, for every $x \in \mathbf{R}^d$, $\tau(x) = x - \overline{x}$. By Theorem 4.2.2, $T_L(pr_h pr_k)$ is a polynomial of degree ≤ 2 and $T_L(pr_h pr_k) - pr_h pr_k$ vanishes on ∂K, i.e., on the subset where the irreducible polynomial $1 - \mathcal{Q} \circ \tau$ vanishes as well. By Hilbert's Nullstellensatz there exists $\gamma_{hk} \in \mathbf{R}$ such that

$$T_L(pr_h pr_k) - pr_h pr_k = \gamma_{hk}(\mathbf{1} - \mathcal{Q} \circ \tau).$$

Therefore, given $x \in \operatorname{int}(K)$,

$$\begin{aligned} 0 = L(T_L(pr_h pr_k))(x) &= L(pr_h pr_k)(x) - \gamma_{hk} L(\mathcal{Q} \circ \tau)(x) \\ &= 2c_{hk} - \gamma_{hk} \sum_{i,j=1}^{d} r_{ij} c_{ij} \end{aligned}$$

and hence

$$\gamma_{hk} = \frac{2c_{hk}}{\sum\limits_{i,j=1}^{d} r_{ij} c_{ij}}$$

and the result follows. □

Remark 4.2.4. Replacing, if necessary, each coefficient c_{ij} of the matrix C in (4.2.20) by $\dfrac{c_{ij}}{\sum_{i,j=1}^{d} r_{ij} c_{ij}}$, $i, j = 1, \dots, d$, we may always assume that

$$\sum_{i,j=1}^{d} r_{ij} c_{ij} = 1. \tag{4.2.25}$$

In this case, the differential operator W_L turns into

$$W_L(u)(x) = \begin{cases} \dfrac{1 - \mathcal{Q}(x - \overline{x})}{2} L(u)(x) & \text{if } x \in \operatorname{int}(K); \\ 0 & \text{if } x \in \partial K \end{cases} \tag{4.2.26}$$

$(u \in \mathscr{C}^2(K), x \in K)$.

We finally mention that, under the same notation of Subsections 3.1.6 and 3.1.7, the differential operators associated with the Markov operators (3.1.31) and (3.1.37) are given, respectively, by

$$W_{U_\lambda} = \lambda W_T + (1 - \lambda) W_S \tag{4.2.27}$$

and

$$W_U = \frac{W_T + W_S}{4}. \tag{4.2.28}$$

4.3 Markov semigroups generated by differential operators associated with Markov operators

The main aim of this section is to show that, under some additional assumptions on the Markov operator T on $\mathscr{C}(K)$, the (abstract) differential operators L_T and W_T are closable in $\mathscr{C}(K)$ and their closures generate Markov semigroups on $\mathscr{C}(K)$. Moreover, these semigroups are obtained as a limit of suitable iterates of Bernstein-Schnabl operators associated with T.

The above mentioned assumptions involve the invariance under T of the spaces of polynomials of degree m, $m \geq 1$. Such a property, which seems to have its own independent interest, is discussed in detail as well.

Given a metrizable convex compact subset K and a Markov operator T on $\mathscr{C}(K)$ satisfying (3.1.2), let $(B_n)_{n\geq 1}$ be the sequence of Bernstein-Schnabl operators associated with T and consider the abstract differential operator L_T defined on $P_\infty(K)$ by (4.1.6). If $K \subset \mathbf{R}^d$, $d \geq 1$, with $\text{int}(K) \neq \emptyset$, we shall also consider the differential operator $(W_T, \mathscr{C}^2(K))$ defined by (4.1.15).

Theorem 4.3.1. *Let D_0 be a dense (in $\mathscr{C}(K)$) subspace of $P_\infty(K)$ and assume that*

$$(\lambda I - L_T)(D_0) \text{ is dense in } \mathscr{C}(K) \text{ for some } \lambda > 0. \tag{4.3.1}$$

Then (L_T, D_0) is closable and its closure generates a Markov semigroup $(T(t))_{t\geq 0}$ on $\mathscr{C}(K)$ such that, if $t \geq 0$ and if $(k(n))_{n\geq 1}$ is a sequence of positive integers satisfying $\lim_{n\to\infty} \frac{k(n)}{n} = t$,

$$T(t)(f) = \lim_{n\to\infty} B_n^{k(n)}(f) \quad \text{uniformly on } K \tag{4.3.2}$$

for every $f \in \mathscr{C}(K)$ (here $B_n^{k(n)}$ denotes the iterate of B_n of order $k(n)$).

Similarly, if $K \subset \mathbf{R}^d$, $d \geq 1$, with $\text{int}(K) \neq \emptyset$, and if D_0 is a dense (in $\mathscr{C}(K)$) subspace of $\mathscr{C}^2(K)$ such that

$$(\lambda I - W_T)(D_0) \text{ is dense in } \mathscr{C}(K) \text{ for some } \lambda > 0, \tag{4.3.3}$$

then (W_T, D_0) is closable and its closure generates a Markov semigroup $(T(t))_{t\geq 0}$ satisfying (4.3.2).

Proof. Combining (4.3.1) or (4.3.3) with Theorem 4.1.2 or Theorem 4.1.5, respectively, and taking Corollary 2.2.3 into account (with $M = 1$ and $\omega = 0$), we infer that (L_T, D_0) (resp., (W_T, D_0)) is closable and its closure generates a contraction C_0-semigroup on $\mathscr{C}(K)$ satisfying (4.3.2). Formula (4.3.2), in turn, implies that each $T(t)$ is a Markov operator and hence the result follows.

□

With the help of the preceding result we shall now determine an important class of convex compact subsets and of Markov operators for which (4.3.2) holds true.

Theorem 4.3.2. *Let K be a metrizable convex compact subset, T a Markov operator satisfying* (3.1.2) *and assume that*

$$T(P_2(K)) \subset A(K), \text{ i.e., } T(h_1h_2) \in A(K) \text{ for every } h_1, h_2 \in A(K). \tag{4.3.4}$$

Then the operator $(L_T, P_\infty(K))$ is closable and its closure $(A_T^*, D(A_T^*))$ generates a Markov semigroup $(T(t))_{t\geq 0}$ on $\mathscr{C}(K)$ satisfying (4.3.2).

Proof. According to Theorem 4.3.1, it suffices to show (4.3.1) with $D_0 = P_\infty(K)$.

Indeed, consider an arbitrary $\lambda > 0$ and a continuous linear functional μ on $\mathscr{C}(K)$ which vanishes on $(\lambda I - L_T)(P_\infty(K))$, i.e.,

$$\mu(u) = \frac{\mu(L_T(u))}{\lambda} \quad \text{for every } u \in P_\infty(K).$$

Therefore, $\mu(h) = 0$ if $h \in A(K)$. If $u = h_1h_2 \in P_2(K)$, taking (4.1.6) and (4.3.4) into account, we get $\mu(u) = -\frac{\mu(u)}{\lambda}$ and hence $\mu(u) = 0$.

Assume now that μ vanishes on $P_m(K)$, $m \geq 2$. Then, if $u = \prod_{i=1}^{m+1} h_i \in P_{m+1}(K)$, again by (4.1.6) and (4.3.4), we obtain

$$\mu(u) = -\frac{m(m-1)}{2\lambda}\mu(u),$$

which implies that $\mu(u) = 0$ as well.

Thus, by induction, we infer that $\mu = 0$ on $P_\infty(K)$ and hence on $\mathscr{C}(K)$, because $P_\infty(K)$ is dense in $\mathscr{C}(K)$.

Therefore $(\lambda I - L_T)(P_\infty(K))$ is dense in $\mathscr{C}(K)$ as well. □

Note that assumption (4.3.4) is satisfied when K is a metrizable Bauer simplex and T is the canonical projection on $\mathscr{C}(K)$ (see Theorem 1.1.12).

Next we show that this is the only case where (4.3.4) can occur.

Theorem 4.3.3. *Assume that there exists a Markov operator T on $\mathscr{C}(K)$ satisfying* (3.1.2) *and* (4.3.4). *Then K is a Bauer simplex and T is the canonical projection associated with it.*

In particular, $T(P_m(K)) \subset A(K)$ for every $m \geq 2$.

Proof. Setting $M_T := \{f \in \mathscr{C}(K) \mid T(f) = f\}$, from (3.1.2) it follows that $A(K) \subset M_T$. Thus, M_T separates the points of K and, by (4.3.4),

$$T(A(K)^2) \subset T(P_2(K)) \subset A(K) \subset M_T.$$

Therefore, by Proposition 1.3.4, T is a projection and $\partial_T K = \partial_H K$, where $H := T(\mathscr{C}(K))$ (see (1.3.4)).

We now proceed by induction to show that $T(P_m(K)) \subset A(K)$ for every $m \geq 2$. Indeed, assume that the inclusion holds true for $m \geq 2$ and fix $u \in P_{m+1}(K)$ having the form $u = \prod_{i=1}^{m+1} h_i$, with $h_1, \dots, h_{m+1} \in A(K)$.

Setting $v := \prod_{i=1}^{m} h_i \in P_m(K)$, we have that $u = vh_{m+1}$ and $T(v) \in A(K)$, so that $h_{m+1}T(v) \in P_2(K)$ and $T(h_{m+1}T(v)) \in A(K)$.

But $h_{m+1}T(v) = h_{m+1}v = u$ on $\partial_T K = \partial_H K$, and hence, by (1.3.10), $T(u) = T(h_{m+1}T(v)) \in A(K)$.

From the above, it follows that $T(P_\infty(K)) \subset A(K)$ and hence, by continuity, $T(\mathscr{C}(K)) = A(K)$. From Theorem 1.1.12 it turns out that K is a Bauer simplex and T is its canonical projection. □

We now turn to finite dimensional settings where other similar generation results can be obtained.

Theorem 4.3.4. *Let K be a convex compact subset of $\mathbf{R}^d$, $d \geq 1$, having non-empty interior and consider a Markov operator T on $\mathscr{C}(K)$ satisfying* (3.1.2). *Furthermore, assume that*

$$T(P_m(K)) \subset P_m(K) \qquad \text{for every } m \geq 2. \tag{4.3.5}$$

Then the differential operator $(W_T, \mathscr{C}^2(K))$ is closable and its closure $(A_T, D(A_T))$ generates a Markov semigroup $(T(t))_{t\geq 0}$ on $\mathscr{C}(K)$ such that, if $t \geq 0$ and $(k(n))_{n\geq 1}$ is a sequence of positive integers satisfying $\lim_{n\to\infty} \frac{k(n)}{n} = t$,

$$T(t)(f) = \lim_{n\to\infty} B_n^{k(n)}(f) \quad \text{uniformly on } K \tag{4.3.6}$$

and

$$\int_0^1 T(t\xi)(f)\,d\xi = \lim_{n\to\infty} \frac{1}{k(n)+1} \sum_{j=0}^{k(n)} B_n^j(f) \quad \text{uniformly on } K \tag{4.3.7}$$

for every $f \in \mathscr{C}(K)$.

Moreover, $P_\infty(K)$ is a core for $(A_T, D(A_T))$.

Proof. Each subspace $P_m(K)$, $m \geq 1$, is finite dimensional and it is invariant under every operator B_n by virtue of Lemma 4.1.1 and assumptions (3.1.2) and (4.3.5). These preliminary remarks along with Theorem 4.1.5 guarantee that all the assumptions of Corollary 2.2.11 are satisfied with $D_0 = \mathscr{C}^2(K)$ and $A_0 = W_T$, and hence the result follows. □

Remarks 4.3.5.

1. If K is a finite dimensional simplex, then its canonical projection T satisfies both (4.3.4) and (4.3.5) and hence Theorems 4.3.2 and 4.3.4 apply. Furthermore, the closures A_T^* and A_T coincide because their generated semigroups are equal by virtue of (4.3.2) and (4.3.6).
2. Under the same assumptions of Theorem 4.3.4 (or Theorem 4.3.3), in the light of Remarks 2.2.4, 1 and 2.2.12, 1, it follows that if $u, v \in \mathscr{C}(K)$ and

$\lim_{n\to\infty} n(B_n(u) - u) = v$ uniformly on K, then $u \in D(A_T)$ (resp., $u \in D(A_T^*)$) and $A_T(u) = v$ (resp., $A_T^*(u) = v$).

In particular, if $\lim_{n\to\infty} n(B_n(u) - u) = 0$ uniformly on K, then $u \in D(A_T)$ (resp., $u \in D(A_T^*)$) and $A_T(u) = 0$ (resp., $A_T^*(u) = 0$).

3. Under the assumptions of Thorems 4.3.1, 4.3.2 and 4.3.4, the relevant Markov semigroup will be often referred to as the *limit semigroup associated with Bernstein-Schnabl operators.*

Examples 4.3.6.

1. Consider the Markov operators T_d *and* S *on the d-dimensional simplex* K_d *and their convex combination* $G := \frac{T_d+S}{2}$ *as in Example 4.2.1, 2. Clearly, the operator* T_d *satisfies* (3.1.2) *and* (4.3.5) *(note that* $T_d(\mathscr{C}(K_d)) \subset A(K_d)$*). Similarly,* S *(and hence* G*) has the same properties because, if* $m_1, \dots, m_d$ *are positive integers, then*

$$S\left(pr_1^{m_1} \cdots pr_d^{m_d}\right) = \begin{cases} pr_2^{m_2} \cdots pr_d^{m_d} & \text{if } m_1 = 0; \\ \left(\mathbf{1} - \sum_{i=2}^{d} pr_i\right)^{m_1-1} pr_1\, pr_2^{m_2} \cdots pr_d^{m_d} & \text{if } m_1 \geq 1. \end{cases}$$

Therefore, Theorem 4.3.4 applies to the differential operators (4.2.10), (4.2.13) *and* (4.2.14).

2. Consider a family $(T_i)_{1\leq i\leq d}$ *of Markov operators on* $\mathscr{C}([0,1])$ *satisfying* (4.2.1) *and* (4.3.5). *Then the tensor product* $T := \bigotimes_{i=1}^{d} T_i$ *on* $\mathscr{C}(Q_d)$ *(see Example 4.2.1, 3) verifies* (4.3.5) *(and* (3.1.2)*) as well because*

$$T\left(pr_1^{m_1} \cdots pr_d^{m_d}\right) = \left(T_1\left(e_1^{m_1}\right) \circ pr_1\right) \cdots \left(T_d\left(e_1^{m_d}\right) \circ pr_d\right)$$

for every positive integers $m_1, \dots, m_d$.

Therefore, Theorem 4.3.4 applies to the differential operator (4.2.11) *and, in particular, to the one defined by* (4.2.14).

The representation formulae (4.3.2) and (4.3.6) are useful tools to investigate several qualitative and quantitative properties of both the semigroups $(T(t))_{t\geq 0}$ (i.e., of the solutions to the initial-boundary value problems associated with their generators) and the transition functions of the corresponding Markov processes (see Subsection 2.3.2).

These aspects will be carefully treated in the next sections. In the last part of the present section we discuss some cases where condition (4.3.5) is satisfied.

To this aim, note that, if T satisfies (4.3.5), then for every $\lambda \in [0,1]$ the operator $U_\lambda := \lambda T + (1-\lambda)I$ satisfies the same property.

We now present a counterexample to (4.3.5).

Example 4.3.7. *Let* $K = K_2$ *be the standard simplex of* $\mathbf{R}^2$ *(see* (1.1.45)*) and consider the Poisson operator* $T_\Delta : \mathscr{C}(K_2) \longrightarrow \mathscr{C}(K_2)$ *associated with the Laplace*

operator

$$\Delta u(x,y) := \frac{\partial^2 u}{\partial x^2}(x,y) + \frac{\partial^2 u}{\partial y^2}(x,y) \tag{4.3.8}$$

($u \in \mathscr{C}^2(\mathrm{int}(K_2)), (x,y) \in \mathrm{int}(K_2)$) *(see Subsection 3.1.4).*

Then $T_\Delta(P_2(K_2)) \not\subset P_2(K_2)$.

Indeed, consider the function $f(x,y) = x^2$ ($(x,y) \in K_2$) and assume that $T_\Delta(f) \in P_2(K_2)$, i.e., there exist $a, b, c, m, n, p \in \mathbf{R}$ such that

$$T_\Delta(f)(x,y) = ax^2 + bxy + cy^2 + mx + ny + p$$

for every $(x,y) \in K_2$.

Since $T_\Delta(f) = f$ on ∂K_2, we have

$$T_\Delta(f)(0,y) = f(0,y) \qquad (y \in [0,1]),$$

$$T_\Delta(f)(x,0) = f(x,0) \qquad (x \in [0,1])$$

and

$$T_\Delta(f)(x,1-x) = f(x,1-x) \qquad (x \in [0,1]).$$

Accordingly, we get $b = c = m = n = p = 0$ and $a = 1$. Thus, $T_\Delta(f) = f$ and this is not possible because f is not harmonic on $\mathrm{int}(K_2)$.

From Theorem 4.2.2 it follows that, if ∂K is an ellipsoid, then several classes of Poisson operators associated with strictly elliptic operators verify (4.3.5).

The next result shows that the inclusion $T(P_2(K)) \subset P_2(K)$ characterizes the ellipsoids between those convex compact subsets of $\mathbf{R}^d$ which are *strictly convex*, i.e., $\partial_e K = \partial K$. In such a case, necessarily $\mathrm{int}(K) \neq \emptyset$ unless K is *trivial*, i.e., K reduces to a singleton.

Theorem 4.3.8. *Given a non-trivial strictly convex compact subset K of $\mathbf{R}^d$, $d \geq 2$, the following statements are equivalent:*

(i) There exists a non-trivial Markov operator T on $\mathscr{C}(K)$, i.e., $T \neq I$, satisfying (3.1.2) *and* (4.3.5).

(ii) There exists a non-trivial Markov operator T on $\mathscr{C}(K)$ satisfying (3.1.2) *such that*

$$T(P_2(K)) \subset P_2(K). \tag{4.3.9}$$

(iii) ∂K is an ellipsoid defined by a quadratic form $Q(x - \overline{x}) := \sum_{i,j=1}^{d} r_{ij}(x_i - \overline{x}_i)(x_j - \overline{x}_j)$ ($x = (x_i)_{1\leq i\leq d} \in \mathbf{R}^d$) with center $\overline{x} = (\overline{x}_i)_{1\leq i\leq d} \in \mathbf{R}^d$ (see (4.2.19)*).*

Moreover, if T is a non-trivial Markov projection on $\mathscr{C}(K)$ satisfying (3.1.2) *and* (4.3.9)*, then one and only one of the following statements holds true:*

(a) *T is the Poisson operator associated with a suitable strictly elliptic differential operator of the form (4.2.20), whose coefficients $(c_{ij})_{\substack{1\le i\le d\\1\le j\le d}}$ are constant and satisfy*

$$\sum_{i,j=1}^{d} r_{ij}c_{ij} = 1. \tag{4.3.10}$$

(b) *For every $x \in \operatorname{int}(K)$ the support $\operatorname{Supp}(\tilde{\mu}_x^T)$ (see (1.3.1)) is contained in an affine hyperplane R_x through x and hence, for every $f \in \mathscr{C}(K)$,*

$$T(f)(x) = \int_{\partial K\cap R_x} f\, d\tilde{\mu}_x^T. \tag{4.3.11}$$

Proof. (i)⇒(ii). It is obvious.

(ii)⇒(iii). Setting $\Phi := \sum_{i=1}^{d} pr_i^2$, from (4.3.9) we infer that $T(\Phi) - \Phi$ is the restriction to K of a polynomial P of degree at most two. Consider the surface

$$\mathcal{S} := \{x \in \mathbf{R}^d \mid P(x) = 0\}.$$

From Theorem 1.3.1 we get

$$\partial_T K = \{x \in K \mid T(\Phi)(x) = \Phi(x)\} = K \cap \mathcal{S}. \tag{1}$$

Now we proceed to show that

$$\partial K = \partial_T K = \mathcal{S}. \tag{2}$$

Indeed, (3.1.4) implies that $\partial K = \partial_e K \subset \partial_T K \subset \mathcal{S}$. Before showing the converse inclusion, we first observe that $\mathcal{S} \neq K$, otherwise, by (1), $\partial_T K = K$ and, as a consequence of Theorem 1.2.10, we should have $T = I$. On account of this preliminary remark, we deduce that $\operatorname{int}(K) \not\subset \mathcal{S}$ and hence we can choose $x_0 \in \operatorname{int}(K) \setminus \mathcal{S} \subset \operatorname{int}(K) \setminus \partial_T K$.

Now, in order to complete the proof of (2), assume, on the contrary, that there exists $\overline{y} \in \mathcal{S} \setminus \partial K$.

Then the straight line R through $\overline{y}$ and x_0 cannot be contained in $\mathcal{S}$ (because $x_0 \notin \mathcal{S}$) and hence, since P is a polynomial of degree at most two, $R \cap \mathcal{S}$ contains at most two points.

On the other hand, $R \cap \mathcal{S}$ contains exactly two points because K is strictly convex, and $R \cap \partial K \subset R \cap \partial_T K \subset R \cap \mathcal{S}$, so that $R \cap \mathcal{S} = R \cap \partial K = R \cap \partial_T K$ and hence $\overline{y} \in \partial K$, a contradiction.

Having now (2) at our disposal, and assuming that

$$P(x) = \sum_{i,j=1}^{d} a_{ij}x_i x_j + \sum_{i=1}^{d} b_i x_i + c$$

$(x = (x_i)_{1\le i\le d} \in \mathbf{R}^d)$, since K is bounded, then necessarily the matrix $(a_{ij})_{\substack{1\le i\le d\\1\le j\le d}}$ is positive-definite or, equivalently, all its eigenvalues are strictly positive. Otherwise,

reasoning by induction on d, it would be possible to find an unbounded sequence of points of ∂K, in contradiction with the fact that K is bounded.

Therefore ∂K is, in fact, an ellipsoid. Now, considering the point $\overline{x} = (\overline{x}_i)_{1\le i\le d}$ whose coordinates are the unique solution to the system

$$2\sum_{j=1}^{d} a_{ij}\xi_j + b_i = 0 \qquad i = 1, \dots, d,$$

and putting $r_{ij} := -\frac{a_{ij}}{c}$ $(i, j = 1, \dots, d)$ and

$$\mathcal{Q}(x-\overline{x}) := \sum_{i,j=1}^{d} r_{ij}(x_i - \overline{x}_i)(x_j - \overline{x}_j)$$

$(x = (x_i)_{1\le i\le d} \in \mathbf{R}^d)$, we get that $P = c(\mathbf{1} - \mathcal{Q})$ and hence

$$K = \{x \in \mathbf{R}^d \mid \mathcal{Q}(x - \overline{x}) \le 1\}.$$

(iii)⇒(i). It is a consequence of Theorem 4.2.2.

In order to show the last part of the statement, consider a Markov projection T on $\mathscr{C}(K)$ satisfying (3.1.2) and (4.3.9). In case (b), for every $x \in \operatorname{int}(K)$, since $\operatorname{Supp}(\tilde{\mu}_x^T) \subset \partial K$ (see (1.3.9) and the preceding formula (2)), we get

$$T(f)(x) = \int_K f\,d\tilde{\mu}_x^T = \int_{\partial K\cap R_x} f\,d\tilde{\mu}_x^T$$

for every $f \in \mathscr{C}(K)$.

Suppose that case (b) does not occur and fix $\overline{z} \in \operatorname{int}(K)$ such that $\operatorname{Supp}(\tilde{\mu}_{\overline{z}}^T)$ is not contained in any affine hyperplane through $\overline{z}$.

Without loss of generality we can assume that ∂K is an ellipsoid with center the origin of $\mathbf{R}^d$, so that we can consider a positive-definite quadratic form $\mathcal{Q}(x) = \sum\limits_{i,j=1}^{d} r_{ij}x_i x_j$ $(x = (x_i)_{1\le i\le d} \in \mathbf{R}^d)$ such that

$$K = \{x \in \mathbf{R}^d \mid \mathcal{Q}(x) \le 1\}.$$

Given $i, j = 1 \dots, d$, since the function $\alpha_{ij} := T(pr_i pr_j) - pr_i pr_j$ is a polynomial of degree at most two which vanishes on $\partial_T K = \partial K = \{x \in \mathbf{R}^d \mid \mathcal{Q}(x) = 1\}$, by Hilbert Nullstellensatz there exists $c_{ij} \in \mathbf{R}$ such that

$$\alpha_{ij} = c_{ij}(\mathbf{1} - \mathcal{Q}). \tag{3}$$

Note that the matrix $(\alpha_{ij}(\overline{z}))_{\substack{1\le i\le d\\1\le j\le d}}$ is positive-definite. Indeed, if $\xi = (\xi_i)_{1\le i\le d} \in \mathbf{R}^d \setminus \{0\}$, then the quantity

$$\sum_{i,j=1}^{d} \alpha_{ij}(\overline{z})\xi_i\xi_j = T\left(\left(\sum_{i=1}^{d} \xi_i(pr_i - pr_i(\overline{z}))\right)^2\right)(x)$$

is strictly positive, otherwise we should have

$$\sum_{i=1}^{d} \xi_i (pr_i - pr_i(\overline{z})) = 0 \quad \text{on } \operatorname{Supp}(\tilde{\mu}_{\overline{z}}^T)$$

i.e.,

$$\operatorname{Supp}(\tilde{\mu}_{\overline{z}}^T) \subset \{y \in \mathbf{R}^d \mid \langle \xi, y - \overline{z} \rangle = 0\},$$

a contradiction (here $\langle \cdot, \cdot \rangle$ denotes the canonical scalar product on $\mathbf{R}^d$).

As a consequence, from (3) it follows that the matrix $(c_{ij})_{\substack{1 \le i \le d \\ 1 \le j \le d}}$ is (symmetric and) positive-definite.

Let us consider the differential operator W_T defined by (4.1.15). From (4.1.11) it follows that, for every $i, j = 1, \dots, d$,

$$W_T(pr_i pr_j + c_{ij}(\mathbf{1} - \mathcal{Q})) = W_T(T(pr_i pr_j)) = 0,$$

so that

$$W_T(\mathcal{Q}) = \sum_{i,j=1}^{d} r_{ij} W_T(pr_i pr_j) = \sum_{i,j=1}^{d} r_{ij} c_{ij} W_T(\mathcal{Q}).$$

On the other hand, since $\mathcal{Q} = \mathbf{1}$ on $\partial K = \partial_T K$, from (1.3.10) it follows that $T(\mathcal{Q}) = \mathbf{1}$ and, by Remark 4.1.3, we have that

$$W_T(\mathcal{Q}) = L_T(\mathcal{Q}) = T(\mathcal{Q}) - \mathcal{Q} = \mathbf{1} - \mathcal{Q}.$$

Thus, $W_T(\mathcal{Q})$ does not vanish on $\operatorname{int}(K)$, so that $\sum\limits_{i,j=1}^{d} r_{ij} c_{ij} = 1$ and the proof is now complete.

□

A special case of the previous result is worth being stated separately.

Corollary 4.3.9. *Given a non-trivial strictly convex compact subset K of $\mathbf{R}^d$, $d \ge 2$, the following statements are equivalent:*

(i) There exists a non-trivial Markov operator T on $\mathscr{C}(K)$ satisfying (3.1.2), such that

$$T(e) - (1 + \lambda)e \in P_1(K) \tag{4.3.12}$$

for some $\lambda \in \mathbf{R}, \lambda \ne 0$, where $e := \sum\limits_{i=1}^{d} pr_i^2$.

(ii) K is a ball with respect to the Euclidean norm $\|\cdot\|_2$ on $\mathbf{R}^d$.

Moreover, if

$$T(e) = (1 + \lambda)e + \sum_{i=1}^{d} b_i x_i + c$$

with $\lambda \in \mathbf{R}, \lambda \ne 0$, and $b_1, \dots, b_d, c \in \mathbf{R}$, then K is the ball of center $\overline{x} = (\overline{x}_i)_{1 \le i \le d} \in \mathbf{R}^d$ and radius r, where

$$\overline{x}_i := -\frac{b_i}{2\lambda} \quad \text{for every } 1 \le i \le d \quad \text{and} \quad r := \sqrt{\|\overline{x}\|_2^2 - \frac{c}{\lambda}}.$$

Proof. (i)⇒(ii). It is enough to apply the same reasoning as in the proof of the implication (iii)⇒(iv) of Theorem 4.3.8 with $P(x) = \sum_{i=1}^{d} \lambda {x_i}^2 + \sum_{i=1}^{d} b_i x_i + c$ ($x = (x_i)_{1 \le i \le d} \in \mathbf{R}^d$).

(ii)⇒(i). Assume that K is the ball of center $\overline{x} = (\overline{x}_i)_{1 \le i \le d} \in \mathbf{R}^d$ and radius r and consider the Poisson operator T_Δ associated with the Laplace operator Δ (see (3.1.21) and (3.1.22)). Then T_Δ is a Markov operator satisfying (3.1.20) and

$$T_\Delta(e) = 2\sum_{i=1}^{d} \overline{x}_i pr_i + r^2 - \| \bar{x} \|_2^2 .$$

□

Remark 4.3.10. In Appendix A.1 we present a complete description of those convex compact subsets K of $\mathbf{R}^2$ such that there exists a Markov projection T on $\mathscr{C}(K)$ satisfying (3.1.2) and (4.3.5).

We proceed further to study condition (4.3.5) in the setting of product spaces.

Consider a finite family $(K_i)_{1 \le i \le d}$ of convex compact subsets having non-empty interior, each contained in some $\mathbf{R}^{s_i}$, $s_i \ge 1$, $i = 1, \dots, d$. For every $i = 1, \dots, d$, let $T_i : \mathscr{C}(K_i) \longrightarrow \mathscr{C}(K_i)$ be a Markov operator satisfying (3.1.2) and (4.3.5). Setting $K := \prod_{i=1}^{d} K_i$ and denoting by T the tensor product of $(T_i)_{1 \le i \le d}$ (see (1.1.30)), then T is a Markov operator on $\mathscr{C}(K)$ which satisfies (3.1.2).

For every $i = 1, \dots, d$, set

$$A_i := A_{T_i} \tag{4.3.13}$$

and, for $j = 1, \dots, d$,

$$A_{i,j} := \begin{cases} A_i & \text{if } j = i; \\ I_{D(A_j)} & \text{if } j \ne i. \end{cases} \tag{4.3.14}$$

Finally, denote by $(T_i(t))_{t \ge 0}$ the Markov semigroup on $\mathscr{C}(K_i)$ generated by $(A_i, D(A_i))$.

Theorem 4.3.11. *The Markov operator* $T := \bigotimes_{i=1}^{d} T_i$ *satisfies* (4.3.5). *Moreover, denoted by* $(A_T, D(A_T))$ *the generator of the semigroup* $(T(t))_{t \ge 0}$ *as in Theorem 4.3.4, then*

(i) $T(t) = \bigotimes_{i=1}^{d} T_i(t)$ *for every* $t \ge 0$.

(ii) The subspace $\bigotimes_{i=1}^{d} D(A_i)$ *is contained in* $D(A_T)$, *it is a core for* $(A_T, D(A_T))$ *and*

$$A_T = \sum_{i=1}^{d} \bigotimes_{j=1}^{d} A_{i,j} \qquad \text{on } \bigotimes_{i=1}^{d} D(A_i)$$

(see Remark 1.1.7).

(iii) $\bigotimes_{i=1}^{d} \mathscr{C}^2(K_i)$ *is a core for* $(A_T, D(A_T))$ *and if* $(u_i)_{1\le i\le d} \in \prod_{i=1}^{d} \mathscr{C}^2(K_i)$, *then*

$$A_T\left(\bigotimes_{i=1}^{d} u_i\right) = \sum_{i=1}^{d} u_1 \otimes \cdots \otimes u_{i-1} \otimes W_{T_i}(u_i) \otimes u_{i+1} \otimes \cdots \otimes u_d.$$

Proof. Note that, given $(u_i)_{1\le i\le d}, (v_i)_{1\le i\le d} \in \prod_{i=1}^{d} A(K_i)$, then, for every $i = 1, \dots, d$,

$$(u_i v_i) \circ pr_i = (u_i \circ pr_i)(v_i \circ pr_i) \in P_2(K)$$

and

$$T((u_i \circ pr_i)(v_i \circ pr_i)) = T_i(u_i v_i) \circ pr_i \in P_2(K).$$

Moreover, for $i, j = 1, \dots, d$, $i \neq j$,

$$T((u_i \circ pr_i)(v_j \circ pr_j)) = (T_i(u_i) \circ pr_i)(T_j(v_j) \circ pr_j) = (u_i \circ pr_i)(u_j \circ pr_j) \in P_2(K).$$

On the other hand, for every $u \in A(K)$, there exist $\alpha_0, \alpha_1, \dots, \alpha_d \in \mathbf{R}$ and $(u_i)_{1\le i\le d} \in \prod_{i=1}^{d} A(K_i)$ such that $u = \alpha_0 + \sum_{i=1}^{d} \alpha_i (u_i \circ pr_i)$. Therefore, on account of the preceding identities, it follows that $T(P_2(K)) \subset P_2(K)$.

By induction, it is now easy to show that $T(P_m(K)) \subset P_m(K)$ for every $m \ge 2$. According to Theorem 4.3.4 we can consider the Markov semigroup $(T(t))_{t\ge 0}$ on $\mathscr{C}(K)$, along with its generator $(A_T, D(A_T))$.

Looking to the family of generators $((A_i, D(A_i)))_{1\le i\le d}$ defined by (4.3.13), from [40, Section A-I-3.7, p. 23] it follows that the operator $\sum_{i=1}^{d} \bigotimes_{j=1}^{d} A_{i,j}$ defined on $\bigotimes_{i=1}^{d} D(A_i)$ (see Remark 1.1.7) is closable on $\mathscr{C}(K)$ and its closure $(B, D(B))$ generates a C_0-semigroup $(S(t))_{t\ge 0}$ on $\mathscr{C}(K)$ given by

$$S(t) = \bigotimes_{i=1}^{d} T_i(t) \qquad (t \ge 0).$$

Moreover, $\bigotimes_{i=1}^{d} D(A_i)$ is a core for $(B, D(B))$.

We now proceed to show that

$$\bigotimes_{i=1}^{d} \mathscr{C}^2(K_i) \subset D(A_T) \text{ and } A_T = B \text{ on } \bigotimes_{i=1}^{d} \mathscr{C}^2(K_i). \tag{1}$$

Indeed, given $(u_i)_{1\le i\le d} \in \bigotimes_{i=1}^{d} \mathscr{C}^2(K_i)$ and considered the sequence $(B_n)_{n\ge 1}$ of Bernstein-Schnabl operators associated with T, then, on account of (3.1.26) and

Theorem 4.1.5 for $u = \bigotimes_{i=1}^{d} u_i$, we get

$$\lim_{n\to\infty} n(B_n(u) - u) = \lim_{n\to\infty} n\left(\bigotimes_{i=1}^{d} B_{n,i}(u_i) - \bigotimes_{i=1}^{d} u_i\right)$$
$$= \lim_{n\to\infty} n\left(\sum_{i=1}^{d} B_{n,1}(u_1)\otimes\cdots\otimes B_{n,i-1}(u_{i-1})\otimes(B_{n,i}(u_i)-u_i)\otimes u_{i+1}\otimes\cdots\otimes u_d\right)$$
$$= \sum_{i=1}^{d} \lim_{n\to\infty} B_{n,1}(u_1)\otimes\cdots\otimes B_{n,i-1}(u_{i-1})\otimes\left[n(B_{n,i}(u_i)-u_i)\right]\otimes u_{i+1}\otimes\cdots\otimes u_d$$
$$= \sum_{i=1}^{d} u_1\otimes\cdots\otimes u_{i-1}\otimes W_{T_i}(u_i)\otimes u_{i+1}\otimes\cdots\otimes u_d = B(u).$$

Therefore, by Remark 4.3.5, 2, $u \in D(A_T)$ and $A_T(u) = B(u)$.

On the other hand, for every $m \geq 1$,

$$P_m(K) \subset \bigcup_{m_1+\ldots+m_d \leq m} \bigotimes_{i=1}^{d} P_{m_i}(K_i) \subset \bigotimes_{i=1}^{d} \mathscr{C}^2(K_i),$$

which implies that $P_\infty(K) \subset \bigotimes_{i=1}^{d} \mathscr{C}^2(K_i) \subset \bigotimes_{i=1}^{d} D(A_i)$ and $A_T = B$ on $P_\infty(K)$ by virtue of (1). Since $P_\infty(K)$ is a core for $(A_T, D(A_T))$ (see Theorem 4.3.4), from Proposition 2.1.7, (a) and (g), it follows that $(B, D(B)) = (A_T, D(A_T))$ and hence, in particular, $T(t) = S(t)$ for every $t \geq 0$ and the proof is now complete. □

The special case where $K_i = [0,1]$ for every $i = 1, \ldots, d$ is worth being studied separately. Consider, therefore, a family $(T_i)_{1\leq i\leq d}$ of Markov operators on $\mathscr{C}([0,1])$ satisfying (3.1.2) and (4.3.5), and set $T := \bigotimes_{i=1}^{d} T_i : \mathscr{C}(Q_d) \longrightarrow \mathscr{C}(Q_d)$. We already pointed out in (4.2.15) that

$$W_T(u)(x) = \frac{1}{2}\sum_{i=1}^{d} \alpha_i(x)\frac{\partial^2 u}{\partial x_i^2}(x)$$

($u \in \mathscr{C}^2(Q_d), x = (x_i)_{1\leq i\leq d} \in Q_d$), where each α_i is given by (4.2.16).

In particular, for $T = S_d$ (see (3.1.28)), we have (see (4.2.18))

$$W_{S_d}(u)(x) = \frac{1}{2}\sum_{i=1}^{d} x_i(1-x_i)\frac{\partial^2 u}{\partial x_i^2}(x)$$

($u \in \mathscr{C}^2(Q_d), x = (x_i)_{1\leq i\leq d} \in Q_d$).

Corollary 4.3.12. *Under the above assumptions, the operator T (in particular, the operator S_d) maps $P_m(Q_d)$ into $P_m(Q_d)$ for every $m \geq 1$.*

Therefore, the differential operator $(W_T, \mathscr{C}^2(Q_d))$ (in particular, the differential operator $(W_{S_d}, \mathscr{C}^2(Q_d))$) is closable and its closure generates a Markov

semigroup $(T(t))_{t\geq 0}$ *on* $\mathscr{C}(Q_d)$ *satisfying all the properties stated in Theorems 4.3.4 and 4.3.11.*

We end this section by discussing property (4.3.5) in the framework of Subsection 3.1.7.

Theorem 4.3.13. *Let* K *be a convex compact subset of* $\mathbf{R}^d$, $d \geq 1$, *having non-empty interior, and consider two Markov operators* S *and* T *on* $\mathscr{C}(K)$ *satisfying* (3.1.2) *and* (4.3.5). *Then the convex convolution product* U *of* S *and* T *defined by* (3.1.37) *maps* $P_m(K)$ *into* $P_m(K)$ *for each* $m \geq 1$.

Therefore, the differential operator $W_U = \frac{W_S+W_T}{4}$ *(see* (4.2.28)*) defined on* $\mathscr{C}^2(K)$ *is closable and its closure generates a Markov semigroup on* $\mathscr{C}(K)$ *satisfying all the properties stated in Theorem 4.3.4.*

Proof. The case $m = 1$ being obvious, we can assume $m \geq 2$. Consider $h_1, \dots, h_m \in A(K)$ and $x_1, x_2 \in K$. Denote by $F(m, 2)$ the set of all mappings $\sigma : \{1, \dots, m\} \longrightarrow \{1, 2\}$ and, for $\sigma \in F(m, 2)$, set $R_1^\sigma := \{i = 1, \dots, m \mid \sigma(i) = 1\}$ and $R_2^\sigma := \{i = 1, \dots, m \mid \sigma(i) = 2\}$. Therefore (see the proof of Lemma 4.1.1),

$$\left(\prod_{i=1}^{m} h_i\right)\left(\frac{x_1+x_2}{2}\right) = \frac{1}{2^m} \sum_{\sigma\in F(m,2)} \prod_{i\in R_1^\sigma} h_i(x_1) \prod_{i\in R_2^\sigma} h_i(x_2),$$

where the product $\prod_{i\in R_k^\sigma} h_i$ is, by convention, equal to 1 if $R_k^\sigma = \emptyset$ for some $k = 1, 2$.

Then, from (3.1.37), it follows that

$$U\left(\prod_{i=1}^{m} h_i\right) = \frac{1}{2^m} \sum_{\sigma\in F(m,2)} S\left(\prod_{i\in R_1^\sigma} h_i\right) T\left(\prod_{i\in R_2^\sigma} h_i\right) \in P_m(K),$$

because of the assumptions on S and T and the fact that $\mathrm{card}R_1^\sigma + \mathrm{card}R_2^\sigma = m$. □

Remark 4.3.14. From Theorem 4.3.13 it turns out that the sum $W_S + W_T = 4W_U$, defined on $\mathscr{C}^2(K)$, is closable and its closure generates a Markov semigroup $(T(t))_{t\geq 0}$, which is the rescaled semigroup with parameter 4 (see Examples 2.1.4, 4 and 2.1.6, 4) of the semigroup generated by the closure of $(W_U, \mathscr{C}^2(K))$.

This result is not trivial because, in general, the investigation of the generation property of the sum of two generators is a delicate problem (see, e.g., [93, Chapter III, Section 1]).

However, the sum $W_S + W_T$ is also equal to $2W_{\frac{S+T}{2}}$ and $\frac{S+T}{2}$ is a Markov operator on $\mathscr{C}(K)$ satisfying (3.1.2) and (4.3.5). Therefore, the semigroup $(T(t))_{t\geq 0}$ also coincides with the rescaled semigroup with parameter 2 generated by the closure of $\left(W_{\frac{S+T}{2}}, \mathscr{C}^2(K)\right)$. Thus it can be represented as in (4.3.6) and (4.3.7) in terms of iterates of Bernstein-Schnabl operators (3.1.38) or of Bernstein-Schnabl operators (3.1.33), with $\lambda = \frac{1}{2}$.

4.4 Preservation properties and asymptotic behaviour

We have seen that in the setting of Section 4.3 the semigroups $(T(t))_{t\geq 0}$ generated by differential operators associated with Markov operators can be represented as suitable limits of Bernstein-Schnabl operators (see Theorem 4.3.1).

On the other hand, several presevation properties of the Bernstein-Schnabl operators have been investigated in Chapter 3.

The aim of this section is to show that some of these preservation properties are inherited by the corresponding semigroups. As a consequence, in the light of Theorem 2.1.15, it will be possible to describe qualitative properties of the solutions to the associated abstract Cauchy problems.

Clearly, any closed subspace of $\mathscr{C}(K)$ which is invariant under each B_n, is invariant under each operator $T(t)$ $(t \geq 0)$. The following result states it more explicitly.

Proposition 4.4.1. *Consider the semigroup $(T(t))_{t\geq 0}$ described in Theorem 4.3.1. Let U be a closed (with respect to the uniform norm) subset of $\mathscr{C}(K)$. If*

$$B_n(U) \subset U \qquad (n \geq 1), \tag{4.4.1}$$

then

$$T(t)(f) \in U \quad \text{for every } f \in U \text{ and } t \geq 0. \tag{4.4.2}$$

Proof. By virtue of (4.4.1), $B_n^m(f) \in U$ for every $f \in U$ and $n, m \geq 1$. Since U is closed, (4.4.2) is a consequence of (4.3.2). □

Turning back to Theorem 4.3.1, the generator $(A_T, D(A_T))$ of the semigroup $(T(t))_{t\geq 0}$ is the closure of (L_T, D_0) (resp., (W_T, D_0)) (see (4.1.9) and (4.1.15)). Consider the abstract Cauchy problem associated with $(A_T, D(A_T))$ and to the initial datum $u_0 \in D(A_T)$ (see (2.1.12))

$$\begin{cases} \dfrac{du}{dt}(t) = A_T(u(t)) & t \geq 0; \\[2ex] u(0) = u_0. \end{cases} \tag{4.4.3}$$

The following result is a reformulation of Proposition 4.4.1 in terms of solutions to abstract Cauchy problems.

Proposition 4.4.2. *Under the same assumptions of Proposition 4.4.1, consider the abstract Cauchy problem (4.4.3) with initial datum u_0 and suppose that $u_0 \in U \cap D(A_T)$. Let $u : [0, +\infty[\longrightarrow D(A_T)$ be the (unique) solution of (4.4.3). Then*

$$u(t) \in U \quad \text{for every } t \geq 0. \tag{4.4.4}$$

Proof. With the above notation, Theorem 2.1.15 guarantees that $u(t) = T(t)(u_0)$ for every $t \geq 0$. According to (4.4.2), $T(t)(u_0) \in U$ for every $t \geq 0$ and this concludes the proof. □

Example 4.4.3. *Let $(B_n)_{n\geq 1}$ be the sequence of Bernstein-Schnabl operators associated with a Markov operator T satisfying* (3.3.2) *with $c = 1$, i.e.,*

$$T(\mathrm{Lip}(1,1)) \subset \mathrm{Lip}(1,1). \tag{4.4.5}$$

Fix $M \geq 0$ and $\alpha \in]0,1]$. Then $U := \mathrm{Lip}(M,\alpha)$ is closed with respect to the uniform convergence and, according to (3.3.12)*, it satisfies* (4.4.1)*.*

Examples of Markov operators satisfying (4.4.5) *are presented in Section 3.3; see, e.g.,*

- *The canonical projection T_d (see* (1.1.48)*) defined on the d-dimensional simplex K_d (endowed with the l_1-metric) (see* (3.3.14)*);*
- *The operator T on $\mathscr{C}([0,1])$ defined by* (3.1.12)*;*
- *The operator T on $\mathscr{C}([0,1])$ defined by* (3.1.14)*, with $b \in \mathrm{Lip}(2M,1)$, $0 \leq M \leq 1$;*
- *The operator S_d defined by* (3.1.28) *on the d-dimensional hypercube Q_d, provided that Q_d is endowed with the l_1-metric.*

Other examples, involving the convexity, will be presented in the sequel. But first, given $f : [0,+\infty[\longrightarrow \mathscr{C}(K)$, as in (2.1.15), let us consider the inhomogeneous abstract Cauchy problem

$$\begin{cases} \dfrac{du}{dt}(t) = A_T(u(t)) + f(t) & t \geq 0; \\ \\ u(0) = u_0 & u_0 \in D(A_T). \end{cases} \tag{4.4.6}$$

We recall that, by means of Theorem 2.1.17, under suitable regularity assumptions on the function f, for every $u_0 \in D(A_T)$, problem (4.4.6) has a unique classical solution $u : [0,+\infty[\longrightarrow \mathscr{C}(K)$ given by

$$u(t) = T(t)(u_0) + \int_0^t T(t-s)(f)(s)\,ds \qquad (t \geq 0). \tag{4.4.7}$$

Now we are in a position to state the next result.

Theorem 4.4.4. *Consider the semigroup $(T(t))_{t\geq 0}$ described in Theorem 4.3.1. Let U be a closed convex cone of $\mathscr{C}(K)$ satisfying* (4.4.1)*. Consider a function $f : [0,+\infty[\longrightarrow \mathscr{C}(K)$ satisfying either assumption (i) or (ii) of Theorem 2.1.17. Assume further that $f(t) \in U$ for every $t \geq 0$ and fix $u_0 \in U \cap D(A_T)$. Then, denoted by $u : [0,+\infty[\longrightarrow D(A_T)$ the unique solution to the inhomogeneous abstract Cauchy problem associated with $(A_T, D(A_T))$ and with initial datum u_0, we have that*

$$u(t) \in U \qquad \textit{for every } t \geq 0. \tag{4.4.8}$$

Proof. Let $t \geq 0$ be fixed. According to Proposition 4.4.1, $T(t)(u_0) \in U$ and $T(t-s)(f)(s) \in U$ for every $s \in [0,t]$. By representing $\int_0^t T(t-s)(f)(s)\,ds$ as a limit of Riemann sums, we get $\int_0^t T(t-s)(f)(s)\,ds \in U$. Now (4.4.8) is a consequence of (2.1.16). □

Let $(A_T, D(A_T))$ be, as before, the generator of $(T(t))_{t\geq 0}$ and let B be a bounded operator on $\mathscr{C}(K)$; then, according to Theorem 2.1.13, $(A_T + B, D(A_T))$ is the generator of a C_0-semigroup $(S(t))_{t\geq 0}$ on $\mathscr{C}(K)$. Moreover,

$$S(t) = \sum_{n=0}^{\infty} S_n(t), \tag{4.4.9}$$

the series being uniformly convergent on compact subintervals of $[0, +\infty[$, where, for every $t \geq 0$, $S_0(t) := T(t)$ and, for any $n \geq 0$,

$$S_{n+1}(t)(f) := \int_0^t T(t-s)BS_n(s)(f)\,ds \qquad (f \in \mathscr{C}(K)). \tag{4.4.10}$$

Consider the *perturbed abstract Cauchy problem*

$$\begin{cases} \dfrac{du}{dt}(t) = A_T(u(t)) + B(u)(t) & t \geq 0; \\ u(0) = u_0 & u_0 \in D(A_T). \end{cases} \tag{4.4.11}$$

Taking Theorem 2.1.13 and Theorem 2.1.15 into account, its unique solution is given by $u(t) = S(t)(u_0) = \sum_{n=0}^{\infty} S_n(t)(u_0)$ for every $t \geq 0$.

On this purpose, the following result holds true.

Theorem 4.4.5. *Let U be a closed convex cone of $\mathscr{C}(K)$ satisfying* (4.4.1). *Moreover, suppose that $B(U) \subset U$. Then $S(t)(U) \subset U$ for every $t \geq 0$. In particular, if $u_0 \in U \cap D(A_T)$, the unique solution $u : [0, +\infty[\longrightarrow D(A_T)$ to* (4.4.11) *with initial datum u_0 satisfies $u(t) \in U$ for every $t \geq 0$.*

Proof. According to Theorem 4.4.1, $T(s)(U) \subset U$ for every $s \geq 0$. In particular, $S_0(t)(U) = T(t)(U) \subset U$ for any $t \geq 0$.

Suppose that $S_n(t)(U) \subset U$ for a certain $n \geq 0$.

If $f \in U$, then $T(t-s)BS_n(s)f \in U$ for every $s \in [0, t]$. By representing $\int_0^t T(t-s)BS_n(s)f\,ds$ as a limit of Riemann sums, we get

$$\int_0^t T(t-s)BS_n(s)f\,ds \in U$$

and, therefore, $S_{n+1}(t)(U) \subset U$ for any $t \geq 0$. Thus, $S_n(t)(U) \subset U$ for any $n \geq 0$ and $t \geq 0$. Now (4.4.9) shows that $S(t)(U) \subset U$ for every $t \geq 0$.

Finally, fix $u_0 \in U$; then the solution of (4.4.11) is $u(t) = S(t)(u_0)$ and this completes the proof. □

Examples 4.4.6.

1. *The set U of all the convex functions $f \in \mathscr{C}([0, 1])$ is a closed convex cone satisfying* (4.4.1) *with respect to the operators B_n associated with the Markov operators studied in Examples 3.4.5, 3.4.6 and 3.4.8.*

2. *The set U of all the T-axially convex functions on a Bauer simplex is a closed convex cone satisfying* (4.4.1) *with respect to the Bernstein-Schnabl operators associated with its canonical projection T (see Theorem 3.5.9).*

3. *Other examples, involving cones of polyhedrally convex functions and cones of subharmonic functions can be found in [182], [57] and [176].*

4. *Let K_d be the canonical simplex in $\mathbf{R}^d$ (see* (1.1.45)*) and let T_d be the canonical projection defined by* (1.1.48). *Let $(B_n)_{n\geq 1}$ be the sequence of Bernstein-Schnabl operators on $\mathscr{C}(K_d)$ associated with T_d (see* (3.1.18)*). The differential operator W_{T_d} associated with T_d is given by* (4.2.10).

Let $\mathscr{C}_0(K_d)$ be the set consisting of all functions $f \in \mathscr{C}(K_d)$ which vanish on the vertices of K_d. Moreover, set

$$H(K_d) := \{W_{T_d}(g) \mid g \in \mathscr{C}_0(K_d) \cap \mathscr{C}^2(K_d)\}. \tag{4.4.12}$$

It was proved in [172] that W_{T_d} acts bijectively between $\mathscr{C}_0(K_d) \cap \mathscr{C}^2(K_d)$ and $H(K_d)$. Moreover, according to [172, Corollary 5.3],

$$\lim_{n\to\infty} \frac{1}{n} \sum_{k=0}^{\infty} B_n^k(h) = -W_{T_d}^{-1}(h) \qquad \textit{for every } h \in H(K_d). \tag{4.4.13}$$

Thus, for each $h \in H(K_d)$, the unique solution $g \in \mathscr{C}_0(K_d) \cap \mathscr{C}^2(K_d)$ to the equation

$$\frac{1}{2} \sum_{i,j=1}^{d} x_i(\delta_{ij} - x_j) \frac{\partial^2 g}{\partial x_i \partial x_j}(x) = -h \tag{4.4.14}$$

can be represented as

$$g = \lim_{n\to\infty} \frac{1}{n} \sum_{k=0}^{\infty} B_n^k(h). \tag{4.4.15}$$

Let U be a closed convex cone satisfying (4.4.1) *(e.g., the cone of non-negative functions, the cone of all T_d-axially convex functions, the cone of polyhedrally convex functions (see [182])).*

If $h \in H(K_d)$, then the solution to (4.4.14)*, given by* (4.4.15)*, belongs to $H(K_d)$.*

In the sequel, we shall investigate some cases where the asymptotic behaviour of the semigroup $(T(t))_{t\geq 0}$ described in Theorems 4.3.1, 4.3.2 and 4.3.4 can be exactly described.

Theorem 4.4.7. *Consider a metrizable convex compact subset K of some locally convex space X and a Markov operator T on $\mathscr{C}(K)$ satisfying* (3.1.2). *Suppose that one of the assumptions of Theorems 4.3.1, 4.3.2 or 4.3.4 is satisfied, so that there exists a Markov semigroup $(T(t))_{t\geq 0}$ on $\mathscr{C}(K)$ satisfying* (4.3.2) *or* (4.3.6). *Furthermore, assume that there exists a Markov projection T_0 on $\mathscr{C}(K)$ verifying* (3.1.2) *and* (3.2.10) *and such that $T \circ T_0 = T_0$ and $\partial_T K \subset \partial_{T_0} K$ (see* (1.3.4)*). Then*

$$\lim_{t\to+\infty} T(t)(f) = T_0(f) \qquad \textit{uniformly on } K \tag{4.4.16}$$

for every $f \in \mathscr{C}(K)$.

In particular, $\lim_{t\to+\infty} T(t)(f) = 0$ *if and only if* $f = 0$ *on* $\partial_{T_0} K$.

Proof. Denote by H_0 the range $T_0(\mathscr{C}(K))$ of T_0. We shall apply Theorem 1.3.8, so that we first fix a sequence $(h_j)_{j\geq 1}$ in $A(K)$ separating the points of K and such that the series $\Phi := \sum_{j=1}^{\infty} h_j^2$ is uniformly convergent on K (see also Proposition 1.3.4). Let $(B_n)_{n\geq 1}$ be the sequence of Bernstein-Schnabl operators associated with T whose iterates converge to the semigroup $(T(t))_{t\geq 0}$ according to (4.3.2) or (4.3.6).

Because of the assumptions on T and T_0, we then get that $B_n(h) = h$ for every $h \in H_0$ (see also the proof of Theorem 3.2.5).

Taking (3.2.4) into account, for every $n \geq 1$,

$$B_n(\Phi) = \sum_{j=1}^{\infty} B_n(h_j^2) = \sum_{j=1}^{\infty} \left(\frac{1}{n}T(h_j^2) + \frac{n-1}{n}h_j^2\right) = \frac{1}{n}T(\Phi) + \frac{n-1}{n}\Phi.$$

Now, according to (1.3.12), we have to evaluate the quantities

$$a_{n,p} := \max_{x\in K} \varphi_{n,p}(x)$$

$(n \geq 1, p \geq 1)$, where

$$\varphi_{n,p} := T_0(\Phi) - \Phi - pn(B_n(\Phi) - \Phi) = T_0(\Phi) - \Phi - p(T(\Phi) - \Phi).$$

Thus, $a_{n,p}$ and $\varphi_{n,p}$ are independent on $n \geq 1$ and, from now on, we shall denote them by a_p and φ_p, respectively.

Note that $a_p \geq 0$ for every $p \geq 1$, because the function φ_p vanishes on $\partial_T K$ and the sequence $(a_p)_{p\geq 1}$ is decreasing. Thus, there exists $a := \lim_{p\to\infty} a_p = \inf_{p\geq 1} a_p \geq 0$.

In order to complete the proof it is enough to show that $a = 0$. Assume, on the contrary, that $a > 0$ and set $F_p := \{x \in K \mid a \leq \varphi_p(x)\}$ for every $p \geq 1$. Every F_p is closed and, definitely, $\bigcap_{p\geq 1} F_p = \emptyset$; otherwise there would exists $x \in K$ such that $a \leq \varphi_p(x)$ for every $p \geq 1$, so that $x \notin \partial_T K$. Accordingly, $\lim_{p\to\infty} p(T(\Phi)(x) - \Phi(x)) = +\infty$ and $\lim_{p\to\infty} \varphi_p(x) = -\infty$, which is not possible.

But, if $\bigcap_{p\geq 1} F_p = \emptyset$, because of the compactness of K, there exists $p \geq 1$ such that $F_p = \emptyset$, which implies that $a_p < a$, a contradiction. This finishes the proof.

The last statement is a consequence of (4.4.16) and (1.3.10). □

Remark 4.4.8. Theorem 4.4.7 applies in particular to $T = T_0$ or $T = \lambda T_0 + (1-\lambda)I$ $(\lambda \in \mathscr{C}(K), 0 \leq \lambda \leq 1)$ provided the Markov projection T_0 satisfies one of the assumptions of Theorems 4.3.1, 4.3.2 or 4.3.4. In this special case Theorem 4.4.7 has been already obtained in [8] (see also [18, Theorem 6.2.6, formula (6.2.20)]).

As a direct application of Theorem 4.4.7 consider the canonical projection T_d on the d-dimensional simplex K_d (see (1.1.48) and (1.1.45)) as well as the Poisson

operator $T_L : \mathscr{C}(K) \longrightarrow \mathscr{C}(K)$, K being an ellipsoid of the form (4.2.19), associated with a strictly elliptic differential operator L with constant coefficients as in (3.1.19) and (4.2.20).

Corollary 4.4.9. *Under the above assumptions, let T be a Markov operator on $\mathscr{C}(K_d)$ (resp., on $\mathscr{C}(K)$) such that $\partial_T K_d \subset \partial_e K_d$ and $T(h) = h$ for every $h \in T_d(\mathscr{C}(K_d)) = A(K_d)$ (resp., $\partial_T K \subset \partial K$ and $T(h) = h$ for every $h \in T_L(\mathscr{C}(K))$) (see Corollary 3.2.8). Furthermore, assume that either (4.3.3) or (4.3.5) holds true and let $(T(t))_{t\geq 0}$ be the relevant Markov semigroup generated by the closure of $(W_T, \mathscr{C}^2(K_d))$ (resp., $(W_T, \mathscr{C}^2(K))$). Then, for every $f \in \mathscr{C}(K_d)$,*

$$\lim_{t\to+\infty} T(t)(f) = T_d(f) \quad \textit{uniformly on } K_d \tag{4.4.17}$$

(resp., for every $f \in \mathscr{C}(K)$,

$$\lim_{t\to+\infty} T(t)(f) = T_L(f) \quad \textit{uniformly on } K). \tag{4.4.18}$$

Another field where the representation formula (4.3.2) can be usefully applied is concerned with Markov processes. Indeed, consider the Markov semigroup $(T(t))_{t\geq 0}$ on $\mathscr{C}(K)$ described in Theorem 4.3.1, and its associated normal Markov process $(\Omega, \mathscr{U}, (P^x)_{x\in K}, (Z_t)_{0\leq t\leq +\infty})$ with state space K (see Theorem 2.3.17). Then, according to (2.3.10), we have

$$T(t)(f)(x) = \int_K f\, dP_t(x,\cdot) = \int_K f \circ Z_t\, dP^x, \tag{4.4.19}$$

where $(P_t)_{t\geq 0}$ is the normal uniformly stochastically continuous Feller transition function associated with $(T(t))_{t\geq 0}$ as in Theorem 2.3.17.

From (4.3.2) it follows that, for every $x \in K$, the measure $P_t(x,\cdot)$ can be represented as a weak limit of a sequence of suitable measures (see [18, pp. 443-444, pp. 662-465]). Moreover, under the assumptions of Theorem 4.4.7, for each $x \in K$, (4.4.16) entails

$$\lim_{t\to+\infty} P_t(x,\cdot) = \tilde{\mu}_x^{T_0} \quad \text{weakly}, \tag{4.4.20}$$

where $\tilde{\mu}_x^{T_0}$ denotes the Borel measure on K defined by (3.1.1).

In particular, if $B \in B_K$ and $\partial B \cap \partial_{H_0} K = \emptyset$, then

$$\lim_{t\to+\infty} P_t(x,B) = 0 \tag{4.4.21}$$

for every $x \in K$, where $H_0 := T_0(\mathscr{C}(K))$.

In the sequel we suppose that K is a compact subset of $\mathbf{R}^d$ and, for each $t \geq 0$, we denote by $Z_t^1, \dots, Z_t^d$ the components of the random variable Z_t.

For each $x \in K$ we denote the expected value of Z_t with respect to P^x by

$$E_x(Z_t) := (E_x(Z_t^1), \dots, E_x(Z_t^d)),$$

and the variance of Z_t by

$$\mathrm{Var}_x(Z_t) := \sum_{i=1}^{d} \mathrm{Var}_x(Z_t^i).$$

After these preliminaries, we can state the following result.

Theorem 4.4.10. *Under the assumptions of Theorem 4.3.4, for every $x \in K$ and $t \geq 0$, the following properties hold true:*
(i) $E_x(Z_t) = x$.
Moreover, if T is a Markov projection satisfying (3.2.10), *then*
(ii) $\mathrm{Var}_x(Z_t^i) = (1 - e^{-t})\alpha_{ii}(x)$ *for every* $i = 1, \dots, d$ *and* $\mathrm{Var}_x(Z_t) = (1 - e^{-t}) \sum_{i=1}^{d} \alpha_{ii}(x) = (1 - e^{-t})(T(e)(x) - e(x))$, *where the α_{ii}'s are the coefficients of the differential operator W_T (see* (4.1.15) *and* (4.1.16)*) and* $e(x) := \sum_{i=1}^{d} x_i^2$ $(x = (x_1, \dots, x_d) \in K)$.
(iii) $\lim_{t \to +\infty} \mathrm{Var}_x(Z_t^i) = \alpha_{ii}(x)$ *uniformly with respect to $x \in K$, and therefore* $\lim_{t \to +\infty} \mathrm{Var}_x(Z_t) = \sum_{i=1}^{d} \alpha_{ii}(x) = T(e)(x) - e(x)$.

Proof. From (3.1.2), (3.1.3) and (4.3.6) it follows that $T(t)(pr_i) = pr_i$ for every $i = 1, \dots, d$. Now

$$E_x(Z_t^i) = E_x(pr_i(Z_t)) = T(t)(pr_i)(x) = x_i,$$

so that $E_x(Z_t) = x$.

To prove (ii), we use (3.2.4) and (3.2.10) in order to deduce that

$$B_n^m(pr_i^2) = \left(1 - \frac{1}{n}\right)^m pr_i^2 + \left(1 - \left(1 - \frac{1}{n}\right)^m\right) T(pr_i^2),$$

for all $n, m \geq 1$ and $i = 1, \dots, d$.

Then, from (4.3.6) we get

$$T(t)(pr_i^2) = e^{-t} pr_i^2 + (1 - e^{-t}) T(pr_i^2),$$

and so

$$\begin{aligned}\mathrm{Var}_x(Z_t^i) &= E_x((Z_t^i)^2) - pr_i^2(x) = T(t)(pr_i^2)(x) - pr_i^2(x)\\ &= (1 - e^{-t})(T(pr_i^2)(x) - pr_i^2(x)) = (1 - e^{-t})\alpha_{ii}(x).\end{aligned}$$

Finally, (iii) is a consequence of (ii). □

We may interpret (4.4.21) and Theorem 4.4.10 imagining a particle which, starting at a position $x \in K$, moves continuously and tends to go away; then, almost surely, it reaches $\partial_T K$, where a viscosity phenomenon occurs.

Other probabilistic applications may be found in [18, pp. 461-466].

We finally point out that, as in many other problems of constructive approximation, from a numerical point of view it would be important to estimate the quantities

$$\|T(t)(f) - B_n^{k(n)}(f)\|_\infty \qquad (f \in \mathscr{C}(K)) \tag{4.4.22}$$

($t \geq 0,\ k(n)/n \to t$), on the light of Theorems 4.3.1, 4.3.2, 4.3.4 and 2.1.15.

This is a challenging problem and only very few results, in the special setting of the unit interval and for the classical Bernstein operators, have been obtained. We refer to Appendix A.2 where such partial results can be found with the hope that they can be a good source for tackling the problem in other settings and, hopefully, in the full generality.

4.5 The special case of the unit interval

We conclude the chapter by discussing all questions treated in the previous sections in the framework of the unit interval, where more complete results can be obtained.

Among other things, we shall see that, given a Markov operator T on $\mathscr{C}([0,1])$ such that $T(e_1) = e_1$ and $T(e_2) > e_2$ on $]0,1[$, then the relevant differential operator W_T is always closable and its closure generates a Markov semigroup on $\mathscr{C}([0,1])$. Moreover, the domain $D(A_T)$ of the closure A_T is explicitly determined.

Note that T does not necessarily leave invariant all spaces of polynomials of degree m, $m \geq 1$.

Furthermore, this semigroup can be approximated by means of suitable iterates of the Bernstein-Schnabl operators associated with T, obtaining in this way a useful tool to get some quantitative and qualitative information about it by means of the study of the operators B_n.

Finally, we state some further asymptotic formulae for the B_n's and we determine both their saturation class and the Favard class of the relevant Markov semigroup.

For the reader's convenience we divide the section in several subsections.

4.5.1 Degenerate differential operators on $[0,1]$

We start by considering an arbitrary function $\alpha \in \mathscr{C}(]0,1[)$ such that $0 < \alpha(x)$ for every $0 < x < 1$. No further conditions are imposed at the end-points 0 and 1.

Consider the differential operator $A : D_V(A) \to \mathscr{C}([0,1])$ defined by

$$A(u)(x) := \begin{cases} \dfrac{\alpha(x)}{2}u''(x) & \text{if } 0 < x < 1; \\ \\ 0 & \text{if } x = 0, 1 \end{cases} \tag{4.5.1}$$

on the linear subspace $D_V(A)$ of all functions $u \in \mathscr{C}([0,1]) \cap \mathscr{C}^2(]0,1[)$ such that

$$\lim_{x \to 0^+} \alpha(x)u''(x) = \lim_{x \to 1^-} \alpha(x)u''(x) = 0. \tag{4.5.2}$$

The boundary conditions (4.5.2), which are incorporated in the domain $D_V(A)$, are often referred to as Ventcel's boundary conditions.

As a first step we get the following generation result.

Theorem 4.5.1. *The operator $(A, D_V(A))$ generates a Markov semigroup on $\mathscr{C}([0,1])$.*

Proof. According to the Feller theory (see Subsection 2.3.3 and Theorem 2.3.19), it suffices to show that 0 and 1 are not entrance end-points. Note that, in this context, choosing $x_0 := 1/2$, then $W = 1$, $Q(x) = \frac{x-1/2}{\alpha(x)}$ and $R(x) = \int_{1/2}^{x} \frac{1}{\alpha(s)}\,ds$

$(0 < x < 1)$. On account of Remark 2.3.18, if $Q \in L^1(0,1/2)$, then $\dfrac{1}{\alpha} \in L^1(0,1/2)$ and hence $R \in L^1(0,1/2)$. Therefore, 0 is not an entrance end-point.

The same reasoning may be used by replacing 0 with 1. □

Under some additional assumptions on α, we shall show that $\mathscr{C}^2([0,1])$ is a core for the operator $(A, D_V(A))$.

To this end first consider the auxiliary operator $(B, D_V(B))$ defined by

$$B(u)(x) := \begin{cases} \dfrac{x(1-x)}{2}u''(x) & \text{if } 0 < x < 1; \\ 0 & \text{if } x = 0, 1, \end{cases} \tag{4.5.3}$$

for every u belonging to

$$D_V(B) := \left\{ u \in \mathscr{C}([0,1]) \cap \mathscr{C}^2(]0,1[) \,|\, \text{there exist} \lim_{\substack{x\to 0^+\\ x\to 1^-}} x(1-x)u''(x) = 0 \right\}.$$

Theorem 4.5.2. *The operator $(B, D_V(B))$ is the generator of a Markov semigroup on $\mathscr{C}([0,1])$ and the subspace $\mathscr{C}^2([0,1])$ is a core for it.*

Proof. The first part of the statement obviously follows from Theorem 4.5.1.

As regards the second part of the statement, given $u \in D_V(B)$, we first observe that

$$\lim_{x\to 0^+} xu'(x) = 0, \quad \lim_{x\to 0^+} xu''(x) = 0 \tag{1}$$

and

$$\lim_{x\to 1^-} (1-x)u'(x) = 0, \quad \lim_{x\to 1^-} (1-x)u''(x) = 0. \tag{2}$$

Actually, setting $M := \sup\limits_{0<x<1} x(1-x)|u''(x)|$, for every $0 < x \le 1/2$ we get

$$\left| \int_x^{\frac{1}{2}} u''(t)\,dt \right| \le \int_x^{\frac{1}{2}} \frac{M}{t(1-t)}\,dt \le 2M \int_x^{\frac{1}{2}} \frac{1}{t}\,dt = 2M\left(\log\left(\frac{1}{2}\right) - \log x\right),$$

i.e.,

$$\left| u'\left(\frac{1}{2}\right) - u'(x) \right| \le 2M\left(\log\left(\frac{1}{2}\right) - \log x\right).$$

Therefore

$$x|u'(x)| \le x\left(\left|u'\left(\frac{1}{2}\right)\right| + 2M\log\left(\frac{1}{2}\right)\right) - 2Mx\log x$$

and hence the first formula of (1) follows. The second one is obvious on account of the Ventcel's condition imposed on 0.

Formula (2) can be proved in a similar manner.

For every $n \ge 1$ and $0 \le x \le 1$ consider

$$u_n(x) := \begin{cases} u\left(\frac{1}{n}\right) + u'\left(\frac{1}{n}\right)\left(x - \frac{1}{n}\right) + \frac{1}{2}u''\left(\frac{1}{n}\right)\left(x - \frac{1}{n}\right)^2 & 0 \le x \le \frac{1}{n}; \\ u(x) & 1/n \le x \le 1 - \frac{1}{n}; \\ u\left(1 - \frac{1}{n}\right) + u'\left(1 - \frac{1}{n}\right)\left(x - 1 + \frac{1}{n}\right) & \\ +\frac{1}{2}u''\left(1 - \frac{1}{n}\right)\left(x - 1 + \frac{1}{n}\right)^2 & 1 - \frac{1}{n} \le x \le 1. \end{cases}$$

Then $u_n \in \mathscr{C}^2([0,1])$. We shall show that $u_n \to u$ and $B(u_n) \to B(u)$ uniformly on $[0,1]$, so that the result will follow.

Indeed, given $\varepsilon > 0$, since both u and $B(u)$ are uniformly continuous, there exists $\delta > 0$ such that for every $x, y \in [0,1]$, $|x - y| \le \delta$,

$$|u(x) - u(y)| \le \frac{\varepsilon}{3} \text{ and } |B(u)(x) - B(u)(y)| \le \frac{\varepsilon}{3}.$$

Moreover, formulae (1) and (2) imply that there exists $\nu \in \mathbf{N}$, $\nu \ge 1/\delta$, such that, for every $n \ge \nu$,

$$\frac{1}{n}\left|u'\left(\frac{1}{n}\right)\right| \le \frac{\varepsilon}{3}, \quad \frac{1}{n}\left|u'\left(1 - \frac{1}{n}\right)\right| \le \frac{\varepsilon}{3},$$
$$\frac{1}{n}\left|u''\left(\frac{1}{n}\right)\right| \le \frac{\varepsilon}{3}, \quad \frac{1}{n}\left|u''\left(1 - \frac{1}{n}\right)\right| \le \frac{\varepsilon}{3}.$$

Accordingly, it is not difficult to show that

$$|u_n(x) - u(x)| \le \varepsilon \quad \text{and} \quad |B(u_n)(x) - B(u)(x)| \le \varepsilon$$

for every $n \ge \nu$ and $x \in [0,1]$ and hence the result follows. □

In some special cases we shall see that $\mathscr{C}^2([0,1])$ is also a core for $(A, D_V(A))$.

Assume, indeed, that

$$\sup_{0<x<1} \frac{\alpha(x)}{x(1-x)} < +\infty \tag{4.5.4}$$

i.e., there exists $c > 0$ such that

$$\alpha(x) \le cx(1-x) \qquad (0 < x < 1). \tag{4.5.5}$$

Thus, $\lim\limits_{x\to 0^+} \alpha(x) = \lim\limits_{x\to 1^-} \alpha(x) = 0$.

In order to state the next result we need some more preliminaries regarding the Banach lattice

$$\mathscr{C}_0(]0,1[) := \left\{ f \in \mathscr{C}(]0,1[) \mid \lim_{\substack{x\to 0^+ \\ x\to 1^-}} f(x) = 0 \right\}$$
$$\cong \{ f \in \mathscr{C}([0,1]) \mid f(0) = f(1) = 0 \}.$$

We recall that a Feller semigroup on $\mathscr{C}_0(]0,1[)$ is a C_0-semigroup of positive linear contractions on $\mathscr{C}_0(]0,1[)$ (see the final part of Section 2.3).

Proposition 4.5.3. *The operators*

$$D_V(A_0) := D_V(A) \cap \mathscr{C}_0(]0,1[), \; A_0 := A|_{D_V(A_0)} : D_V(A_0) \longrightarrow \mathscr{C}_0(]0,1[)$$

and

$$D_V(B_0) := D_V(B) \cap \mathscr{C}_0(]0,1[), \; B_0 := B|_{D_V(B_0)} : D_V(B_0) \longrightarrow \mathscr{C}_0(]0,1[)$$

are the generators of some Feller semigroups on $\mathscr{C}_0(]0,1[)$.

Moreover, $\mathscr{C}^2([0,1]) \cap \mathscr{C}_0(]0,1[)$ *is a core for* $(B_0, D_V(B_0))$.

Proof. The first part of the statement is a direct consequence of Corollary 2.3.21. By (4.5.5), $D_V(B_0) \subset D_V(B)$. Therefore, if $u \in D_V(B_0)$, since $\mathscr{C}^2([0,1])$ is a core for $(B, D_V(B))$, then there exists a sequence $(u_n)_{n\geq 1}$ in $\mathscr{C}^2([0,1])$ such that $u_n \to u$ and $B(u_n) \to B(u)$ uniformly on $[0,1]$ (the functions $B(u_n)$ and $B(u)$ are extended to 0 at the boundary points 0 and 1). In particular, $u_n(0) \to 0$ and $u_n(1) \to 0$. For any $n \geq 1$ set $v_n(x) := u_n(x) - xu_n(1) - (1-x)u_n(0)$; then $v_n \in \mathscr{C}^2([0,1]) \cap \mathscr{C}_0(]0,1[)$, $v_n \to u$ uniformly on $[0,1]$ and finally, taking (4.5.3) into account, $B(v_n) = B(u_n) \to B(u)$ uniformly on $[0,1]$. □

At this point we are able to show the before announced result.

Theorem 4.5.4. *Under assumption* (4.5.4), $\mathscr{C}^2([0,1]) \cap \mathscr{C}_0(]0,1[)$ *is a core for* $(A_0, D_V(A_0))$ *and* $\mathscr{C}^2([0,1])$ *is a core for* $(A, D_V(A))$.

Proof. From Proposition 4.5.3 and Theorem 2.3.13 it follows that the operator $(\lambda B_0, D_V(B_0))$ is closable and its closure $(C, D(C))$ is the generator of a Feller semigroup on $\mathscr{C}_0(]0,1[)$, where λ is the continuous bounded function on $]0,1[$ defined by

$$\lambda(x) := \frac{\alpha(x)}{x(1-x)} \qquad (0 < x < 1). \tag{1}$$

By referring to the definition of closure and, once again, to the final part of Proposition 4.5.3, it is not difficult to show that $\mathscr{C}^2([0,1]) \cap \mathscr{C}_0(]0,1[)$ is a core for $(C, D(C))$. But $A_0 = C$ on $\mathscr{C}^2([0,1]) \cap \mathscr{C}_0(]0,1[)$ and hence $(A_0, D_V(A_0)) = (C, D(C))$ by virtue of Proposition 2.1.7, part (g). This reasoning proves the first part of the statement.

Now, let $u \in D_V(A)$ and $\varepsilon > 0$ and consider the function $v(x) := u(x) - xu(1) - (1-x)u(0)$ $(0 \leq x \leq 1)$. Since $v \in D_V(A_0)$, there exists $v^* \in \mathscr{C}^2([0,1]) \cap \mathscr{C}_0(]0,1[)$ such that $\|v - v^*\|_\infty \leq \varepsilon$ and $\|A_0(v) - A_0(v^*)\|_\infty \leq \varepsilon$. Therefore, setting $w(x) = v^*(x) + xu(1) + (1-x)u(0)$ $(0 \leq x \leq 1)$, clearly $w \in \mathscr{C}^2([0,1])$, $\|u - w\|_\infty \leq \varepsilon$ and $\|A(u) - A(w)\|_\infty = \|A_0(v) - A_0(v^*)\|_\infty \leq \varepsilon$. □

Coming back to the general theme of this book, consider a Markov operator T on $\mathscr{C}([0,1])$ satisfying

$$T(e_1) = e_1 \tag{4.5.6}$$

and

$$x^2 < T(e_2)(x) \qquad \text{for every } 0 < x < 1. \tag{4.5.7}$$

For every $u \in \mathscr{C}^2([0,1])$ and $x \in [0,1]$, let W_T be the elliptic second-order differential operator associated with T which is defined as

$$W_T(u)(x) = \frac{\alpha(x)}{2} u''(x), \tag{4.5.8}$$

where

$$\alpha(x) := T(e_2)(x) - x^2 \qquad (0 \le x \le 1) \tag{4.5.9}$$

(see Section 4.2).

Then $\alpha \in \mathscr{C}([0,1])$, $\alpha(0) = \alpha(1) = 0$ and

$$0 < \alpha(x) \le x(1-x) \quad \text{for every } 0 < x < 1. \tag{4.5.10}$$

Remark 4.5.5. Conversely, if $\alpha \in \mathscr{C}([0,1])$ satisfies (4.5.10), then there always exists a Markov operator T on $\mathscr{C}([0,1])$ satisfying (4.5.6) such that $\alpha = T(e_2) - e_2$. Indeed, it is enough to consider the function $\lambda :]0,1[\longrightarrow \mathbf{R}$ defined as in formula (1) of the proof of Theorem 4.5.4 and the Markov operator T on $\mathscr{C}([0,1])$ defined by (4.2.9). Such an operator will be referred to as an *admissible Markov operator for* α (or *for* $(A, D_V(A))$).

Corollary 4.5.6. *The following statements hold true:*

(1) If T is a Markov operator on $\mathscr{C}([0,1])$ satisfying (4.5.6)-(4.5.7), *then the differential operator $(A, D_V(A))$ defined by* (4.5.1)-(4.5.2), *where α is defined by* (4.5.9), *is the generator of a Markov semigroup on $\mathscr{C}([0,1])$, $\mathscr{C}^2([0,1])$ is a core for it and it coincides with the closure $(A_T, D(A_T))$ of $(W_T, \mathscr{C}^2([0,1]))$ (see* (4.5.8)*).*

(2) If $(A, D_V(A))$ is a differential operator of the form (4.5.1)-(4.5.2) *with α satisfying* (4.5.5), *then there exists an admissible Markov operator T on $\mathscr{C}([0,1])$ for $\frac{\alpha}{c}$ and $(A, D_V(A))$ coincides with the closure of $(c\,W_T, \mathscr{C}^2([0,1]))$, i.e., with $(c\,A_T, D(A_T))$, where, once again, $(A_T, D(A_T))$ denotes the closure of $(W_T, \mathscr{C}^2([0,1]))$.*

Proof. (1). The first part of the statement directly follows from Theorems 4.5.1 and 4.5.4. As regards the last part, since $A = W_T$ on $\mathscr{C}^2([0,1])$, then the operator $(W_T, \mathscr{C}^2([0,1]))$ is closable and its closure $(A_T, D(A_T))$ satisfies $D(A_T) \subset D_V(A)$ and $A = A_T$ on $D(A_T)$. Hence, the result follows by applying property (g) of Proposition 2.1.7.

(2). It is a consequence of (1) and of Remark 4.5.5. □

4.5.2 Approximation properties by means of Bernstein-Schnabl operators

In this subsection we deepen the analysis on the Markov semigroup generated by $(A, D_V(A))$ by approximationg it in terms of suitable iterates of Bernstein-Schnabl operators on $[0, 1]$. Moreover, we shall see that, thanks to such an approximation formula, it is also possible to infer some preservation properties of the semigroup by similar ones held by the approximating operators.

We shall assume that the strictly positive continuous function α satisfies (4.5.5) so that Theorems 4.5.1 and 4.5.4 as well as Corollary 4.5.6 may be applied. Moreover, to our purposes, it is not limitative to assume that $c = 1$, i.e.,

$$0 < \alpha(x) \leq x(1-x) \qquad (0 < x < 1). \tag{4.5.11}$$

Indeed, setting $\tilde{\alpha} := \dfrac{\alpha}{c}$ and denoting by $(\widetilde{A}, D_V(\widetilde{A}))$ the relevant differential operator defined as in (4.5.1)-(4.5.2), clearly $D_V(A) = D_V(\widetilde{A})$ and $A = c\,\widetilde{A}$.

Furthermore, considering the Markov semigroups $(T(t))_{t\geq 0}$ and $(\widetilde{T}(t))_{t\geq 0}$ generated by $(A, D_V(A))$ and $(\widetilde{A}, D_V(\widetilde{A}))$, respectively, it turns out that

$$T(t) = \widetilde{T}(ct) \qquad \text{for every } t \geq 0, \tag{4.5.12}$$

and hence all the approximation formulae and the preservation properties we shall establish for $(\widetilde{T}(t))_{t\geq 0}$ can be naturally transferred to $(T(t))_{t\geq 0}$ as well.

So, from now on we shall assume that $\alpha \in \mathscr{C}(]0, 1[)$ satisfies (4.5.11). According to Corollary 4.5.6, part (2), and Remark 4.5.5, consider an admissible Markov operator T on $\mathscr{C}([0, 1])$ for α.

Let $(B_n)_{n\geq 1}$ be the sequence of Bernstein-Schnabl operators associated with T (see (3.1.7)). From Theorem 1.3.1, $\partial_T([0, 1]) = \{x \in [0, 1] \,|\, T(e_2)(x) = e_2(x)\}$ and hence, by (4.5.11) and (3.2.1), we have

$$B_n(f)(x) = f(x) \quad \text{for } x = 0, 1; \tag{4.5.13}$$

in particular the B_n's map the space $\mathscr{C}_0(]0, 1[)$ into itself.

We have now prepared all the necessary tools to easily get the approximation result for our semigroup.

Theorem 4.5.7. *Under assumption* (4.5.11)*, let* $(A, D_V(A))$ *be the operator defined by* (4.5.1)-(4.5.2) *and denote by* $(T(t))_{t\geq 0}$ *the relevant Markov semigroup on* $\mathscr{C}([0, 1])$*. Furthermore, consider an admissible Markov operator* T *for* $(A, D_V(A))$ *and the relevant sequence* $(B_n)_{n\geq 1}$ *of Bernstein-Schnabl operators.*

Then, for any $t \geq 0$*, for every sequence* $(k(n))_{n\geq 1}$ *of positive integers such that* $k(n)/n \to t$ *and for every* $f \in \mathscr{C}([0, 1])$*,*

$$T(t)(f) = \lim_{n\to\infty} B_n^{k(n)}(f) \quad \textit{uniformly on } [0, 1]. \tag{4.5.14}$$

Proof. By Theorem 4.1.5, for every $u \in \mathscr{C}^2([0,1]) \subset D_V(A)$,

$$\lim_{n\to\infty} n(B_n(u) - u) = A(u) \quad \text{uniformly on } [0,1]$$

and $\mathscr{C}^2([0,1])$ is a core for $(A, D_V(A))$ (see Theorem 4.5.4). Finally, for every $n \geq 1$ and $p \geq 1$, $\|B_n^p\| = 1$, since $B_n^p(\mathbf{1}) = \mathbf{1}$. Therefore, formula (4.5.14) easily follows from Remark 2.2.4, 2. □

Remark 4.5.8. When $\alpha(x) := x(1-x)$, then the approximating Bernstein-Schnabl operators are the classical Bernstein operators (3.1.11).

The main significance of Theorem 4.5.7 lies in the possibility to numerically approximate and to infer qualitative properties of the semigroup $(T(t))_{t\geq 0}$ and hence of the solutions to the initial-boundary value problem associated with $(A, D_V(A))$

$$\begin{cases} \dfrac{\partial u}{\partial t}(x,t) = \dfrac{\alpha(x)}{2}\dfrac{\partial^2 u}{\partial x^2}(x,t) & 0 < x < 1,\, t \geq 0; \\[2ex] u(x,0) = u_0(x) & 0 \leq x \leq 1,\, u_0 \in D_V(A); \\[2ex] \lim\limits_{\substack{x\to 0^+ \\ x\to 1^-}} \alpha(x)\dfrac{\partial^2 u}{\partial x^2}(x,t) = 0 & t \geq 0, \end{cases} \tag{4.5.15}$$

because of the formula

$$u(x,t) = T(t)u_0(x) \qquad (0 \leq x \leq 1,\, t \geq 0). \tag{4.5.16}$$

Such a problem is strictly related to a stochastic model from genetics which is involved in the study of the fluctuations of gene frequency under the influence of mutation and selection (see, e.g., [18, pp. 465-466] for more details).

We also point out that the semigroup involved in Theorem 4.5.1 is the transition semigroup of some normal Markov process, having $[0,1]$ as state space, with absorbing barriers at 0 and 1 and with mean instantaneous velocity 0 and variance instantaneous velocity $\alpha(x)$ at position $x \in [0,1]$ (see [195, Section I.4]).

From formula (4.5.14) it is then possible to infer several useful information on such a Markov process (see Section 4.4 and [18, pp. 461-465]).

For some results concerning numerical estimates of the quantities

$$\|T(t)(f) - B_n^{k(n)}(f)\|_\infty \quad (n \to \infty)$$

($f \in \mathscr{C}([0,1])$, $t \geq 0$, $k(n)/n \to t$) we refer to Appendix A.2.

Before passing on to illustrate some spatial regularity properties of the solutions to problems (4.5.15), we only point out that it is also possible to consider other type of differential problems like (4.4.6) and (4.4.11). For the sake of brevity we omit the details.

4.5.3 Preservation properties and asymptotic behaviour

In this subsection we discuss some preservation properties of the limit semigroup, which have their counterparts in terms of spatial properties of the solutions to problem (4.5.15) because of formula (4.5.16).

We begin to prove that the B_n's (associated with an arbitrary Markov operator T on $\mathscr{C}([0,1])$ satisfying $T(e_1) = e_1$) preserve the class of the increasing continuous functions. To this end we recall the following auxiliary functions defined by (3.3.4)-(3.3.7) in a more general context. Suppose $n \geq 2$ and consider $f \in \mathscr{C}([0,1])$. For every $x_1, \dots, x_{n-1} \in [0,1]$ and $x \in [0,1]$, set

$$f_{x_1,\dots,x_{n-1}}(t) := f\left(\frac{x_1 + \dots + x_{n-1} + t}{n}\right) \qquad (0 \leq t \leq 1) \tag{4.5.17}$$

and

$$f^x_{x_1,\dots,x_{n-2}}(t) := T\left(f_{x_1,\dots,x_{n-2},t}\right)(x) \qquad (0 \leq t \leq 1). \tag{4.5.18}$$

Moreover, for every $k = 3, \dots, n-1$,

$$f^x_{x_1,\dots,x_{n-k}}(t) := T(f^x_{x_1,\dots,x_{n-k},t})(x) \quad (0 \leq t \leq 1). \tag{4.5.19}$$

Finally, set

$$f^x(t) := T(f^x_t)(x) \quad (0 \leq t \leq 1). \tag{4.5.20}$$

Theorem 4.5.9. *Assume that the Markov operator T maps increasing continuous functions into increasing continuous functions. Then, for every $n \geq 1$, B_n maps increasing continuous functions into increasing continuous functions too.*

Proof. If $n = 1$ the statement is true since $B_1 = T$. Assuming $n \geq 2$, let f be an increasing continuous function on $[0,1]$, $x \in [0,1]$ and consider the auxiliary functions defined in (4.5.17)-(4.5.20). By finite induction it is easy to prove that $f_{x_1,\dots,x_{n-1}}$ and $f^x_{x_1,\dots,x_h}$ are increasing for any $h = 1, \dots, n-2$. Moreover, observe that f^x is increasing as well. Now let $y \in [0,1]$, $x < y$ and note that

$$T(f^x_{x_1,\dots,x_h})(x) \leq T(f^x_{x_1,\dots,x_h})(y) \text{ for every } h = 1, \dots, n-2, \tag{1}$$

$$T(f^x)(x) \leq T(f^x)(y) \tag{2}$$

and

$$T(f_{x_1,\dots,x_{n-1}})(x) \leq T(f_{x_1,\dots,x_{n-1}})(y). \tag{3}$$

By (1.3.1), (3.1.7) and (3), we get

$$\begin{aligned} B_n(f)(x) &= \int_0^1 \cdots \int_0^1 \left(\int_0^1 f_{x_1,\dots,x_{n-1}}(x_n)\, d\tilde{\mu}^T_x(x_n)\right) d\tilde{\mu}^T_x(x_1) \cdots d\tilde{\mu}^T_x(x_{n-1}) \\ &= \int_0^1 \cdots \int_0^1 T(f_{x_1,\dots,x_{n-1}})(x)\, d\tilde{\mu}^T_x(x_1) \cdots d\tilde{\mu}^T_x(x_{n-1}) \\ &\leq \int_0^1 \cdots \int_0^1 T(f_{x_1,\dots,x_{n-1}})(y)\, d\tilde{\mu}^T_x(x_1) \cdots d\tilde{\mu}^T_x(x_{n-1}). \end{aligned}$$

Now, since $T(f_{x_1,\dots,x_{n-1}})(y) = f^y_{x_1,\dots,x_{n-2}}(x_{n-1})$, by (1), we obtain

$$B_n(f)(x) \le \int_0^1 \cdots \int_0^1 \left(\int_0^1 f^y_{x_1,\dots,x_{n-2}}(x_{n-1})\, d\tilde{\mu}^T_x(x_{n-1}) \right) d\tilde{\mu}^T_x(x_1) \cdots d\tilde{\mu}^T_x(x_{n-2})$$
$$= \int_0^1 \cdots \int_0^1 T(f^y_{x_1,\dots,x_{n-2}})(x)\, d\tilde{\mu}^T_x(x_1) \cdots d\tilde{\mu}^T_x(x_{n-2})$$
$$\le \int_0^1 \cdots \int_0^1 T(f^y_{x_1,\dots,x_{n-2}})(y)\, d\tilde{\mu}^T_x(x_1) \cdots d\tilde{\mu}^T_x(x_{n-2})$$

and so on. Finally,

$$B_n(f)(x) \le \int_0^1 T(f^y_{x_1})(x)\, d\tilde{\mu}^T_x(x_1) \le \int_0^1 T(f^y_{x_1})(y)\, d\tilde{\mu}^T_x(x_1)$$
$$= \int_0^1 f^y(x_1)\, d\tilde{\mu}^T_x(x_1) = T(f^y)(x) \le T(f^y)(y),$$

because of (2).

On the other hand, $B_n(f)(y) = T(f^y)(y)$ since

$$B_n(f)(y) = \int_0^1 \cdots \int_0^1 \left(\int_0^1 f_{x_1,\dots,x_{n-1}}(x_n)\, d\tilde{\mu}^T_y(x_n) \right) d\tilde{\mu}^T_y(x_1) \cdots d\tilde{\mu}^T_y(x_{n-1})$$
$$= \int_0^1 \cdots \int_0^1 T(f_{x_1,\dots,x_{n-1}})(y)\, d\tilde{\mu}^T_y(x_1) \cdots d\tilde{\mu}^T_y(x_{n-1})$$
$$= \int_0^1 \cdots \int_0^1 \left(\int_0^1 f^y_{x_1,\dots,x_{n-2}}(x_{n-1})\, d\tilde{\mu}^T_y(x_{n-1}) \right) d\tilde{\mu}^T_y(x_1) \cdots d\tilde{\mu}^T_y(x_{n-1})$$
$$= \int_0^1 \cdots \int_0^1 T(f^y_{x_1,\dots,x_{n-2}})(y)\, d\tilde{\mu}^T_y(x_1) \cdots d\tilde{\mu}^T_y(x_{n-2})$$
$$= \dots = \int_0^1 T(f^y_{x_1})(y)\, d\tilde{\mu}^T_y(x_1) = \int_0^1 f^y(x_1)\, d\tilde{\mu}^T_y(x_1) = T(f^y)(y)$$

and hence the proof is complete. □

Theorem 4.5.10. *Consider the Markov semigroup $(T(t))_{t\ge 0}$ generated by the operator $(A, D_V(A))$ and the approximating sequence $(B_n)_{n\ge 1}$ of Bernstein-Schnabl operators associated with an admissible Markov operator T on $\mathscr{C}([0,1])$ as in Theorem 4.5.7. Then*

(1) $T(t)(f)(x) = f(x)$ for every $f \in \mathscr{C}([0,1])$, $t \ge 0$ and $x = 0, 1$.

(2) If the operator T maps continuous increasing functions into (continuous) increasing functions, then each $T(t)$ maps continuous increasing functions into increasing functions.

(3) If $T(\mathrm{Lip}(1,1)) \subset \mathrm{Lip}(1,1)$ then, for every $M > 0$, $0 < \alpha \le 1$ and $t \ge 0$, $T(t)(\mathrm{Lip}(M,\alpha)) \subset \mathrm{Lip}(M,\alpha)$.

(4) If $f \in \mathscr{C}([0,1])$ the following propositions are equivalent:
   (i) f is convex.
   (ii) $B_{n+1}(f) \le B_n(f)$ for every $n \ge 1$.

(iii) $f \le B_n(f)$ *for every* $n \ge 1$.

(iv) $f \le T(t)(f)$ *for any* $t \ge 0$.

(5) *Under assumptions* (c_1) *and* (c_2) *of Theorem 3.4.3, if* $f \in \mathscr{C}([0,1])$ *is convex, then each* $T(t)(f)$ *is convex* $(t \ge 0)$ *and the family* $(T(t)(f))_{t\ge 0}$ *is increasing with respect to the parameter* $t \ge 0$.

(6) *For every* $f \in \mathscr{C}([0,1])$,

$$\lim_{t\to+\infty} T(t)(f) = T_1(f) \qquad \text{uniformly on } [0,1], \tag{4.5.21}$$

where T_1 *denotes the canonical projection defined by* (1.1.49). *Therefore,* $\lim_{t\to+\infty} T(t)(f) = 0$ *uniformly on* $[0,1]$ *if and only if* $f(0) = f(1) = 0$.

Proof. Statement (1) follows from (4.5.13). Statements (2) and (3) follow from Theorem 4.5.9 and Corollary 3.3.2, respectively.

We pass on to prove property (4).

(i)⇒(ii). The implication follows from Theorem 3.5.2.

(ii)⇒(iii). For every $n, m \ge 1$, we have $B_{n+m}(f) \le B_n(f)$; therefore, letting $m \to \infty$, statement (iii) follows from Theorem 3.2.1.

(iii)⇒(iv). It is a consequence of Theorem 4.5.7.

(iv)⇒(i). Set, for every $r > 0$,

$$u(r) := \frac{1}{r}\int_0^r T(s)(f)\,ds \in D_V(A).$$

Then $A(u)(r) = \frac{1}{r}(T(r)(f) - f) \ge 0$, so that $u(r)$ is convex for every $r > 0$. Accordingly, $f = \lim_{r\to 0^+} u(r)$ is convex too.

The first part of (5) is a consequence of Theorem 4.4.1 and Example 4.4.6, 2. As regards the second part, first take $u \in D_V(A)$, u convex. Then for every $t \ge 0$, $T(t)(u) \in D_V(A)$ and $T(t)(u)$ is convex too. Therefore $AT(t)(u) \ge 0$ and hence $\frac{d}{dt}T(t)(u) = AT(t)(u) \ge 0$ $(t \ge 0)$, so that $(T(t)(u))_{t\ge 0}$ is increasing. Consider again the function $u(r)$ $(r > 0)$ as above. Then $\lim_{r\to 0^+} u(r) = f$ uniformly on $[0,1]$ and each $u(r)$ is convex. So, for $0 \le s < t$, we get $T(s)(u(r)) \le T(t)(u(r))$ and hence $T(s)(f) \le T(t)(f)$, because of the continuity of the operators $T(s)$ and $T(t)$, and this completes the proof of (5).

In order to obtain property (6) it is enough to apply Theorem 4.4.7 to the Markov operator T and the Markov projection T_1 on $\mathscr{C}([0,1])$. In this case, $T_1(\mathscr{C}([0,1])) = A([0,1]) = \operatorname{span}(\{\mathbf{1}, e_1\})$ and $\partial_{T_1}[0,1] = \{0,1\}$ so that it is easily seen that all the assumptions of Theorem 4.4.7 are satisfied and hence (6) follows. □

4.5.4 The saturation class of Bernstein-Schnabl operators and the Favard class of their limit semigroups

In this last subsection we shall determine both the saturation class of Bernstein-Schnabl operators on the unit interval and the Favard class of the relevant Markov semigroup. Moreover, we shall describe the classes of all continuous functions verifying a pointwise or a uniform asymptotic formula. Among other things, the characterization of the Favard class reveals some further "spatial regularity" properties preserved under the evolution governed by the semigroup approximated by Bernstein-Schnabl operators.

We begin by stating a pointwise asymptotic formula for Bernstein-Schnabl operators on $[0,1]$ associated with a Markov operator $T : \mathscr{C}([0,1]) \longrightarrow \mathscr{C}([0,1])$ satisfying (4.5.6) and (4.5.7). Setting α as is (4.5.9), we recall that $\alpha \in \mathscr{C}([0,1])$, $\alpha(0) = \alpha(1) = 0$ and

$$0 < \alpha(x) \le x(1-x) \quad \text{for every } 0 < x < 1.$$

Given $x \in [0,1]$, consider the auxiliary function $\psi_x(t) := t - x$ $(0 \le t \le 1)$. Then, for every $n \ge 1$,

$$B_n(\mathbf{1}) = \mathbf{1}, \qquad B_n(e_1) = e_1, \qquad B_n(\psi_x) = \psi_x \tag{4.5.22}$$

and

$$B_n(\psi_x^2)(x) = \frac{\alpha(x)}{n}, \qquad B_n(\psi_x^4)(x) \le \frac{M^*}{n^2}, \tag{4.5.23}$$

where M^* is a constant independent on $n \ge 1$ (see the last formula in the proof of Theorem 4.1.5).

Moreover, for every $f \in \mathscr{C}([0,1])$, (4.5.13) holds.

From [18, Theorem 5.1.2 and the subsequent Remark] it also follows that, if $f \in \mathscr{C}^1([0,1])$, then, for every $x \in [0,1]$,

$$|B_n(f)(x) - f(x)| \le 2\sqrt{\frac{\alpha(x)}{n}}\,\omega\left(f', \sqrt{\frac{\alpha(x)}{n}}\right), \tag{4.5.24}$$

where ω denotes the usual modulus of continuity (see (1.4.8)).

In particular, if $f' \in \mathrm{Lip}(M,1)$, i.e., f' is Lipschitz continuous and $|f'(x) - f'(y)| \le M|x-y|$ for every $x, y \in [0,1]$, then

$$|B_n(f)(x) - f(x)| \le \frac{2M}{n}\alpha(x). \tag{4.5.25}$$

Theorem 4.1.5 shows that, if $u \in \mathscr{C}^2([0,1])$, then

$$\lim_{n\to\infty} n(B_n(u) - u) = \frac{\alpha}{2}u'' \qquad \text{uniformly on } [0,1] \tag{4.5.26}$$

(see (4.5.8)).

Below we state some further asymptotic formulae. We shall denote by $\mathscr{C}_b^2(]0,1[)$ the linear subspace of all continuous functions on $[0,1]$ which possess a bounded continuous second derivative on $]0,1[$. From formulae (4.5.22) and (4.5.23) and from Proposition 1.5.5 and Theorem 1.5.6 the following result easily follows.

Proposition 4.5.11. *Let $u \in \mathscr{C}([0,1])$. Then*

(i) If u is differentiable in a neighborhood of a point x_0 of $]0,1[$ and if, in addition, it is two times differentiable at it, then

$$\lim_{n\to\infty} n(B_n(u)(x_0) - u(x_0)) = \frac{\alpha(x_0)}{2} u''(x_0).$$

(ii) If $u \in \mathscr{C}_b^2(]0,1[)$, then

$$\lim_{n\to\infty} n(B_n(u) - u) = \frac{\alpha}{2} u''$$

uniformly on each compact subinterval of $]0,1[$.

The previous result can be further refined.

If $u \in \mathscr{C}_b^2(]0,1[)$ we denote by $A(u)$ the continuous function on $[0,1]$ defined by (4.5.1).

Theorem 4.5.12. *If $u \in \mathscr{C}_b^2(]0,1[)$, then*

$$\lim_{n\to\infty} n(B_n(u) - u) = A(u) \qquad \textit{uniformly on } [0,1]. \tag{4.5.27}$$

Proof. On account of (4.5.13) and Proposition 4.5.11, clearly $\lim_{n\to\infty} n(B_n(u)-u) = A(u)$ pointwise on $[0,1]$. Therefore, in order to get the desired result, it is sufficient to show that the sequence $(n(B_n(u) - u))_{n\geq 1}$ is equicontinuous on $[0,1]$. This is certainly true on $]0,1[$ by virtue of Proposition 4.5.11, part (ii). As concerns the end-points 0 and 1, setting $M := \sup_{0<x<1} |u''(x)|$, we then get that $u' \in \mathrm{Lip}(M,1)$. Therefore, by (4.5.25),

$$n|B_n(u)(x) - u(x)| \leq 2M\alpha(x) \qquad (0 \leq x \leq 1).$$

So, given $\varepsilon > 0$, choose $0 < \delta < 1$ such that $\alpha(x) \leq \varepsilon/(2M)$ for every $x \in [0,\delta] \cup [1-\delta, 1]$ and hence

$$n|B_n(u)(x) - u(x)| - n|B_n(u)(0) - u(0)| = n|B_n(u)(x) - u(x)| \leq \varepsilon$$

and

$$n|B_n(u)(x) - u(x)| - n|B_n(u)(1) - u(1)| = n|B_n(u)(x) - u(x)| \leq \varepsilon.$$

□

Consider the differential operator $A : D_V(A) \longrightarrow \mathscr{C}([0,1])$ defined by (4.5.1) with α defined by (4.5.9).

The operator $(A, D_V(A))$ is the generator of a Markov semigroup $(T(t))_{t\geq 0}$ on $\mathscr{C}([0,1])$ (see Theorem 4.5.1). Moreover, for every $t \geq 0$ and for every sequence $(k(n))_{n\geq 1}$ of positive integers such that $k(n)/n \to t$,

$$T(t)(f) = \lim_{n\to\infty} B_n^{k(n)}(f) \qquad \text{uniformly on } [0,1] \tag{4.5.28}$$

$(f \in \mathscr{C}([0,1]))$(see Theorem 4.5.7).

By the next result we determine the *saturation class* of Bernstein-Schnabl operators and the *Favard class* of the relevant semigroup $(T(t))_{t\geq 0}$ which are, respectively, the linear subspaces of all functions $f \in \mathscr{C}([0,1])$ such that

$$\sup_{n\geq 1} n\|B_n(f) - f\|_\infty < +\infty \tag{4.5.29}$$

and

$$\sup_{0<t} \frac{\| T(t)(f) - f \|_\infty}{t} < +\infty \tag{4.5.30}$$

(see, e.g., [61, Section 2.1]). Among other things, the characterization of the Favard class reveals some further "spatial regularity" properties preserved under the evolution governed by the semigroup $(T(t))_{t\geq 0}$.

However, apart from its own interest, the next result will be fruitfully used to obtain a complete characterization of functions verifying the asymptotic formula (4.5.26).

Theorem 4.5.13. *Given $f \in \mathscr{C}([0,1])$ and $M \geq 0$, the following statements are equivalent:*
(i) For every $n \geq 1$ and $x \in [0,1]$,

$$|B_n(f)(x) - f(x)| \leq \frac{M\alpha(x)}{n}.$$

(ii) $f \in \mathscr{C}^1([0,1])$ and $f' \in Lip(M,1)$.
If, in addition, α is concave, then statements (i) and (ii) are also equivalent to
(iii) For every $t \geq 0$ and $x \in [0,1]$

$$|T(t)(f)(x) - f(x)| \leq Mt\alpha(x).$$

Proof. The equivalence (i)⇔(ii) follows from [148, Satz 3.10] taking Proposition 4.5.11 into account.
(ii)⇒(iii). We first point out that, for any $n \geq 1$ and $p \geq 1$,

$$B_n^p(f) - f = \sum_{k=1}^{p} B_n^{k-1}(B_n(f) - f). \tag{1}$$

By assumption and on account of Theorem 3.5.2 we get

$$|B_n^p(f) - f| \leq \sum_{k=1}^{p} B_n^{k-1}(|B_n(f) - f|) \leq \frac{M}{n} \sum_{k=1}^{p} B_n^{k-1}(\alpha) \leq \frac{pM}{n}\alpha$$

for any $n \geq 1$ and $p \geq 1$. So, if $t \geq 0$ and if $k(n)/n \to t$, from the previous estimate and from (4.5.28) the result follows.
(iii)⇒(ii). By definition, if $u \in D_V(A)$, then $\lim_{t\to 0^+} \frac{T(t)(u)-u}{t} = A(u)$ uniformly on $[0,1]$. Therefore, given $x \in [0,1]$, since $\psi_x^2 \in D_V(A)$ and $\psi_x^4 \in D_V(A)$, we get

$$\lim_{t\to 0^+} \frac{T(t)(\psi_x^2)(x)}{t} = \lim_{t\to 0^+} \frac{T(t)(\psi_x^2)(x) - \psi_x^2(x)}{t} = \alpha(x)$$

and, similarly,

$$\lim_{t\to 0^+} \frac{T(t)(\psi_x^4)(x)}{t} = 0.$$

Moreover, by (4.5.2) and (4.5.22), $T(t)(\mathbf{1}) = \mathbf{1}$ and $T(t)(\psi_x) = \psi_x$ for every $t \geq 0$. Therefore, the same proof of part (i) of Proposition 1.5.5 adapted to the family $(T(t))_{t\geq 0}$, shows that, if a function $u \in \mathscr{C}([0,1])$ is differentiable in a neighborhood of a point x_0 of $]0,1[$ and if, in addition, it is two times differentiable at it, then

$$\lim_{t\to 0^+} \frac{T(t)(u)(x_0) - u(x_0)}{t} = \frac{\alpha(x_0)}{2} u''(x_0). \tag{2}$$

Therefore, by applying again Satz 3.10 of [148] to the sequence of positive linear operators $(T(1/n))_{n\geq 1}$, the result follows. □

For $M \geq 0$, denote by $\mathscr{C}^2_{M,\alpha}([0,1])$ the linear subspace of all functions $f \in \mathscr{C}^1([0,1])$ such that f' is absolutely continuous and

$$\alpha(x)|f''(x)| \leq M \text{ a.e. on } [0,1]. \tag{4.5.31}$$

Theorem 4.5.14. *Given $f \in \mathscr{C}([0,1])$ and $M \geq 0$, the following statements are equivalent:*
(i) For every $n \geq 1$,

$$\|B_n(f) - f\|_\infty \leq \frac{M}{n}.$$

(ii) For every $t \geq 0$,

$$\|T(t)(f) - f\|_\infty \leq Mt.$$

(iii) $f \in \mathscr{C}^2_{M,\alpha}([0,1])$.
In particular, for every $t \geq 0$, $T(t)(\mathscr{C}^2_{M,\alpha}([0,1])) \subset \mathscr{C}^2_{M,\alpha}([0,1])$.

Proof. (i)⇒(ii). By appealing to formula (1) of the proof of Theorem 4.5.13, we get

$$\|B_n^p(f) - f\|_\infty \leq \sum_{k=1}^{p} \|B_n^{k-1}\| \|B_n(f) - f\|_\infty \leq \frac{p}{n} M.$$

Now, given $t \geq 0$ and considering a sequence $(k(n))_{n\geq 1}$ of positive integers such that $k(n)/n \to t$, we obtain

$$\|B_n^{k(n)}(f) - f\|_\infty \leq \frac{k(n)}{n} M$$

and hence $\|T(t)(f) - f\| \le Mt$ by (4.5.28).
(ii)⇒(iii). Setting $L_n := T(1/n)$ for each $n \ge 1$, we get

$$\sup_{n\ge1} n\|L_n(f) - f\|_\infty \le M.$$

Moreover, by formula (2) in the proof of Theorem 4.5.13,

$$\lim_{n\to\infty} n(L_n(u)(x_0) - u(x_0)) = \frac{\alpha(x_0)}{2}u''(x_0)$$

for every $x_0 \in]0,1[$ and for every function $u \in \mathscr{C}([0,1])$ which is differentiable in a neighborhood of x_0 and two times differentiable at it. Therefore $f \in \mathscr{C}^2_{M,\alpha}([0,1])$ because of Theorem 4.3 of [134] applied with $\rho = 1$, $\lambda_n = n$, $w_0 = w_1 = 1$ and $w_2 = 1/\alpha$ (see also Section 6 of [134] and, in particular, formulae (6.2) and (3.3)).

The implication (iii)⇒(i) can be proved similarly by applying again Theorem 4.3 of [134] to the sequence $(B_n)_{n\ge1}$ and by taking into account that the remainder $o(1)$ which appears in formula (4.5) of Theorem 4.3 of [134], is zero in this case because the operators B_n leave invariant the affine functions (see the proof of Lemma 4.4 of [134]).

As regards the last part of the statement, given $s, t \ge 0$, we get, for $f \in \mathscr{C}^2_{M,\alpha}([0,1])$,

$$\|T(s)T(t)(f) - T(t)(f)\|_\infty \le \|T(t)\|\|T(s)(f) - f\|_\infty \le \|T(s)(f) - f\|_\infty$$

and hence $T(t)(f) \in \mathscr{C}^2_{M,\alpha}([0,1])$ because of the equivalence (ii)⇔(iii) applied both to f and $T(t)(f)$. □

Corollary 4.5.15. *Denote by* $\mathrm{Sat}((B_n)_{n\ge1})$ *the saturation class of the Bernstein-Schnabl operators associated with a Markov operator satisfying* (4.5.6) *and* (4.5.7) *and by* $\mathrm{Fav}((T(t))_{t\ge0})$ *the Favard class of the relevant Markov semigroup (see* (4.5.29)-(4.5.30)*). Then*

$$\mathrm{Sat}((B_n)_{n\ge1}) = \mathrm{Fav}((T(t))_{t\ge0}) = \bigcup_{M\ge0} \mathscr{C}^2_{M,\alpha}([0,1]), \tag{4.5.32}$$

where α is defined by (4.5.9).

Combining the preceding results, we shall characterize those functions $u \in \mathscr{C}([0,1])$ for which formula (4.5.26) may hold pointwise or uniformly.

Once again, consider a Markov operator T on $\mathscr{C}([0,1])$, the linear operator $(A, D_V(A))$ defined by (4.5.1)-(4.5.2) and the Markov semigroup $(T(t))_{t\ge0}$ generated by it. From Proposition 4.5.11 it also follows that

$$\lim_{n\to\infty} n(B_n(u) - u) = A(u) \qquad \text{pointwise on } [0,1],$$

for every $u \in D_V(A)$ and, finally, the Bernstein-Schnabl operators are positive and contractive. Therefore Theorem 2.3.6 applies.

Theorem 4.5.16. *Given $u \in \mathscr{C}([0,1])$, the following statements are equivalent:*
(i) There exists $\lim_{n\to\infty} n(B_n(u)-u)$ uniformly on $[0,1]$.
(ii) There exists $v \in \mathscr{C}([0,1]$ such that $\lim_{n\to\infty} n(B_n(u)-u) = v$ pointwise on $[0,1]$.
(iii) $u \in D_V(A)$, i.e., $u \in \mathscr{C}^2(]0,1[)$ and

$$\lim_{x\to 0^+} \alpha(x)u''(x) = \lim_{x\to 1^-} \alpha(x)u''(x) = 0.$$

(iv) There exists $w \in \mathscr{C}([0,1]$ such that $\lim_{t\to 0^+} \frac{T(t)(u)-u}{t} = w$ pointwise on $[0,1]$.
Moreover, if one of the previous statements holds true, then $v = w = A(u)$.

In particular, if $\lim_{n\to\infty} n(B_n(u)-u) = 0$ pointwise on $[0,1]$ or if $\lim_{t\to 0^+} \frac{T(t)(u)-u}{t} = 0$ pointwise on $[0,1]$, then u is affine.

Proof. According to Theorem 2.3.6 it suffices to show that (iii)⇒ (i). To this end, consider on the subspace $D_V(A)$ the graph norm

$$\|u\|_A := \|u\|_\infty + \|A(u)\|_\infty \qquad (u \in D_V(A)).$$

Since $(A, D_V(A))$ is closed, $(D_V(A), \|\cdot\|_A)$ is a Banach space.

Moreover, by Theorem 4.5.4, $\mathscr{C}^2([0,1])$ is a core for $(A, D_V(A))$ and hence it is dense in $(D_V(A), \|\cdot\|_A)$. Note that, if $v \in D_V(A)$, then $v \in \mathscr{C}^2_{M,\alpha}([0,1])$, where $M := \sup_{0<x<1} \alpha(x)|v''(x)|$, and hence, by Theorem 4.5.14, $\sup_{n\geq 1} n\|B_n(v)-v\|_\infty < +\infty$. Therefore, the sequence of bounded linear operators $(n(B_n - I))_{n\geq 1}$ (I denoting the identity operator) is pointwise bounded on $D_V(A)$ and hence, by the uniform boundedness principle, it is equibounded. On the other hand, if $v \in \mathscr{C}^2([0,1])$, then

$$\lim_{n\to\infty} n(B_n(v)-v) = A(v) \qquad \text{uniformly on } [0,1]$$

(see Theorem 4.1.5); consequently, by a density argument, we obtain a similar formula for every $u \in D_V(A)$ and so the result follows. □

4.6 Notes and comments

The results of this chapter generalize those documented in [18, Chapter 6 and the references contained in the relevant notes] which are concerned with the (abstract) differential operators (4.1.9)-(4.1.10) and (4.1.15) in the special case where T is a Markov projection. The more general framework of this chapter guarantees to notably enlarge the class of differential operators by including, in particular, those which can be obtained by usual operations with Markov operators such as convex combinations, compositions, tensor products and so on. Moreover, this non-trivial generalization discloses new challenging problems as well.

Theorems 4.1.2 and 4.1.5 are due to the authors. Theorem 4.1.5 also appears in [21, Theorem 4.2] and generalizes a similar result for Bernstein operators due to Voronovskaja ([207]) (see also Section 4.5). For additional comments on asymptotic formulae we refer to the final notes of Chapter 1 of this monograph and the ones of Section 6.2 of [18]. In the setting of Banach spaces, asymptotic formulae for Bernstein-Schnabl operators have been obtained by M. Beśka and K. Dziedziul ([54]) and Altomare and Diomede ([23]).

Section 4.2 has an illustrative character. Corollary 4.2.3 is due to Romito ([176]).

All the results of Sections 4.3 and 4.4 are due to the authors. Some of them also appear in [21] and [22]. The preservation properties (4.3.4) and (4.3.5) have been first investigated by Altomare and Raşa in [34] in the case where T is a Markov projection. Theorem 4.3.8 generalizes Theorem 2.3 of [176].

Both the differential operator $(W_T, \mathscr{C}^2(K))$ defined by (4.1.15) and its closure $(A_T, D(A_T))$ degenerate on the set $\partial_T K$ of the interpolation points of T (and, hence, on the set of the extreme points of K). The relevant initial-boundary value problems (4.4.3), (4.4.6) and (4.4.11) seem to be introduced and studied here for the first time at this level of generality. Multiplicative perturbations of them as well as additive perturbations with first-order differential operators will be considered in the subsequent Chapter 5.

In the case where K is the d-dimensional simplex and T is the canonical projection T_d, then the relevant differential operator given by (4.2.10) arises in the theory of Fleming-Viot processes. They are involved in Wright-Fisher diffusion model in population genetics in which neither mutation, migration nor selection affects. For additional details we refer to the Notes and Comments 2.4.

Differential operators (4.2.23) and (4.2.24) are special cases of degenerate differential operators whose coefficients behave like a power of the distance from the boundary of the domain. For some further results in this respect we refer, e.g., to [195]–[199].

However, in all the above-mentioned particular cases the generation results are obtained via the theory of partial differential equations. Our approach, which is based on Trotter-Schnabl-type results and involves the Bernstein-Schnabl operators, allows us to get not only generation but also several preservation properties of the semigroup itself as showed in Section 4.4. In some particular cases this same approach also allows us to determine the asymptotic behaviour of the semigroup and to describe some properties of the relevant Markov process.

Theorem 4.4.7 generalizes a result obtained by Altomare ([8]; see also [18, Theorem 6.2.6, formula (6.2.20)]) in the case where $T = T_0$. The germ of its proof can be tracked in a paper by Raşa ([171, Lemma 2.1 and Theorem 2.2]) which refers to the unit interval.

For both theoretical and applicative reasons it is important to explicitly determine the domain $D(A_T)$ of the closure A_T. This is a delicate problem which has been solved in particular cases only. In this respect we refer to the papers by Blümlinger ([57, Satz 6] and Romito ([176, Theorem 3.4]) which are concerned with the differential operators (4.2.24) and (4.2.23) in the settings of balls and ellipsoids of $\mathbf{R}^d$, respectively.

Another case where the domain $D(A_T)$ can be explicitly determined concerns the unit interval. All the relevant details are contained in Section 4.5 which, in turn, is mainly based on the papers by Altomare, Leonessa and Raşa ([28]) and Altomare ([10]). Moreover, all the results of Section 4.5 generalize similar ones obtained for Bernstein operators by several authors (we refer to the final notes of Section 6.3 of [18] for more details).

Subsections 4.5.1-4.5.3 are taken from [28] (see also [25]). Statement (6) of Theorem 4.5.10 is due to Raşa ([171]). Theorems 4.5.13 and 4.5.14 were obtained in [10] and they describe the pointwise and the uniform saturation classes of Bernstein-Schnabl operators in $\mathscr{C}([0,1])$.

For Bernstein operators the equivalences (i)⇔(ii) and (i)⇔(iii) in Theorem 4.5.13 are due, respectively, to Lorentz ([131, Theorem 11, p. 102]; see also [84]) and Micchelli ([142, Theorem 3.2]). Moreover, the equivalence (i)⇔(iii) in Theorem 4.5.14 is due to Lorentz and Schumaker ([134, Section 6, formula (6.6)]). Corollary 4.5.15 is taken from [10] and it seems to be new also in the special case $\alpha(x) = x(1-x)$ $(0 \le x \le 1)$.

A determination of both the saturation class of Bernstein-Schnabl operators and the Favard class of the associated Markov semigroup in the general context of this chapter is not yet available and its achievement represents a challenging problem. However, partial results can be found in the papers by Blümlinger ([57, Satz 8]), Romito ([176, Section 5]), Nishishiraho ([151; 153; 156]) and Schnabl ([185; 186]).

Theorem 4.5.16 was established in [10]. For Bernstein operators it was obtained, with different methods, by Berens and Lorentz ([53, Section 4, Remark to Lemma 4]).

5 Perturbed differential operators and modified Bernstein-Schnabl operators

Each progress in mathematics is based on the discovery of stronger tools and easier methods, which at the same time makes it easier to understand earlier methods. By making these stronger tools and easier methods his own, it is possible for the individual researcher to orientate himself in the different branches of mathematics.

David Hilbert

In this chapter we shall deal with multiplicative as well as additive perturbations of the (abstract) differential operators which were introduced and studied in the previous chapter.

In particular, in finite dimensional settings, given a suitable Markov operator T on $\mathscr{C}(K)$, we are concerned with both multiplicative perturbations of the form λW_T ($\lambda \in \mathscr{C}(K)$, λ strictly positive) of the differential operator W_T (see (4.1.15)) and with first-order additive perturbations of it of the form

$$\begin{aligned} V_T(u) &= W_T(u) + \sum_{i=1}^{d} \beta_i \frac{\partial u}{\partial x_i} + \gamma u \\ &= \frac{1}{2} \sum_{i,j=1}^{d} (T(pr_i pr_j) - pr_i pr_j) \frac{\partial^2 u}{\partial x_i \partial x_j} + \sum_{i=1}^{d} \beta_i \frac{\partial u}{\partial x_i} + \gamma u \end{aligned}$$

($u \in \mathscr{C}^2(K)$), where the functions $\beta_1, ..., \beta_d, \gamma \in \mathscr{C}(K)$ satisfy suitable assumptions.

These complete differential operators often occur in a wide class of diffusion models arising from different fields, such as mathematical finance, physics, biology and so on. In particular, a class of them occurs in the theory of Fleming-Viot processes on the d-dimensional simplex.

In this chapter we shall establish some conditions under which the perturbed operators are closable and their closures generate C_0-semigroups on $\mathscr{C}(K)$. Moreover, we shall show that such semigroups may be represented in terms of iterates of suitable modifications of Bernstein-Schnabl operators, which will be introduced and studied in this same chapter.

As in the preceding chapter, the study of some qualitative properties of these modifications allows us to infer some properties of the relevant C_0-semigroups and, hence, some spatial regularity properties of the underlying diffusion problems.

Finally, in Sections 5.7 and 5.8, we consider the special cases of the interval $[0, 1]$ and the d-dimensional simplex and hypercube, where additional results and applications can be achieved.

5.1 Lototsky-Schnabl operators

Motivated by the goal to study multiplicative perturbations of the (abstract) differential operators defined by (4.1.9)-(4.1.10) and (4.1.15), in this section we turn our attention on the special class of Bernstein-Schnabl operators introduced in Subsection 3.1.6, namely the so-called *Lototsky-Schnabl operators.*

More precisely, from now on, let us fix a Markov operator T on $\mathscr{C}(K)$ satisfying (3.1.2), K being a metrizable convex compact subset of some locally convex Hausdorff space X and, as usual, let us denote by $(B_n)_{n\geq 1}$ the sequence of Bernstein-Schnabl operators associated with T, as in (3.1.7). Moreover, we consider $\lambda \in \mathscr{C}(K)$, $0 \leq \lambda \leq 1$, and set

$$U_\lambda := \lambda T + (1-\lambda)I.$$

U_λ is a Markov operator and the Bernstein-Schnabl operators B_{n,U_λ}, $n \geq 1$, associated with U_λ will be referred to as the *Lototsky-Schnabl operators associated with T and λ* and they will be simply denoted by $B_{n,\lambda}$, $n \geq 1$. Hence, if $f \in \mathscr{C}(K)$, $x \in K$ and $n \geq 1$, then

$$B_{n,\lambda}(f)(x) = \begin{cases} \sum_{k=0}^{n} \binom{n}{k} \lambda(x)^k (1-\lambda(x))^{n-k} B_k(f_{x,k/n})(x) & \text{if } x \notin \partial_T K; \\ \\ f(x) & \text{if } x \in \partial_T K \end{cases}$$

(see (3.1.35) and (3.1.32)).

Clearly, when $\lambda = \mathbf{1}$, the $B_{n,\lambda}$'s turn into the Bernstein-Schnabl operators associated with T.

As Bernstein-Schnabl operators, the $B_{n,\lambda}$'s share all the preservation properties presented in Sections 3.3-3.5. Moreover, it is possible to state for them an asymptotic formula such as in Theorem 4.1.2; in addition, in this respect, the following asymptotic formula holds true.

Theorem 5.1.1. *For every $m \geq 1$ and $h_1, \dots, h_m \in A(K)$,*

$$\lim_{n\to\infty} n \left(B_{n,\lambda}\left(\prod_{j=1}^{m} h_j\right) - \prod_{j=1}^{m} h_j \right) = \lambda\Theta_T(h_1, \dots, h_m) \qquad \textit{uniformly on } K, \tag{5.1.1}$$

where

$$\Theta_T(h_1, \dots, h_m) := \begin{cases} 0 & \textit{if } m = 1; \\ \\ T(h_1 h_2) - h_1 h_2 & \textit{if } m = 2; \\ \\ \sum_{1\leq i<j\leq m} (T(h_i h_j) - h_i h_j) \prod_{\substack{r=1 \\ r\neq i,j}}^{m} h_r & \textit{if } m \geq 3. \end{cases}$$

Proof. The proof follows directly by applying Theorem 4.1.2 to the sequence $(B_{n,\lambda})_{n\geq 1}$ associated with the Markov operator U_λ, since it is easy to prove that $\Theta_{U_\lambda}(h_1, \dots, h_m) = \lambda\Theta_T(h_1, \dots, h_m)$ for every $m \geq 1$ and $h_1, \dots, h_m \in A(K)$. □

From Theorem 5.1.1 it also follows that, for every $u \in P_\infty(K)$ (see (4.1.2)), there exists $\lim_{n\to\infty} n(B_{n,\lambda}(u) - u)$ in $\mathscr{C}(K)$ and hence we can define a linear operator $L_\lambda : P_\infty(K) \longrightarrow \mathscr{C}(K)$ by setting

$$L_\lambda(u) := \lim_{n\to\infty} n(B_{n,\lambda}(u) - u) \qquad (u \in P_\infty(K)). \tag{5.1.2}$$

Moreover,

$$L_\lambda = \lambda L_T, \tag{5.1.3}$$

where the operator $L_T : P_\infty(K) \longrightarrow \mathscr{C}(K)$ is defined by

$$L_T(u) := \lim_{n\to\infty} n(B_n(u) - u) \qquad (u \in P_\infty(K))$$

(see (4.1.9)).

Assume now that K is a convex compact subset of $\mathbf{R}^d$ with non-empty interior and consider the elliptic second-order differential operator $W_T : \mathscr{C}^2(K) \longrightarrow \mathscr{C}(K)$ defined by

$$W_T(u) := \frac{1}{2}\sum_{i,j=1}^{d} (T(pr_i pr_j) - pr_i pr_j)\frac{\partial^2 u}{\partial x_i \partial x_j} \qquad (u \in \mathscr{C}^2(K))$$

(see (4.1.15)).

As we have seen in Section 4.1, such an operator is elliptic and it degenerates on $\partial_T K$; moreover, it is involved in an asymptotic formula concerning the operators B_n (see Theorem 4.1.5).

Also with respect to the Lototsky-Schnabl operators $B_{n,\lambda}$ we can state an asymptotic formula related to W_T, as the following result shows.

Theorem 5.1.2. *For every $u \in \mathscr{C}^2(K)$,*

$$\lim_{n\to\infty} n(B_{n,\lambda}(u) - u) = \lambda W_T(u) \qquad \textit{uniformly on } K. \tag{5.1.4}$$

Therefore,

$$L_\lambda = \lambda W_T \qquad \textit{on } P_\infty(K). \tag{5.1.5}$$

Proof. The proof of (5.1.4) is a direct consequence of Theorem 4.1.5 and of the fact that $W_{U_\lambda} = \lambda W_T$ (see (4.1.15)).

Formula (5.1.5) easily follows from (5.1.2)–(5.1.4). □

We are now in a position to state the main result of the section.

Theorem 5.1.3. *Let K be a metrizable convex compact subset of some locally convex space X and fix $\lambda \in \mathscr{C}(K)$, $0 < \lambda(x) \leq 1$ for every $x \in K$. Consider a Markov operator T on $\mathscr{C}(K)$ such that (3.1.2) holds true. Assume that either*

(i) $T(P_2(K)) \subset A(K)$, *i.e.,* K *is a Bauer simplex and* T *is its canonical projection (see Theorem 4.3.3)*

or

(ii) K *is a convex compact subset of* $\mathbf{R}^d$, $d \geq 1$, *having non-empty interior and* $T(P_m(K)) \subset P_m(K)$ *for every* $m \geq 2$.

Moreover, according to the notation of Theorems 4.3.2 and 4.3.4 (see also Remark 4.3.5, 1), set

$$(A, D(A)) := \begin{cases} (A_T^*, D(A_T^*)) & \text{if assumption (i) holds true;} \\ \\ (A_T, D(A_T)) & \text{if assumption (ii) holds true.} \end{cases} \tag{5.1.6}$$

Then the operator $A_\lambda := \lambda A$ *defined on* $D(A_\lambda) = D(A)$ *generates a Markov semigroup* $(T_\lambda(t))_{t\geq 0}$ *on* $\mathscr{C}(K)$. *Moreover, for every* $t \geq 0$ *and for every sequence* $(k(n))_{n\geq 1}$ *of positive integers satisfying* $\lim_{n\to\infty} \frac{k(n)}{n} = t$,

$$T_\lambda(t)(f) = \lim_{n\to\infty} B_{n,\lambda}^{k(n)}(f) \quad \text{uniformly on } K \tag{5.1.7}$$

$(f \in \mathscr{C}(K))$.

Furthermore,

(a) under assumption (i), $P_\infty(K)$ *is a core for* $(A_\lambda, D(A_\lambda))$ *and* $A_\lambda = \lambda L_T$ *on* $P_\infty(K)$, *so that* $(A_\lambda, D(A_\lambda))$ *is the closure of* $(\lambda L_T, P_\infty(K))$ *as well;*

(b) under assumption (ii), $\mathscr{C}^2(K)$ *is a core for* $(A_\lambda, D(A_\lambda))$ *and* $A_\lambda = \lambda W_T$ *on* $\mathscr{C}^2(K)$, *so that* $(A_\lambda, D(A_\lambda))$ *coincides with the closure of* $(\lambda W_T, \mathscr{C}^2(K))$.

Proof. Since, according to Theorems 4.3.2 and 4.3.4, $(A, D(A))$ is the generator of a Markov semigroup and since $0 < \lambda(x) \leq 1$ for every $x \in K$, then the operator $(\lambda A, D(A))$, by Theorem 2.3.11, is the generator of a Markov semigroup $(T_\lambda(t))_{t\geq 0}$ on $\mathscr{C}(K)$.

Consider now the operator $(Z_\lambda, D(Z_\lambda))$ defined by setting

$$D(Z_\lambda) := \left\{ g \in \mathscr{C}(K) \mid \text{there exists } \lim_{n\to\infty} n(B_{n,\lambda}(g) - g) \in \mathscr{C}(K) \right\}$$

and, for every $f \in D(Z_\lambda)$,

$$Z_\lambda(f) := \lim_{n\to\infty} n(B_{n,\lambda}(f) - f).$$

Then, under assumption (i) (resp., (ii)), from Theorem 5.1.1 (resp., Theorem 5.1.2) and (5.1.3) (resp., (5.1.5)) it follows that $P_m(K) \subset D(Z_\lambda)$ and $Z_\lambda = \lambda L_T$ on $P_m(K)$.

Hence $P_\infty(K) \subset D(Z_\lambda)$, so that $D(Z_\lambda)$ is dense in $\mathscr{C}(K)$.

Suppose now that assumption (i) holds true; then Theorem 4.3.2 implies, in particular, that $(I - L_T)(P_\infty(K)) = (I - A)(P_\infty(K))$ is dense in $\mathscr{C}(K)$, so that $P_\infty(K)$ is a core for $(A, D(A))$.

On the other hand, by virtue of Theorem 4.3.4, $P_\infty(K)$ is a core for $(A, D(A))$ also in the case where assumption (ii) holds true.

This implies that $P_\infty(K)$ is a core for $(\lambda A, D(A))$, since, for every $u \in P_\infty(K)$,

$$\|u\|_{\lambda A} = \|u\|_\infty + \|\lambda A(u)\|_\infty \le \|u\|_\infty + \|A(u)\|_\infty = \|u\|_A.$$

Therefore, $(I - \lambda A)(P_\infty(K)) = (I - \lambda L_T)(P_\infty(K))$ is dense in $\mathscr{C}(K)$, so that the range $(I - Z_\lambda)(D(Z_\lambda))$ of $I - Z_\lambda$ is dense in $\mathscr{C}(K)$.

Then, according to Theorem 2.2.1, the operator $(Z_\lambda, D(Z_\lambda))$ is closable and its closure $(\overline{Z_\lambda}, D(\overline{Z_\lambda}))$ generates a Markov semigroup $(S(t))_{t\ge0}$ such that, for every $t \ge 0$ and for every sequence $(k(n))_{n\ge1}$ of positive integers satisfying $\lim_{n\to\infty} \frac{k(n)}{n} = t$, one gets

$$S(t)(f) = \lim_{n\to\infty} B_{n,\lambda}^{k(n)}(f) \quad \text{uniformly on } K$$

for every $f \in \mathscr{C}(K)$.

Moreover, since $\overline{Z_\lambda} = Z_\lambda = \lambda L_T = \lambda A$ on $P_\infty(K)$ and $P_\infty(K)$ is a core for $(\lambda A, D(A))$, we conclude that $\overline{Z_\lambda} = \lambda A$ and, consequently, $S(t) = T_\lambda(t)$ for every $t \ge 0$.

Finally, the last statements (a) and (b) can be easily achieved since, in case (ii), $\mathscr{C}^2(K)$ is a core for $(\lambda A, D(A))$. □

As in Theorem 4.4.7, the existence on K of a Markov projection satisfying suitable assumptions allows us to establish some remarkable results about the asymptotic behaviour of the semigroup $(T_\lambda(t))_{t\ge0}$.

In fact, the following theorem holds true.

Theorem 5.1.4. *Under the assumptions of Theorem 5.1.3, suppose further that there exists a Markov projection T_0 on $\mathscr{C}(K)$ verifying (3.1.2) and (3.2.10) and such that $T \circ T_0 = T_0$ and $\partial_T K \subset \partial_{T_0} K$ (see (1.3.4)). Then*

(i) For every $t \ge 0$,

$$T_\lambda(t)(h) = h \text{ for every } h \in H_0, \tag{5.1.8}$$

where $H_0 := T_0(\mathscr{C}(K))$ is the range of T_0.

(ii) For every $f \in \mathscr{C}(K)$,

$$\lim_{t\to+\infty} T_\lambda(t)(f) = T_0(f). \tag{5.1.9}$$

Consequently, if $f \in \mathscr{C}(K)$, then $\lim_{t\to+\infty} T_\lambda(t)(f) = 0$ if and only if $f = 0$ on $\partial_{H_0} K$.

Proof. Consider the Markov operator

$$U_\lambda := \lambda T + (1 - \lambda)I;$$

taking the assumption on the projection T_0 into account, it is easy to prove that $U_\lambda \circ T_0 = T_0$ and $\partial_{U_\lambda} K \subset \partial_{T_0} K$ (see (1.3.4)).

Moreover, by means of (3.2.10), we get that $B_{n,\lambda}(h) = h$ for every $h \in H_0$ (see the proof of Theorem 3.2.5). Then formula (5.1.8) follows directly from (5.1.7).

To prove (5.1.9), it suffices to apply Theorem 4.4.7. □

Remark 5.1.5. Theorem 5.1.4 naturally holds true if T itself is a Markov projection on $\mathscr{C}(K)$ satisfying one of the assumptions of Theorem 5.1.2; it is enough to consider $T_0 = T$.

Anyway, in this particular case, the result has been already obtained in [8] (see also [18, Theorem 6.2.6, formula (6.2.20)]).

Turning back to the general case where T is a Markov operator on $\mathscr{C}(K)$ satisfying (3.1.2), it is possible to present some preservation properties for the semigroup $(T_\lambda(t))_{t\geq 0}$, by deriving them from the corresponding ones held by the operators $B_{n,\lambda}$ throughout formula (5.1.7). In the following results, we collect some of them.

Theorem 5.1.6. *Under the same assumptions of Theorem 5.1.3, if $f \in \mathscr{C}(K)$ is convex, then, for every $t \geq 0$,*

$$f \leq T_\lambda(t)(f) \leq T(f). \tag{5.1.10}$$

Proof. Fix a convex function $f \in \mathscr{C}(K)$; then, by Theorem 3.5.2, for every $n \geq 2$, we have that

$$f \leq B_{n,\lambda}(f) \leq \lambda T(f) + (1-\lambda)f. \tag{1}$$

Moreover, $f = \sup\{h \in A(K) \mid h \leq f\}$ (see, e.g., [77, Vol. II, Proposition 21.18]), so that, since (3.1.2) holds true, $f \leq T(f)$. Hence, taking (1) and Theorem 5.1.3 into account, (5.1.10) follows. □

Theorem 5.1.7. *If $\lambda \in]0,1]$, under the assumptions of Theorem 5.1.3,*

$$T_\lambda(t)(P_m(K)) \subset P_m(K) \tag{5.1.11}$$

($m \geq 1$) for every $t \geq 0$.

Proof. It is easy to show that, if $\lambda \in]0,1]$ and $T(P_m(K)) \subset P_m(K)$ for every $m \geq 1$, then $U_\lambda(P_m(K)) \subset P_m(K)$ for every $m \geq 1$.

Hence, from (4.1.5), it follows that

$$B_{n,\lambda}(P_m(K)) \subset P_m(K)$$

for every $n, m \geq 1$.

Then (5.1.11) is a consequence of (5.1.7) by virtue of Theorem 5.1.3. □

Theorem 5.1.8. *Let K be a convex compact subset of a metrizable locally convex space X and let us assume that (1.6.23) holds true.*

Let T be a Markov operator on $\mathscr{C}(K)$ such that

$$T(\mathrm{Lip}(1,1)) \subset \mathrm{Lip}(1,1) \tag{5.1.12}$$

(see (1.6.21)).

Further, assume that $\lambda \in]0,1]$. *Then, under the assumptions of Theorem 5.1.3, for every* $t \geq 0$,

$$T_\lambda(t)(\mathrm{Lip}(M,\alpha)) \subset \mathrm{Lip}(M,\alpha) \tag{5.1.13}$$

for every $M > 0$ *and* $\alpha \in]0,1]$.

Proof. Since λ is constant, from (5.1.12) it follows that $U_\lambda(\mathrm{Lip}(1,1)) \subset \mathrm{Lip}(1,1)$; hence, formula (3.3.12) in Corollary 3.3.2 implies that

$$B_{n,\lambda}(\mathrm{Lip}(M,\alpha)) \subset \mathrm{Lip}(M,\alpha)$$

for every $M > 0$ and $\alpha \in]0,1]$.

Accordingly, (5.1.13) follows at once from (5.1.7). □

Theorem 5.1.9. *Under the same assumptions of Theorem 5.1.3, suppose further that*

(c_1) T maps continuous convex functions into (continuous) convex functions.

(c_2) For every convex function $f \in \mathscr{C}(K)$ and for every $x, y \in K$,

$$\Delta(\tilde{f}; x, y) \geq 0$$

(see (3.4.2) and (3.4.3)).

If $\lambda \in]0,1]$, *then* $T_\lambda(t)(f)$ *is convex for every* $t \geq 0$ *and for every convex* $f \in \mathscr{C}(K)$.

Proof. Under the assumptions above, on account of Example 3.4.8 and Theorem 3.4.3, it turns out that $B_{n,\lambda}(f)$ is convex for every $n \geq 1$ and for every convex $f \in \mathscr{C}(K)$. Then, by applying (5.1.7), we obtain the required assertion. □

Theorems 5.1.3, 5.1.4, 5.1.8 and 5.1.9 may be fruitfully employed in order to infer some properties of the solutions to the abstract Cauchy problem associated with $(\lambda A, D(A))$ with initial datum $u_0 \in D(A)$

$$\begin{cases} \dfrac{du}{dt}(t) = \lambda A(u(t)) \quad t \geq 0; \\ \\ u(0) = u_0, \end{cases} \tag{5.1.14}$$

which are given by $u(t) := T_\lambda(t)(u_0)$ $(t \geq 0)$. We omit the details for the sake of brevity.

5.2 A modification of Bernstein-Schnabl operators

In this and in the following sections, by adapting an idea first developed in [14] (see also [35]), we introduce and study a further modification of the Bernstein-Schnabl operators associated with a Markov operator T on $\mathscr{C}(K)$ satisfying (3.1.2).

So, from now on we fix a convex compact subset K of some locally convex space X and a Markov operator $T : \mathscr{C}(K) \longrightarrow \mathscr{C}(K)$ such that

$$T(h) = h \qquad \text{for every } h \in A(K).$$

Let $(B_n)_{n\geq 1}$ be the sequence of Bernstein-Schnabl operators associated with the operator T. Thus, for every $f \in \mathscr{C}(K)$, $x \in K$ and $n \geq 1$,

$$B_n(f)(x) := \int_K \cdots \int_K f\left(\frac{x_1 + \dots + x_n}{n}\right) d\tilde{\mu}_x^T(x_1) \cdots d\tilde{\mu}_x^T(x_n),$$

$(\tilde{\mu}_x^T)_{x\in K}$ being the continuous selection of probability Borel measures associated with T via Riesz representation theorem (see Theorem 1.1.1). In other words, for every $x \in K$ and $f \in \mathscr{C}(K)$,

$$T(f)(x) = \int_K f \, d\tilde{\mu}_x^T.$$

Moreover, we fix $\beta \in \mathscr{C}(K, X)$ and $\gamma \in \mathscr{C}(K)$ and we assume that there exists $n_0 \geq 1$ such that

$$x + \frac{\beta(x)}{n} \in K \qquad \text{and} \qquad 1 + \frac{\gamma(x)}{n} \geq 0, \tag{5.2.1}$$

for every $x \in K$ and $n \geq n_0$.

If K is a convex compact subset of some Euclidean space $\mathbf{R}^d$, $d \geq 1$ and $\beta \in \mathscr{C}(K, \mathbf{R}^d)$, its components will be denoted by $\beta_1, \dots, \beta_d \in \mathscr{C}(K, \mathbf{R})$, i.e.,

$$\beta(x) := (\beta_1(x), \dots, \beta_d(x)) \qquad (x \in K). \tag{5.2.2}$$

Below, we present some concrete settings in which (5.2.1) holds true.

Examples 5.2.1.

1. *If $\beta(x) := -\rho x + \rho c$ ($x \in K$), with $\rho \geq 0$ and $c \in K$, then (5.2.1) is satisfied for $n \geq \max\{\|\gamma\|_\infty, \rho\}$.*
2. *If $K = [0,1]^d$ is the hypercube of $\mathbf{R}^d$ and if, for every $i = 1, \dots, d$,*

$$\theta_i := \sup_{\substack{x\in K \\ pr_i(x)\neq 0}} \frac{-\beta_i(x)}{pr_i(x)} < +\infty \qquad \text{and} \qquad \eta_i := \sup_{\substack{x\in K \\ pr_i(x)\neq 1}} \frac{\beta_i(x)}{1 - pr_i(x)} < +\infty,$$

then (5.2.1) holds true whenever $n \geq \max\{-\theta_1, \dots, -\theta_d, \eta_1, \dots, \eta_d, \|\gamma\|_\infty\}$ (see (5.2.2)).

3. *If K is the canonical simplex of $\mathbf{R}^d$, then (5.2.1) is satisfied for $n \geq \max\{-\theta_1, \dots, -\theta_d, M, \|\gamma\|_\infty\}$, where the θ_i's, defined as above, are supposed to be finite and*

$$M := \sup_{\substack{x\in K \\ pr_1(x)+\dots+pr_d(x)<1}} \frac{\beta_1(x) + \dots + \beta_d(x)}{1 - pr_1(x) - \dots - pr_d(x)} < +\infty.$$

For every $n \geq n_0$, consider the positive linear operator $M_n : \mathscr{C}(K) \longrightarrow \mathscr{C}(K)$ defined by setting

$$M_n(f) := B_n\left(\left(\mathbf{1} + \frac{\gamma}{n}\right) \times \left(f \circ \left(id + \frac{\beta}{n}\right)\right)\right) \tag{5.2.3}$$

for every $f \in \mathscr{C}(K)$, where $id(y) := y$ for every $y \in K$.

More precisely, for every $f \in \mathscr{C}(K)$ and $x \in K$,

$$M_n(f)(x) = \int_K \cdots \int_K \left(1 + \frac{1}{n}\gamma\left(\frac{x_1 + \dots + x_n}{n}\right)\right)$$
$$\times f\left(\frac{x_1 + \dots + x_n}{n} + \frac{1}{n}\beta\left(\frac{x_1 + \dots + x_n}{n}\right)\right) d\mu_x^T(x_1) \cdots d\mu_x^T(x_n).$$

The sequence $(M_n)_{n\geq n_0}$ was introduced and studied in [19] (see also [69]) in the special case where the relevant operators B_n are the Bernstein-Schnabl operators associated with a positive projection T, satisfying suitable assumptions.

It is easy to show that, for every $n \geq n_0$,

$$\|M_n\| \leq 1 + \frac{\|\gamma\|_\infty}{n} \tag{5.2.4}$$

and, hence, the sequence $(M_n)_{n\geq n_0}$ is equibounded.

In particular, if $\gamma(x) = 0$ for every $x \in K$, then

$$\|M_n\| = 1 \qquad \text{for every } n \geq n_0. \tag{5.2.5}$$

According to (3.1.6), it is possible to express also each operator M_n in an integral form.

Indeed, for any $n \geq n_0$, consider the mapping $\pi_n : K^n \longrightarrow K$ defined by (3.1.5), i.e.,

$$\pi_n(x_1, \dots, x_n) := \frac{x_1 + \dots + x_n}{n}$$

($(x_1, \dots, x_n) \in K^n$) and, for each $x \in K$, set $\tilde{\nu}_{x,n}^T := \pi_n(\tilde{\mu}_x^T \otimes \cdots \otimes \tilde{\mu}_x^T)$, the image measure under π_n of the tensor product of $\tilde{\mu}_x^T$ with itself n-times, so that, for any $f \in \mathscr{C}(K)$,

$$M_n(f)(x) = \int_K \left(1 + \frac{\gamma}{n}\right) \times f \circ \left(id + \frac{\beta}{n}\right) \, d\tilde{\nu}_{x,n}^T.$$

Then, denoted by $\tilde{\lambda}_{x,n}^T$ the image measure under $id + \frac{\beta}{n}$ of the measure with density $\mathbf{1} + \frac{\gamma}{n}$ relative to $\tilde{\nu}_{x,n}^T$, we finally obtain

$$M_n(f)(x) = \int_K f \, d\tilde{\lambda}_{x,n}^T \qquad (f \in \mathscr{C}(K), x \in K). \tag{5.2.6}$$

A further expression of the operators M_n can be obtained in terms of tensor product of operators (see (1.1.30)).

As usual, let $T^{(n)} : \mathscr{C}(K^n) \longrightarrow \mathscr{C}(K^n)$ be the Markov operator defined as the tensor product of T with itself n times, i.e., $T^{(n)} := \underbrace{T \otimes \cdots \otimes T}_{n \text{ times}}$.

Then, for every $f \in \mathscr{C}(K)$,

$$M_n(f) = T^{(n)}\left(\left(\left(\mathbf{1} + \frac{\gamma}{n}\right) \times f\right) \circ \pi_n\right) \circ \left(i_n + \frac{\beta}{n}\right), \tag{5.2.7}$$

where the function i_n is defined by (3.1.9).

In the particular settings presented in Section 3.1, clearly, for every $n \geq n_0$, the operators M_n can be explicitly described starting from the corresponding B_n's, just by replacing in the defining sum the term $f\left(\frac{x_1+\ldots+x_n}{n}\right)$ with

$$\left(1 + \frac{1}{n}\gamma\left(\frac{x_1 + \ldots + x_n}{n}\right)\right) f\left(\frac{x_1 + \ldots + x_n}{n} + \frac{1}{n}\beta\left(\frac{x_1 + \ldots + x_n}{n}\right)\right).$$

For the sake of simplicity, we describe some of them only in the particular cases of K being a convex compact subset in $\mathbf{R}$ or in $\mathbf{R}^2$.

Examples 5.2.2.

1. *From now on, let us fix $\beta, \gamma \in \mathscr{C}([0,1])$ satisfying* (5.2.1).

If we consider the Markov projection $T_1 : \mathscr{C}([0,1]) \longrightarrow \mathscr{C}([0,1])$ defined by (1.1.49), *then, according to* (3.1.11),

$$M_n(f)(x) = \sum_{k=0}^{n} \binom{n}{k} \left(1 + \frac{1}{n}\gamma\left(\frac{k}{n}\right)\right) f\left(\frac{k}{n} + \frac{1}{n}\beta\left(\frac{k}{n}\right)\right) x^k (1-x)^{n-k}$$

$(n \geq n_0,\ f \in \mathscr{C}([0,1]), 0 \leq x \leq 1)$.

Moreover, if T is the Markov operator on $\mathscr{C}([0,1])$ defined by (3.1.12) *and the Bernstein-Schnabl operators associated with T are defined by* (3.1.13), *then the corresponding sequence $(M_n)_{n\geq n_0}$ is defined by setting, for every $n \geq n_0$, $f \in \mathscr{C}([0,1])$ and $x \in [a_k, a_{k+1}]$ $(k = 0, \ldots, d-1)$,*

$$M_n(f)(x) = \frac{1}{(a_{k+1} - a_k)^n} \sum_{h=0}^{n} \binom{n}{h} (x - a_k)^h (a_{k+1} - x)^{n-h}$$
$$\times \left(1 + \frac{1}{n}\gamma(b_{k,h,n})\right) f\left(b_{k,h,n} + \frac{1}{n}\beta(b_{k,h,n})\right),$$

where, for every $h = 0, \ldots, n$, $b_{k,h,n} := \frac{h}{n}a_{k+1} + \frac{n-h}{n}a_k$.

Finally, if T is the Markov operator defined by (3.1.14), *then, according to* (3.1.16), *the relevant sequence $(M_n)_{n\geq n_0}$ is defined by setting*

$$M_n(f)(x) = \sum_{h=0}^{n} \sum_{k=0}^{n-h} \binom{n}{h} \binom{n-h}{k} \left(1 - x - \frac{b(x)}{2}\right)^{n-h-k} b(x)^h \left(x - \frac{b(x)}{2}\right)^k$$
$$\times \left(1 + \frac{1}{n}\gamma\left(\frac{h+2k}{2n}\right)\right) f\left(\frac{h+2k}{2n} + \frac{1}{n}\beta\left(\frac{h+2k}{2n}\right)\right),$$

for every $f \in \mathscr{C}([0,1])$, $x \in [0,1]$ *and* $n \geq n_0$.

2. *Let* K_2 *be the canonical simplex in* $\mathbf{R}^2$ *(see (1.1.45)) and consider* $\beta_1, \beta_2, \gamma \in \mathscr{C}(K_2)$ *satisfying (5.2.1). Moreover, let* T_2 *be the canonical projection on* K_2 *(see (1.1.48)) and consider the sequence* $(B_n)_{n\geq 1}$ *of Bernstein operators associated with* T_2 *(see (3.1.18)); then*

$$\begin{aligned} M_n(f)(x,y) = &\sum_{\substack{h\geq 0, k\geq 0\\ h+k\leq n}} \frac{n!}{h!\,k!\,(n-h-k)!}\left(1+\frac{1}{n}\gamma\left(\frac{h}{n},\frac{k}{n}\right)\right) \\ &\times f\left(\frac{h}{n}+\frac{1}{n}\beta_1\left(\frac{h}{n},\frac{k}{n}\right),\frac{k}{n}+\frac{1}{n}\beta_2\left(\frac{h}{n},\frac{k}{n}\right)\right)x^h y^k (1-x-y)^{n-h-k} \end{aligned} \tag{5.2.8}$$

$(n \geq n_0, f \in \mathscr{C}(K_2), (x,y) \in K_2)$.

3. *Let* $Q_2 := [0,1]^2$ *be the unit square in* $\mathbf{R}^2$ *and fix* $\beta_1, \beta_2, \gamma \in \mathscr{C}(Q_2)$ *satisfying (5.2.1). Consider the projection* S_2 *on* $\mathscr{C}(Q_2)$ *defined by (3.1.28) and its associated Bernstein-Schnabl operators* B_n *(see (3.1.30)); then*

$$\begin{aligned} M_n(f)(x,y) = &\sum_{h,k=0}^{n} \binom{n}{h}\binom{n}{k}\left(1+\frac{1}{n}\gamma\left(\frac{h}{n},\frac{k}{n}\right)\right) \\ &\times f\left(\frac{h}{n}+\frac{1}{n}\beta_1\left(\frac{h}{n},\frac{k}{n}\right),\frac{k}{n}+\frac{1}{n}\beta_2\left(\frac{h}{n},\frac{k}{n}\right)\right)x^h(1-x)^{n-h}y^k(1-y)^{n-k} \end{aligned} \tag{5.2.9}$$

$(f \in \mathscr{C}(Q_2), (x,y) \in Q_2)$.

4. *Let* K *be the closed ball of center the origin of* $\mathbf{R}^d$ *and radius 1 and consider* $\beta_1, \dots, \beta_d, \gamma \in \mathscr{C}(K)$ *such that (5.2.1) holds true. Moreover, let* T_Δ *be the Poisson operator associated with the Laplace operator* Δ *(see (3.1.21)) and defined explicitly by (3.1.22). The Bernstein-Schnabl operators associated with* T_Δ *are defined by (3.1.23) and (3.1.24). Therefore, for every* $f \in \mathscr{C}(K)$, $x \in K$ *and* $n \geq n_0$, *we get*

$$M_n(f)(x) = \left(1+\frac{\gamma(x)}{n}\right) f\left(x+\frac{\beta(x)}{n}\right) \qquad \text{if } x \in \partial K$$

and, if $x \in \operatorname{int}(K)$,

$$\begin{aligned} M_n(f)(x) = &\left(\frac{1-\|x\|_2^2}{\sigma_p}\right)^n \int_{\partial K} \cdots \int_{\partial K} \left(1+\frac{1}{n}\gamma\left(\frac{x_1+\dots+x_n}{n}\right)\right) \\ &\times \frac{f\left(\frac{x_1+\dots+x_n}{n}+\frac{1}{n}\beta\left(\frac{x_1+\dots+x_n}{n}\right)\right)}{\|x_1-x\|_2^d \cdots \|x_n-x\|_2^d}\, d\sigma(x_1)\ \cdots\ d\sigma(x_n). \end{aligned}$$

5.3 Approximation properties

In this section we prove that the sequence $(M_n)_{n\geq n_0}$ (see (5.2.3)) is an approximation process on $\mathscr{C}(K)$ and we give some estimates of the rate of convergence.

From now on we shall assume that (1.6.23) holds true, i.e., the distance ρ on K is induced by some distance ρ_0 on X such that, for every $x, y, z \in X$ and $k \in [0, 1]$,

$$\rho_0(x+z, y+z) = \rho_0(x, y) \text{ and } \rho_0(kx, 0) \leq k\rho_0(x, 0).$$

We explicitly notice that from the previous properties we get

$$\rho_0(kx, ky) \leq k\rho_0(x, y)$$

(see (3.3.1)) for every $k \in [0, 1]$ and $x, y \in X$.

Under these assumptions, the following approximation result holds true.

Proposition 5.3.1. *For every $f \in \mathscr{C}(K)$,*

$$\lim_{n\to\infty} f \circ \left(id + \frac{\beta}{n}\right) = f \qquad \textit{uniformly on } K \tag{5.3.1}$$

and

$$\lim_{n\to\infty} M_n(f) = f \qquad \textit{uniformly on } K. \tag{5.3.2}$$

Proof. We begin by showing (5.3.1).

Indeed, set

$$M_\beta := \sup_{x\in K} \rho_0(\beta(x), 0).$$

If $f \in \mathscr{C}(K)$ and $\varepsilon > 0$ are given, there exists $\delta > 0$ such that, for any $x, y \in K$, $\rho_0(x, y) \leq \delta$, we have that $|f(x) - f(y)| \leq \varepsilon$.

By (1.6.23), for every $x \in K$, we have

$$\rho_0\left(x, x + \frac{\beta(x)}{n}\right) = \rho_0\left(\frac{\beta(x)}{n}, 0\right) \leq \frac{1}{n} M_\beta,$$

and so we can choose $\nu \geq 1$ such that, for every $x \in K$ and for every $n \geq \nu$,

$$\rho_0\left(x, x + \frac{\beta(x)}{n}\right) \leq \delta.$$

Hence, for every $x \in K$ and $n \geq \nu$,

$$\left|f\left(x + \frac{\beta(x)}{n}\right) - f(x)\right| \leq \varepsilon$$

and (5.3.1) holds true.

We pass on proving (5.3.2). Fix now $f \in \mathscr{C}(K)$ and $x \in K$; then

$$|M_n(f)(x) - f(x)| \le \left|B_n\left(f \circ \left(id + \frac{\beta}{n}\right)\right)(x) - B_n(f)(x)\right|$$
$$+ |B_n(f)(x) - f(x)| + \frac{1}{n}\left|B_n\left(\gamma f \circ \left(id + \frac{\beta}{n}\right)\right)(x)\right|$$
$$\le \left\|f \circ \left(id + \frac{\beta}{n}\right) - f\right\|_\infty + \|B_n(f) - f\|_\infty + \frac{1}{n}\|\gamma f\|_\infty,$$

so that, thanks to (5.3.1) and Proposition 3.2.1, this completes the proof. □

In order to provide an estimate of the rate of both pointwise and uniform convergence of the approximating sequence $(M_n)_{n \ge n_0}$, we shall again appeal to the modulus of continuity $\omega_\rho(f, \delta)$ defined by (1.6.20).

In particular, we show that the estimates of Proposition 3.2.2 about the rate of convergence of Bernstein-Schnabl operators by means of the total modulus of smoothness $\Omega(f, \delta)$ (see (1.6.10)) may be useful in order to evaluate the rate of convergence of the sequence $(M_n)_{n \ge n_0}$.

Proposition 5.3.2. *If $f \in \mathscr{C}(K)$, then, for every $n \ge n_0$ and $x \in K$,*

$$|M_n(f)(x) - f(x)| \le 2\Omega\left(\left(\mathbf{1} + \frac{\gamma}{n}\right) \times \left(f \circ \left(id + \frac{\beta}{n}\right)\right), \frac{1}{\sqrt{n}}\right)$$
$$+ \omega_\rho\left(f, \frac{\rho_0(\beta(x), 0)}{n}\right) + \frac{1}{n}\left|\gamma(x) f\left(x + \frac{\beta(x)}{n}\right)\right|.$$

In particular, for every $n \ge n_0$,

$$\|M_n(f) - f\|_\infty \le 2\Omega\left(\left(\mathbf{1} + \frac{\gamma}{n}\right) \times \left(f \circ \left(id + \frac{\beta}{n}\right)\right), \frac{1}{\sqrt{n}}\right)$$
$$+ \omega_\rho\left(f, \frac{M_\beta}{n}\right) + \frac{1}{n}\|\gamma\|_\infty \|f\|_\infty,$$

where $M_\beta := \sup_{x \in K} \rho_0(\beta(x), 0)$.

Proof. It is enough to show the pointwise estimate only. Fix $f \in \mathscr{C}(K)$, $x \in K$ and $n \ge n_0$; preliminarily we notice that, by virtue of (1.6.23),

$$\left|f \circ \left(id + \frac{\beta}{n}\right)(x) - f(x)\right| \le \omega_\rho\left(f, \frac{\rho_0(\beta(x), 0)}{n}\right) \le \omega_\rho\left(f, \frac{M_\beta}{n}\right).$$

Thus,

$$|M_n(f)(x) - f(x)|$$
$$\le \left|B_n\left(\left(\mathbf{1} + \frac{\gamma}{n}\right) \times \left(f \circ \left(id + \frac{\beta}{n}\right)\right)\right)(x) - \left(\mathbf{1} + \frac{\gamma}{n}\right) \times \left(f \circ \left(id + \frac{\beta}{n}\right)\right)(x)\right|$$

$$
\begin{aligned}
&+\left|\left(\mathbf{1}+\frac{\gamma}{n}\right)\times\left(f\circ\left(id+\frac{\beta}{n}\right)\right)(x)-f(x)\right| \\
&\leq\left|B_n\left(\left(\mathbf{1}+\frac{\gamma}{n}\right)\times\left(f\circ\left(id+\frac{\beta}{n}\right)\right)\right)(x)-\left(\mathbf{1}+\frac{\gamma}{n}\right)\times\left(f\circ\left(id+\frac{\beta}{n}\right)\right)(x)\right| \\
&+\omega_\rho\left(f,\frac{\rho_0(\beta(x),0)}{n}\right)+\frac{1}{n}|\gamma(x)|\left|f\left(x+\frac{\beta(x)}{n}\right)\right|.
\end{aligned}
$$

Hence, taking (3.2.6) into account, the result easily follows. □

Assume now that K is a convex compact subset of $\mathbf{R}^d$, $d \geq 1$. Then, by means of (3.2.7), it is possible to estimate the rate of pointwise as well as uniform convergence of the sequence $(M_n(f))_{n\geq n_0}$ to f $(f \in \mathscr{C}(K))$ with respect to the first modulus of continuity.

Proposition 5.3.3. *Let $f \in \mathscr{C}(K)$ and set $r(K) := \max_{x\in K} \|x\|_2$. Assume that there exists $M \geq 0$ such that, for every $x, y \in K$,*

$$
\|\beta(x)-\beta(y)\|_2 \leq M\|x-y\|_2. \tag{5.3.3}
$$

Then, for every $n \geq n_0$,

$$
\begin{aligned}
\|M_n(f)-f\|_\infty &\leq 2\left(1+\frac{\|\gamma\|_\infty}{n}\right)\left(2+\frac{M}{n}\right)\omega\left(f,\frac{r(K)}{\sqrt{n}}\right) \\
&+2\frac{\|f\|_\infty}{n}\omega\left(\gamma,\frac{r(K)}{\sqrt{n}}\right)+\omega\left(f,\frac{\|\beta\|_\infty}{n}\right)+\frac{1}{n}\|\gamma\|_\infty\|f\|_\infty.
\end{aligned}
$$

Proof. For every $f \in \mathscr{C}(K)$, $n \geq n_0$ and $x \in K$, as we have seen in the proof of Proposition 5.3.2,

$$
\begin{aligned}
&|M_n(f)(x)-f(x)| \\
\leq\ &\left|B_n\left(\left(\mathbf{1}+\frac{\gamma}{n}\right)\times\left(f\circ\left(id+\frac{\beta}{n}\right)\right)\right)(x)-\left(\mathbf{1}+\frac{\gamma}{n}\right)\times\left(f\circ\left(id+\frac{\beta}{n}\right)\right)(x)\right| \\
+\ &\omega\left(f,\frac{\|\beta(x)\|_2}{n}\right)+\frac{1}{n}|\gamma(x)|\left|f\left(x+\frac{\beta(x)}{n}\right)\right|.
\end{aligned}
$$

Hence, taking estimate (3.2.7) into account, the result easily follows since, by virtue of (5.3.3) and of Lemma 1.6.1, for any $n \geq n_0$ and $\delta > 0$, we have

$$
\begin{aligned}
&\omega\left(\left(\mathbf{1}+\frac{\gamma}{n}\right)\times\left(f\circ\left(id+\frac{\beta}{n}\right)\right),\delta\right)\leq\left(1+\frac{\|\gamma\|_\infty}{n}\right)\omega\left(f,\left(1+\frac{M}{n}\right)\delta\right) \\
&+\frac{\|f\|_\infty}{n}\omega(\gamma,\delta)\leq\left(1+\frac{\|\gamma\|_\infty}{n}\right)\left(2+\frac{M}{n}\right)\omega(f,\delta)+\frac{\|f\|_\infty}{n}\omega(\gamma,\delta).
\end{aligned}
$$

□

We now provide a further estimate of the rate of convergence by means of the second modulus of smoothness.

Proposition 5.3.4. *Let K be a compact convex subset of $\mathbf{R}^d$, $d \geq 1$. There exists $P > 0$ (depending on K, d and on the functions β and γ) such that, for every $f \in \mathscr{C}(K)$ and $n \geq n_0$,*

$$\|M_n(f) - f\|_\infty \leq \max\{\|\gamma\|_\infty + 2, C\} \left(\frac{P}{n} \|f\|_\infty + \omega_2 \left(f, \sqrt{\frac{P}{n}} \right) \right),$$

where C is an absolute constant that only depends on d and K.

Proof. Since $\|M_n\| \leq 1 + \frac{\|\gamma\|_\infty}{n} \leq 1 + \|\gamma\|_\infty$ for every $n \geq n_0$, we may apply Theorem 1.6.3; hence, for every $n \geq n_0$, let us evaluate

$$\begin{aligned} \lambda_\infty &= \max\{\|M_n(\mathbf{1}) - \mathbf{1}\|_\infty, \|M_n(pr_1) - pr_1\|_\infty, \dots, \\ &\|M_n(pr_d) - pr_d\|_\infty, \|M_n(e) - e\|_\infty\}, \end{aligned}$$

where the function e is defined by (1.6.4).

In particular,

$$\|M_n(\mathbf{1}) - \mathbf{1}\|_\infty \leq \frac{1}{n} \|\gamma\|_\infty.$$

Moreover, we recall that $\beta = (\beta_1, \dots, \beta_d)$, with $\beta_i \in \mathscr{C}(K)$ $(i = 1, \dots, d)$.

Since $pr_i \circ \left(id + \frac{\beta}{n} \right) = pr_i + \frac{\beta_i}{n}$ for every $i = 1, \dots, d$, we get

$$\begin{aligned} \|M_n(pr_i) - pr_i\|_\infty &= \left\| \frac{1}{n^2} B_n(\gamma\beta_i) + \frac{1}{n} B_n(\beta_i) + \frac{1}{n} B_n(\gamma pr_i) \right\|_\infty \\ &\leq \frac{1}{n} (\|\gamma\|_\infty (\|\beta_i\|_\infty + r(K)) + \|\beta_i\|_\infty), \end{aligned}$$

where $r(K) := \max_{x \in K} \|x\|_2$ and, taking also Proposition 3.2.3 into account,

$$\begin{aligned} &\|M_n(e) - e\|_\infty \leq \|B_n(e) - e\|_\infty \\ &+ \frac{1}{n} \sum_{i=1}^d \left\| B_n(2pr_i\beta_i + \gamma pr_i^2) + \frac{1}{n} B_n(\beta_i^2 + 2\gamma\beta_i pr_i) + \frac{1}{n^2} B_n(\gamma\beta_i^2) \right\|_\infty \\ &\leq \frac{1}{n} \left(r(K)^2 + 2r(K) \sum_{i=1}^d \|\beta_i\|_\infty + \|\gamma\|_\infty r(K)^2 + \sum_{i=1}^d \|\beta_i^2\|_\infty \right. \\ &\left. + 2\|\gamma\|_\infty r(K) \sum_{i=1}^d \|\beta_i\|_\infty + \|\gamma\|_\infty \sum_{i=1}^d \|\beta_i^2\|_\infty \right). \end{aligned}$$

Now, choosing in a suitable way the constant P and applying Theorem 1.6.3, we get the required assertion. □

5.4 Preservation properties

In this section we discuss some preservation properties of the operators M_n, which may be inferred from the relevant ones held by the corresponding Bernstein-Schnabl operators.

First of all, we prove that they preserve Lipschitz-continuous functions.

From now on, if K is a metrizable compact convex subset of some locally convex space X, we shall assume again that (1.6.23) holds true.

Moreover, in accordance with (1.6.21), we set

$$\mathbb{L}ip_K(M,\alpha) := \{g \in \mathscr{C}(K,X) \mid \rho(g(x),g(y)) \leq M\rho(x,y)^\alpha \text{ for every } x,y \in K\}. \tag{5.4.1}$$

A function $f \in \mathbb{L}\mathrm{ip}_K(M,\alpha)$ is said to be *K-Hölder continuous with constant M and exponent α.*

Then the following result holds.

Proposition 5.4.1. *Let K be a convex compact subset of some locally convex space X satisfying* (1.6.23) *and consider a Markov operator T on $\mathscr{C}(K)$ satisfying* (3.1.2). *Suppose that condition* (3.3.2) *is fulfilled, i.e., $T(\mathrm{Lip}(1,1)) \subset \mathrm{Lip}(c,1)$ for some $c \geq 1$, and that condition* (5.2.1) *holds true.*

If $\beta \in \mathbb{L}\mathrm{ip}_K(C,1)$ and $\gamma \in \mathrm{Lip}(N,1)$ for some $C,N \geq 0$, then, for every $f \in \mathrm{Lip}(M,1)$ and $n \geq n_0$,

$$M_n(f) \in \mathrm{Lip}\left(cM\left(1+\frac{\|\gamma\|_\infty}{n}\right)\left(1+\frac{C}{n}\right)+c\|f\|_\infty\frac{N}{n},1\right).$$

Moreover, if γ is constant, for any $n \geq n_0$,

$$M_n(f) \in \mathrm{Lip}\left(cM\left(1+\frac{|\gamma|}{n}\right)\left(1+\frac{C}{n}\right),1\right).$$

In particular, if $\gamma = 0$ and β is constant, then, for every $n \geq n_0$,

$$M_n(f) \in \mathrm{Lip}\,(cM,1)\,.$$

Proof. It is easy to check that, taking (1.6.23) and (3.3.1) into account, for any $n \geq n_0$,

$$\left(1+\frac{\gamma}{n}\right)f \circ \left(id+\frac{\beta}{n}\right) \in \mathrm{Lip}\left(M\left(1+\frac{\|\gamma\|_\infty}{n}\right)\left(1+\frac{C}{n}\right)+\|f\|_\infty\frac{N}{n},1\right)$$

and hence the result follows at once from (3.3.2) and Proposition 3.3.8. □

Remark 5.4.2. We explicitly mention that, if K is a convex compact subset of some $\mathbf{R}^d$, $d \geq 1$, if $\beta = (\beta_1,\dots,\beta_d)$ and if there exists $C \geq 0$ such that $\beta_i \in \mathrm{Lip}\left(\frac{C}{\sqrt{d}},1\right)$ for every $i = 1,\dots,d$, then $\beta \in \mathbb{L}\mathrm{ip}_K(C,1)$.

Finally, taking (3.3.2) and Corollary 3.3.2 into account, the next proposition easily follows.

Proposition 5.4.3. *Let K be a convex compact subset of some locally convex space X satisfying* (1.6.23) *and consider a Markov operator T on $\mathscr{C}(K)$ satisfying* (3.1.2)*. Suppose that condition* (3.3.2) *is fulfilled, i.e., $T(\mathrm{Lip}(1,1)) \subset \mathrm{Lip}(c,1)$ for some $c \geq 1$. Assume, further, that $\beta \in \mathbb{L}\mathrm{ip}_K(C,1)$ for some $C \geq 0$.*

Under condition (5.2.1)*, for every $f \in \mathscr{C}(K)$, $\delta > 0$ and $n \geq n_0$,*

$$\omega(M_n(f),\delta) \leq (1+c)\left(1+\frac{\|\gamma\|_\infty}{n}\right)\,\omega\left(f,\left(1+\frac{C}{n}\right)\delta\right)+\frac{(1+c)\|f\|_\infty}{n}\,\omega(\gamma,\delta).$$

Moreover, if γ is constant, then, for any $n \geq n_0$,

$$\omega(M_n(f),\delta) \leq (1+c)\left(1+\frac{|\gamma|}{n}\right)\,\omega\left(f,\left(1+\frac{C}{n}\right)\delta\right).$$

In particular, if $\gamma = 0$ and β is a constant function, then, for every $n \geq n_0$,

$$\omega(M_n(f),\delta) \leq (1+c)\,\omega(f,\delta).$$

Finally, if $c = 1$,

$$\omega(M_n(f),\delta) \leq 2\,\omega(f,\delta)$$

and

$$M_n(\mathrm{Lip}(M,\alpha)) \subset \mathrm{Lip}(M,\alpha)$$

for every $M > 0$ and $\alpha \in]0,1]$.

Below we state some further preservation properties of the operators M_n's.

First of all, we are interested in stating sufficient conditions in order that they preserve convexity.

Theorem 5.4.4. *Let K be a convex compact subset of some locally convex space X and consider a Markov operator T on $\mathscr{C}(K)$ satisfying* (3.1.2)*. Assume that*
(a) T maps continuous convex functions into (continuous) convex functions.
(b) For every convex function $f \in \mathscr{C}(K)$ and for every $x, y \in K$,

$$\Delta(\tilde{f};x,y) \geq 0$$

(see (3.4.2) *and* (3.4.3)*).*
Assume that γ is constant and that (5.2.1) *holds true. If $f \in \mathscr{C}(K)$ and if $f \circ \left(id + \frac{\beta}{n}\right)$ is convex for every $n \geq n_0$, then $M_n(f)$ is convex for any $n \geq n_0$.*

In particular, the result holds true whenever β is affine and f is convex.

Proof. Because of Theorem 3.4.3, under assumptions (a) and (b), the B_n's map continuous convex functions into continuous convex functions. This, together with (5.2.3), completes the proof by also recalling that a function $g \in \mathscr{C}(K)$ is convex if and only if $g\left(\frac{x+y}{2}\right) \leq \frac{1}{2}(g(x)+g(y))$ for every $x, y \in K$. □

As we have seen in Theorem 3.5.2, if T satisfies (3.1.2) and $f \in \mathscr{C}(K)$ is T-axially convex (see Definition 3.5.1), then

$$f \leq B_n(f) \leq T(f)$$

for any $n \geq 1$.

By means of this last result, we present some estimates about the operators M_n.

Proposition 5.4.5. *Let K be a convex compact subset of some locally convex space X and consider a Markov operator T on $\mathscr{C}(K)$ satisfying* (3.1.2)*. Under condition* (5.2.1)*, assume further that γ is constant. If $f \in \mathscr{C}(K)$ and if $f \circ \left(id + \frac{\beta}{n}\right)$ is T-axially convex for every $n \geq n_0$, then*

$$\left(\mathbf{1} + \frac{\gamma}{n}\right)\left(f \circ \left(id + \frac{\beta}{n}\right)\right) \leq M_n(f) \leq \left(\mathbf{1} + \frac{\gamma}{n}\right) T\left(f \circ \left(id + \frac{\beta}{n}\right)\right) \tag{5.4.2}$$

$(n \geq n_0)$.

Examples 5.4.6.

1. *Consider a metrizable Bauer simplex K and its canonical projection T. If β is affine and $f \in \mathscr{C}(K)$ is T-axially convex function (or T-convex (see Theorem 3.5.9)), then $f \circ \left(id + \frac{\beta}{n}\right)$ is T-axially convex (and, hence, T-convex). Consequently,* (5.4.2) *holds true for every constant function γ.*

Moreover, again by virtue of Theorem 3.5.9, each $M_n(f)$ is T-axially convex $(n \geq n_0)$.

2. *If K is a convex compact subset of $\mathbf{R}^d$, $d \geq 1$, then inequalities* (5.4.2) *hold true if all the components $\beta_1, \dots, \beta_d$ of β are convex and f is convex and increasing with respect to the partial ordering $\leq_d$ on $\mathbf{R}^d$ defined by $x \leq_d y$ if $pr_i(x) \leq pr_i(y)$ for every $i = 1, \dots, d$ $(x, y \in \mathbf{R}^d)$.*

If this is the case and if, moreover, $\beta_i \geq 0$ for every $i = 1, \dots, d$ and $\gamma \geq 0$, then

$$f \leq M_n(f) \leq (\mathbf{1} + \gamma) T\left(f \circ (id + \beta)\right).$$

5.5 Asymptotic formulae

The main aim of this section is to establish some asymptotic formulae for modified Bernstein-Schnabl operators.

We first assume that K is a metrizable convex compact subset of a locally convex space X satisfying (1.6.23). We fix a Markov operator $T : \mathscr{C}(K) \longrightarrow \mathscr{C}(K)$ satisfying (3.1.2) and the relevant sequence $(B_n)_{n\geq 1}$ of Bernstein-Schnabl operators.

Moreover, we consider $\beta \in \mathscr{C}(K, X)$ such that $\beta(K) \subset K$ and $\gamma \in \mathscr{C}(K)$ satisfying (5.2.1).

Further, we assume that $0 \in K$ (here $\mathbf{0}$ denotes the null vector in X) and we consider the space

$$A_0(K) := \{h \in A(K) \mid h(\mathbf{0}) = 0\}. \tag{5.5.1}$$

We point out that, if $0 \in K$, then $\lambda x \in K$ for every $x \in K$ and $0 \leq \lambda \leq 1$.

Moreover, any function in $A_0(K)$ verifies the following simple properties.

Proposition 5.5.1. *Consider $h \in A_0(K)$, $x, y \in K$ and $\lambda \geq 0$.*
(1) If $\lambda x \in K$, then $h(\lambda x) = \lambda h(x)$.
(2) If $x + y \in K$, then $h(x + y) = h(x) + h(y)$.

Proof. (1). If $0 \leq \lambda \leq 1$, then

$$h(\lambda x) = h(\lambda x + (1 - \lambda)\mathbf{0}) = \lambda h(x) + (1 - \lambda)h(\mathbf{0}) = \lambda h(x).$$

If $\lambda > 1$, then

$$h(x) = h\left(\frac{1}{\lambda}\lambda x\right) = \frac{1}{\lambda}h(\lambda x)$$

and this completes the proof.

(2). We have

$$h(x + y) = h\left(2\frac{x + y}{2}\right) = 2h\left(\frac{x + y}{2}\right) = 2\frac{h(x) + h(y)}{2} = h(x) + h(y). \qquad \square$$

Set

$$A_1(K) := A_0(K) \cup \{\mathbf{1}\}.$$

We remark that

$$A(K) = \operatorname{span}(A_1(K)),$$

so that, for every $m \geq 2$, we may infer that

$$P_m(K) := \operatorname{span}\left(\left\{\prod_{i=1}^{m} h_i \mid h_1, \dots, h_m \in A_1(K)\right\}\right)$$

(see (4.1.1)).

To present our next result about an asymptotic formula for the operators M_n, for every $m \geq 1$ and $h_1, \dots, h_m \in A_0(K)$ we set

$$B^*_\beta(\mathbf{1}) = 0 \quad \text{and } B^*_\beta(h_1, \dots, h_m) := \begin{cases} h_1 \circ \beta & \text{if } m = 1; \\ \sum_{i=1}^{m} (h_i \circ \beta) \prod_{\substack{j=1 \\ j \neq i}}^{m} h_j & \text{if } m \geq 2. \end{cases} \tag{5.5.2}$$

Moreover, for every $m, p \geq 1$, $1 \leq p \leq m$, set

$$N_m(p) := \{(i_1, \dots, i_p) \in \{1, \dots, m\}^p \mid i_r \neq i_s \text{ for } r \neq s\} \tag{5.5.3}$$

and

$$\begin{aligned} \widetilde{N}_m := \{((i_1, \dots, i_p), (j_1, \dots, j_{m-p})) \in N_m(p) \times N_m(m-p) \mid i_h \neq j_k \\ \text{for every } h = 1, \dots, p, \text{ and } k = 1, \dots, m-p\}. \end{aligned} \tag{5.5.4}$$

Then the following result holds true.

Proposition 5.5.2. *For every $m \geq 1$ and $h_1, \dots, h_m \in A_1(K)$,*

$$\lim_{n\to\infty} n \left(\prod_{i=1}^{m} h_i \circ \left(id + \frac{\beta}{n} \right) - \prod_{i=1}^{m} h_i \right) = B^*_\beta(h_1, \dots, h_m).$$

Proof. Since $B^*_\beta(\mathbf{1}) = 0$, the result is trivial for $h = \mathbf{1}$; it is also a consequence of easy calculations if $m = 1, 2$ and $h_1, h_2 \in A_0(K)$.

Fix now some $m \geq 3$ and $h_1, \dots, h_m \in A_0(K)$. Since, by virtue of Proposition 5.5.1, for any $n \geq n_0$,

$$\begin{aligned} \prod_{i=1}^{m} h_i \circ \left(id + \frac{\beta}{n} \right) &= \prod_{i=1}^{m} \left(h_i + \frac{h_i \circ \beta}{n} \right) = \prod_{i=1}^{m} h_i + \frac{1}{n} \sum_{i=1}^{m} (h_i \circ \beta) \prod_{\substack{j=1 \\ j \neq i}}^{m} h_j \\ &+ \sum_{p=1}^{m-2} \frac{1}{n^{m-p}} \sum_{((i_1, \dots, i_p), (j_1, \dots, j_{m-p})) \in \widetilde{N}_m} h_{i_1} \cdots h_{i_p} (h_{j_1} \circ \beta) \cdots (h_{j_{m-p}} \circ \beta) \\ &+ \frac{1}{n^m} \prod_{i=1}^{m} (h_i \circ \beta), \end{aligned}$$

we obtain

$$\begin{aligned} n \left[\left(\prod_{i=1}^{m} h_i \right) \circ \left(id + \frac{\beta}{n} \right) - \prod_{i=1}^{m} h_i \right] &= B^*_\beta(h_1, \dots, h_m) \\ &+ \sum_{p=1}^{m-2} \frac{1}{n^{m-p-1}} \sum_{((i_1, \dots, i_p), (j_1, \dots, j_{m-p})) \in \widetilde{N}_m} h_{i_1} \cdots h_{i_p} (h_{j_1} \circ \beta) \cdots (h_{j_{m-p}} \circ \beta) \\ &+ \frac{1}{n^{m-1}} \prod_{i=1}^{m} (h_i \circ \beta). \end{aligned}$$

This completes the proof. □

We are now ready to state an asymptotic formula for the operators M_n. It involves the operator B^*_β defined by (5.5.2), and the operator Θ_T (see (4.1.6)) introduced in Chapter 4 in association with Bernstein-Schnabl operators B_n.

Theorem 5.5.3. *Fix $m \geq 1$ and $h_1, \dots, h_m \in A_1(K)$; then*

$$\begin{aligned}\lim_{n\to\infty} n\left(M_n\left(\prod_{i=1}^m h_i\right) - \prod_{i=1}^m h_i\right)\\ = \Theta_T(h_1,\dots h_m) + B^*_\beta(h_1,\dots,h_m) + \gamma\prod_{i=1}^m h_i\end{aligned} \tag{5.5.5}$$

uniformly on K (see (4.1.6) *and* (5.5.2)*).*

Proof. The result is straightforward for $m = 1$ or, if $m > 1$, if $h_1 = \dots = h_m = \mathbf{1}$, taking (5.5.2) and (4.1.6) into account.

Assume, instead, that $m \geq 2$ and fix $h_1, \dots, h_m \in A_0(K)$. We preliminarily show that

$$\lim_{n\to\infty} n\left(B_n\left(\prod_{i=1}^m h_i\circ\left(id+\frac{\beta}{n}\right)\right) - \prod_{i=1}^m h_i\circ\left(id+\frac{\beta}{n}\right)\right) = \Theta_T(h_1,\dots,h_m), \tag{1}$$

$$\lim_{n\to\infty}\left(B_n\left(\gamma\prod_{i=1}^m h_i\circ\left(id+\frac{\beta}{n}\right)\right) - \gamma\prod_{i=1}^m h_i\circ\left(id+\frac{\beta}{n}\right)\right) = 0 \tag{2}$$

and

$$\lim_{n\to\infty} n\left(\left(1+\frac{\gamma}{n}\right)\prod_{i=1}^m h_i\circ\left(id+\frac{\beta}{n}\right) - \prod_{i=1}^m h_i\right) = B^*_\beta(h_1,\dots,h_m) + \gamma\prod_{i=1}^m h_i, \tag{3}$$

the three limits being uniform on K.

Starting to prove (1), the calculation made in the preceding proposition leads to

$$\begin{aligned}& n\left(B_n\left(\prod_{i=1}^m h_i\circ\left(id+\frac{\beta}{n}\right)\right) - \prod_{i=1}^m h_i\circ\left(id+\frac{\beta}{n}\right)\right)\\ &= n\left[B_n\left(\prod_{i=1}^m h_i\right) - \prod_{i=1}^m h_i\right]\\ &+ \sum_{p=1}^{m-1}\frac{1}{n^{m-p-1}}\sum_{((i_1,\dots,i_p),(j_1,\dots,j_{m-p}))\in\widetilde{N}_m}\left[B_n(h_{i_1}\cdots h_{i_p}(h_{j_1}\circ\beta)\cdots(h_{j_{m-p}}\circ\beta))\right.\\ &\left.- h_{i_1}\cdots h_{i_p}(h_{j_1}\circ\beta)\cdots(h_{j_{m-p}}\circ\beta)\right] + \frac{1}{n^{m-1}}\left[B_n\left(\prod_{i=1}^m(h_i\circ\beta)\right) - \prod_{i=1}^m(h_i\circ\beta)\right];\end{aligned}$$

taking Theorem 4.1.2 and (3.2.5) into account, the result easily follows.

To prove (2), simply observe that

$$B_n\left(\gamma\prod_{i=1}^m h_i\circ\left(id+\frac{\beta}{n}\right)\right)-\gamma\prod_{i=1}^m h_i\circ\left(id+\frac{\beta}{n}\right)$$
$$=B_n\left(\gamma\prod_{i=1}^m h_i\circ\left(id+\frac{\beta}{n}\right)-\gamma\prod_{i=1}^m h_i\right)$$
$$+\left(B_n\left(\gamma\prod_{i=1}^m h_i\right)-\gamma\prod_{i=1}^m h_i\right)+\left(\gamma\prod_{i=1}^m h_i-\gamma\prod_{i=1}^m h_i\circ\left(id+\frac{\beta}{n}\right)\right),$$

where each term tends uniformly to zero by virtue of (5.3.1) and (3.2.5).

Finally, from (5.3.1) and Proposition 5.5.2, we obtain (3).

Hence, since

$$n\left(M_n\left(\prod_{i=1}^m h_i\right)-\prod_{i=1}^m h_i\right)$$
$$=n\left(B_n\left(\left(\mathbf{1}+\frac{\gamma}{n}\right)\times\left(\prod_{i=1}^m h_i\circ\left(id+\frac{\beta}{n}\right)\right)\right)-\prod_{i=1}^m h_i\right)$$
$$=n\left(B_n\left(\prod_{i=1}^m h_i\circ\left(id+\frac{\beta}{n}\right)\right)-\prod_{i=1}^m h_i\circ\left(id+\frac{\beta}{n}\right)\right)$$
$$+\left(B_n\left(\gamma\prod_{i=1}^m h_i\circ\left(id+\frac{\beta}{n}\right)\right)-\gamma\prod_{i=1}^m h_i\circ\left(id+\frac{\beta}{n}\right)\right)$$
$$+n\left(\left(\mathbf{1}+\frac{\gamma}{n}\right)\times\left(\prod_{i=1}^m h_i\circ\left(id+\frac{\beta}{n}\right)\right)-\prod_{i=1}^m h_i\right),$$

by using (1)–(3), (5.5.5) follows at once. □

From Theorem 5.5.3 it also follows that, for every $u\in P_\infty(K)$, there exists $\lim_{n\to\infty} n(M_n(u)-u)$ in $\mathscr{C}(K)$ and hence we can consider the linear operator $N_T: P_\infty(K)\longrightarrow\mathscr{C}(K)$ defined by

$$N_T(u):=\lim_{n\to\infty} n(M_n(u)-u)\qquad (u\in P_\infty(K)). \tag{5.5.6}$$

Thus, if $h_1,\dots,h_m\in A_1(K)$, $m\geq 1$, then

$$N_T\left(\prod_{i=1}^m h_i\right)=\Theta_T(h_1,\dots,h_m)+B_\beta^*(h_1,\dots,h_m)+\gamma\prod_{i=1}^m h_i. \tag{5.5.7}$$

In other words, considering the linear operator $B_\beta:=N_T-L_T-\gamma I$ from $P_\infty(K)$ into $\mathscr{C}(K)$ (see (4.1.9)-(4.1.10)), we have

$$B_\beta\left(\prod_{i=1}^m h_i\right)=B_\beta^*(h_1,\dots,h_m) \tag{5.5.8}$$

$(h_1,\dots,h_m\in A_1(K), m\geq 1)$ and

$$N_T=L_T+B_\beta+\gamma I. \tag{5.5.9}$$

Remark 5.5.4. We stress that the assumptions $\beta : K \longrightarrow K$ and $0 \in K$, together with the construction of $B^*_\beta(h_1, \dots, h_m)$ on $A_1(K)$ and the relevant asymptotic formula (5.5.5), are needed only if there exists $x \in K$ such that $\beta(x) \neq 0$.

If β is constant of constant value 0, a careful inspection of the proof of Theorem 5.5.3 shows that the assumptions above are no longer necessary.

We present another situation where the assumptions of Theorem 5.5.3 may be relaxed.

When K is a convex compact subset of some Euclidean space $\mathbf{R}^d$, $d \geq 1$ (which not necessarily contains the null vector) with non-empty interior, then it is possible to state a similar asymptotic formula for functions $u \in \mathscr{C}^2(K)$.

To this end, we introduce a first-order additive perturbation of the differential operator (4.1.15), namely

$$V_T(u)(x) := \frac{1}{2}\sum_{i,j=1}^{d} \alpha_{ij}(x)\frac{\partial^2 u}{\partial x_i \partial x_j}(x) + \sum_{i=1}^{d} \beta_i(x)\frac{\partial u}{\partial x_i}(x) + \gamma(x)u(x) \qquad (5.5.10)$$

($u \in \mathscr{C}^2(K)$, $x \in K$), where the coefficients α_{ij} are defined by (4.1.16) and $\beta = (\beta_1, \dots, \beta_d)$, $\beta_i \in \mathscr{C}(K)$ $(i = 1, \dots, d)$, and γ satisfy (5.2.1).

In order to state the above mentioned asymptotic formula, we start by presenting some further technical properties of the sequence $(B_n)_{n\geq 1}$.

Proposition 5.5.5. *If $f \in \mathscr{C}(K)$ and $i, j = 1, \dots, d$, then*
(a) $\lim_{n\to\infty} B_n(f \times (pr_i \circ \Psi_x))(x) = 0$ uniformly w.r.t. $x \in K$;
(b) $\lim_{n\to\infty} B_n(f \times (pr_i \circ \Psi_x)(pr_j \circ \Psi_x))(x) = 0$ uniformly w.r.t. $x \in K$;
here, for every $x \in K$ the function Ψ_x is defined by (1.5.1), i.e., $\Psi_x(y) = y - x$ $(y \in K)$.

Proof. (a) Set $r(K) := \max_{x\in K} \|x\|_2$ and consider $i = 1, \dots, d$ and $f \in \mathscr{C}(K)$; then, $f \times (pr_i \circ \Psi_x) = fpr_i - x_i f$ for every $x = (x_1, \dots, x_d) \in K$.

Accordingly, for every $n \geq 1$,

$$B_n(f \times (pr_i \circ \Psi_x)) - f \times (pr_i \circ \Psi_x) = B_n(fpr_i) - fpr_i - x_i(B_n(f) - f).$$

Hence, for every $x \in K$,

$$\begin{aligned}
|B_n(f \times (pr_i \circ \Psi_x))(x)| &= |B_n(f \times (pr_i \circ \Psi_x))(x) - f(x)(pr_i \circ \Psi_x)(x)| \\
&\leq \|B_n(f \times (pr_i \circ \Psi_x)) - f \times (pr_i \circ \Psi_x)\|_\infty \\
&\leq \|B_n(fpr_i) - fpr_i\|_\infty + r(K)\|B_n(f) - f\|_\infty
\end{aligned}$$

and, taking Proposition 3.2.1 into account, this completes the proof.

(b) For every $f \in \mathscr{C}(K)$, $x = (x_1, \dots, x_d) \in K$ and $i, j = i, \dots, d$,

$$f \times (pr_i \circ \Psi_x)(pr_j \circ \Psi_x) = fpr_ipr_j - x_j fpr_i - x_i fpr_j + x_i x_j f;$$

accordingly,

$$
\begin{aligned}
&B_n(f \times (pr_i \circ \Psi_x)(pr_j \circ \Psi_x)) - f \times (pr_i \circ \Psi_x)(pr_j \circ \Psi_x)\\
&= B_n(f pr_i pr_j) - f pr_i pr_j - x_j(B_n(f pr_i) - f pr_i)\\
&- x_i(B_n(f pr_j) - f pr_j) + x_i x_j (B_n(f) - f).
\end{aligned}
$$

Then,

$$
\begin{aligned}
&|B_n(f \times (pr_i \circ \Psi_x)(pr_j \circ \Psi_x))(x)|\\
&= |B_n(f \times (pr_i \circ \Psi_x)(pr_j \circ \Psi_x))(x) - f(x)(pr_i \circ \Psi_x)(x)(pr_j \circ \Psi_x)(x)|\\
&\le \|B_n(f pr_i pr_j) - f pr_i pr_j\|_\infty + r(K)\|B_n(f pr_i) - f pr_i\|_\infty\\
&+ r(K)\|B_n(f pr_j) - f pr_j\|_\infty + (r(K))^2 \|B_n(f) - f\|_\infty,
\end{aligned}
$$

and the result now follows from Proposition 3.2.1. □

Now we are in a position to state the following theorem.

Theorem 5.5.6. *For every $u \in \mathscr{C}^2(K)$,*

$$\lim_{n\to\infty} n(M_n(u) - u) = V_T(u) \tag{5.5.11}$$

uniformly on K, where the operator $(V_T, \mathscr{C}^2(K))$ is defined by (5.5.10).

Therefore, if $0 \in K$, then, for every $u \in P_\infty(K)$,

$$N_T(u) = V_T(u) \qquad \text{and} \qquad B_\beta(u) = \sum_{i=1}^{d} \beta_i \frac{\partial u}{\partial x_i}. \tag{5.5.12}$$

Proof. According to Theorem 1.5.2, to prove (5.5.11) it is sufficient to show that

(i) $\lim_{n\to\infty} [n(M_n(\mathbf{1})(x) - 1) - \gamma(x)] = 0$ uniformly w.r.t. $x \in K$;

(ii) $\lim_{n\to\infty} [nM_n(pr_i \circ \Psi_x)(x) - \beta_i(x)] = 0$ uniformly w.r.t. $x \in K$ for every $i = 1, \dots, d$;

(iii) $\lim_{n\to\infty} [nM_n((pr_i \circ \Psi_x)(pr_j \circ \Psi_x))(x) - \alpha_{ij}(x)] = 0$ uniformly w.r.t. $x \in K$ for every $i, j = 1, \dots, d$;

(iv) $\sup_{\substack{n \ge n_0\\ x\in K}} nM_n(\Phi_x^2)(x) < +\infty$;

(v) $\lim_{n\to\infty} nM_n(\Phi_x^4)(x) = 0$ uniformly w.r.t. $x \in K$,

where the functions Ψ_x and Φ_x are defined by (1.5.1) and (1.5.2), respectively.

We begin to prove (i). Indeed, for every $x \in K$ and $n \ge n_0$,

$$n(M_n(\mathbf{1})(x) - 1) - \gamma(x) = B_n(\gamma)(x) - \gamma(x),$$

so that the result follows from Proposition 3.2.1.

We now proceed to verify (ii). To this end, fix $i = 1, \dots, d$. Then, for every $x \in K$, the function $pr_i \circ \Psi_x \in A(K)$ and therefore, according to (3.2.2), $B_n(pr_i \circ \Psi_x)(x) = 0$ for every $n \ge 1$.

Moreover, an easy calculation shows that, for every $x \in K$ and $n \geq n_0$,

$$(pr_i \circ \Psi_x) \circ \left(id + \frac{\beta}{n}\right) = pr_i \circ \left(\Psi_x + \frac{\beta}{n}\right);$$

hence

$$\begin{aligned}
&M_n(pr_i \circ \Psi_x)(x) \\
&= B_n\left(pr_i \circ \left(\Psi_x + \frac{\beta}{n}\right)\right)(x) + \frac{1}{n}B_n\left(\gamma\left(pr_i \circ \left(\Psi_x + \frac{\beta}{n}\right)\right)\right)(x) \\
&= \frac{1}{n}B_n(\beta_i)(x) + \frac{1}{n}B_n(\gamma(pr_i \circ \Psi_x))(x) + \frac{1}{n^2}B_n(\gamma\beta_i)(x).
\end{aligned}$$

Accordingly,

$$\begin{aligned}
&|nM_n(pr_i \circ \Psi_x)(x) - \beta_i(x)| \\
&\leq |B_n(\beta_i)(x) - \beta_i(x)| + |B_n(\gamma(pr_i \circ \Psi_x))(x)| + \frac{1}{n}|B_n(\gamma\beta_i)(x)|.
\end{aligned}$$

In order to complete the proof of (ii), it now suffices to apply Proposition 3.2.1, Proposition 5.5.5 and the fact that the sequence $(B_n(\gamma\beta_i))_{n\geq 1}$ is equibounded.

To show condition (iii), we first note that, for every $i, j = 1, \ldots, d$, $x \in K$ and $n \geq n_0$,

$$\begin{aligned}
&((pr_i \circ \Psi_x)(pr_j \circ \Psi_x)) \circ \left(id + \frac{\beta}{n}\right) \\
&= (pr_i \circ \Psi_x)(pr_j \circ \Psi_x) + \frac{\beta_i}{n} pr_j \circ \Psi_x + \frac{\beta_j}{n} pr_i \circ \Psi_x + \frac{\beta_i\beta_j}{n^2}.
\end{aligned}$$

Therefore, taking (3.2.3) into account,

$$\begin{aligned}
&|nM_n((pr_i \circ \Psi_x)(pr_j \circ \Psi_x))(x) - \alpha_{ij}(x)| \\
&\leq |B_n(\beta_i \times (pr_j \circ \Psi_x))(x)| + |B_n(\beta_j \times (pr_i \circ \Psi_x))(x)| + \frac{1}{n}|B_n(\beta_i\beta_j)(x)| \\
&+ |B_n(\gamma \times (pr_i \circ \Psi_x)(pr_j \circ \Psi_x))(x)| + \frac{1}{n}|B_n(\gamma\beta_i \times (pr_j \circ \Psi_x))(x)| \\
&+ \frac{1}{n}|B_n(\gamma\beta_j \times (pr_i \circ \Psi_x))(x)| + \frac{1}{n^2}|B_n(\gamma\beta_i\beta_j)(x)|.
\end{aligned}$$

Hence, by using Proposition 5.5.5 and the equiboundedness of the sequences $(B_n(\beta_i\beta_j))_{n\geq 1}$ and $(B_n(\gamma\beta_i\beta_j))_{n\geq 1}$, we get the required assertion.

We proceed to verify conditions (iv) and (v); to this aim, preliminarily notice that

$$r_0 := \sup_{\substack{n\geq 1 \\ x\in K}} nB_n(\Phi_x^2)(x) < +\infty \quad \text{and} \quad \lim_{n\to\infty} nB_n(\Phi_x^4)(x) = 0 \tag{1}$$

uniformly w.r.t. $x \in K$ (see the proof of Theorem 4.1.5). Moreover, if $q \geq 2$, $x, y \in K$ and $n \geq n_0$, then

$$\Phi_x^q \circ \left(id + \frac{\beta}{n}\right)(y) = \left\|y + \frac{\beta(y)}{n} - x\right\|_2^q \leq 2^{q-1}\left(\|y - x\|_2^q + \frac{\|\beta(y)\|_2^q}{n^q}\right),$$

so that

$$\Phi_x^q \circ \left(id + \frac{\beta}{n}\right) \leq 2^{q-1}\left(\Phi_x^q + \frac{M_q}{n^q}\mathbf{1}\right),$$

where $M_q := \sup_{y \in K} \|\beta(y)\|_2^q$.

Therefore,

$$nM_n(\Phi_x^2)(x) \leq 2\left(nB_n(\Phi_x^2)(x) + \frac{M_2}{n}\right) + 2B_n\left(|\gamma|\left(\Phi_x^2 + \frac{M_2}{n^2}\mathbf{1}\right)\right)(x)$$

$$\leq 2\left(1 + \frac{\|\gamma\|_\infty}{n}\right)\left(nB_n(\Phi_x)^2(x) + \frac{M_2}{n}\right) \leq 2(1 + \|\gamma\|_\infty)(r_0 + M_2).$$

Analogously,

$$nM_n(\Phi_x^4)(x)$$

$$\leq 8\left(nB_n(\Phi_x^4)(x) + \frac{M_4}{n^3}\right) + 8\left(B_n(|\gamma|\Phi_x^4)(x) + \frac{M_4}{n^4}B_n(|\gamma|)(x)\right)$$

$$\leq 8\left(1 + \frac{\|\gamma\|_\infty}{n}\right)\left(nB_n(\Phi_x)^4(x) + \frac{M_4}{n^3}\right)$$

and hence part (iv) and (v) easily follow from (1); this completes the proof of (5.5.11).

Finally, (5.5.12) follows directly from Theorem 5.5.3. □

5.6 Modified Bernstein-Schnabl operators and first-order perturbations

This section is devoted to discuss the pre-announced generation results for the additive perturbations N_T and V_T of the operators L_T and W_T (see (5.5.9) and (5.5.10)). Moreover, we shall show that the relevant Feller semigroups can be approximated in terms of iterates of modified Bernstein-Schnabl operators.

In the finite-dimensional case, this will allow us to approximate the solutions to suitable diffusion problems associated with the operator $(V_T, \mathscr{C}^2(K))$ and, in particular, to deduce some of their spatial regularity properties, by means of the relevant ones held by the operators M_n.

Turning back to the more general setting, from now on we assume that K is a metrizable convex compact subset of some locally convex space X satisfying (1.6.23) and such that $0 \in K$. Moreover, we suppose that there exist $a, b \in \mathbf{R}$ and $z \in K$ such that $0 < a \leq 1$, $b \geq 0$, $bz \in K$ and

$$ax + bz \in K \text{ for every } x \in K; \tag{5.6.1}$$

we point out that, if $0 \leq b \leq 1$, then automatically $bz \in K$; on the other hand, if $a + b \leq 1$, then (5.6.1) is satisfied (since $ax \in K$ for every $x \in K$).

We consider the function $\beta : K \longrightarrow K$ defined by setting

$$\beta(x) := ax + bz \qquad (x \in K). \tag{5.6.2}$$

After these preliminaries, we are ready to state the following result.

Theorem 5.6.1. *Under the preceding hypotheses, let T be a Markov operator satisfying* (3.1.2) *and assume that*

$$T(h_1 h_2) \in A_0(K) \text{ if } h_1, h_2 \in A(K) \text{ and } h_1 \text{ or } h_2 \text{ belongs to } A_0(K) \tag{5.6.3}$$

(see (5.5.1)*).*

Consider $\gamma \in \mathscr{C}(K)$, the function β defined by (5.6.2) *and assume that* (5.2.1) *holds true. Moreover, consider the modified Bernstein-Schnabl operators $(M_n)_{n \geq n_0}$ associated with T, γ and β (see* (5.2.3)*) and the linear operator N_T defined by* (5.5.6) *(see also* (5.5.9)*).*

Then the operator $(N_T, P_\infty(K))$ is closable and its closure $(B_T^*, D(B_T^*))$ generates a positive C_0-semigroup $(T(t))_{t \geq 0}$ on $\mathscr{C}(K)$ such that

$$\|T(t)\| \leq \exp(\|\gamma\|_\infty t) \qquad (t \geq 0). \tag{5.6.4}$$

If $\gamma \leq 0$, then $(T(t))_{t \geq 0}$ is, in fact, a Feller semigroup.

Moreover,

(a) If $t \geq 0$ and if $(k(n))_{n \geq 1}$ is a sequence of positive integers such that $\lim_{n \to \infty} \frac{k(n)}{n} = t$, then

$$\lim_{n \to \infty} M_n^{k(n)}(f) = T(t)(f) \quad \textit{uniformly on } K \tag{5.6.5}$$

for every $f \in \mathscr{C}(K)$, where each $M_n^{k(n)}$ denotes the iterate of M_n of order $k(n)$.

(b) The generator $(B_T^*, D(B_T^*))$ coincides with the closure of the linear operator $Z : D(Z) \to \mathscr{C}(K)$ defined by

$$Z(f) := \lim_{n\to\infty} n(M_n(f) - f) \tag{5.6.6}$$

for every $f \in D(Z)$, where

$$D(Z) := \left\{g \in \mathscr{C}(K) \mid \lim_{n\to\infty} n(M_n(g) - g) \text{ exists in } \mathscr{C}(K)\right\}. \tag{5.6.7}$$

(c) $P_\infty(K)$ is a core for $(B_T^*, D(B_T^*))$.

Proof. Let us consider the positive linear contractions $L_n : \mathscr{C}(K) \longrightarrow \mathscr{C}(K)$ defined by setting

$$L_n := B_n\left(f \circ \left(id + \frac{\beta}{n}\right)\right) \qquad (f \in \mathscr{C}(K),\, n \geq n_0).$$

Moreover, we look at the linear operator $(\Omega, D(\Omega))$ with domain

$$D(\Omega) := \left\{g \in \mathscr{C}(K) \mid \lim_{n\to\infty} n(L_n(g) - g) \text{ exists in } \mathscr{C}(K)\right\},$$

defined by

$$\Omega(f) := \lim_{n\to\infty} n(L_n(f) - f)$$

for every $f \in D(\Omega)$.

By applying Theorem 5.5.3 and the subsequent formula (5.5.6) for $\gamma = 0$, we get that $P_\infty(K) \subset D(\Omega)$ and $\Omega(u) = N_T(u) - \gamma u =: \widetilde{N_T}(u)$ for every $u \in P_\infty(K)$.

We pass to prove that, if $\lambda > 0$ and $\lambda \neq a$, then the range $(\lambda I - \Omega)(D(\Omega))$ of $\lambda I - \Omega$ is dense in $\mathscr{C}(K)$. To this end, since $P_\infty(K)$ is dense in $\mathscr{C}(K)$ it suffices to show that

$$\overline{(\lambda I - \Omega)(P_\infty(K))} = \mathscr{C}(K) \tag{1}$$

with respect to $\|\cdot\|_\infty$.

Consider a continuous linear functional $\mu : \mathscr{C}(K) \longrightarrow \mathbf{R}$ such that $\mu = 0$ on $(\lambda I - \Omega)(P_\infty(K))$. By a consequence of Hahn-Banach theorem, (1) will be proved if we show that $\mu = 0$ and, to this end, it suffices to prove that $\mu = 0$ on $P_\infty(K)$.

Indeed,

$$\mu(\mathbf{1}) = \frac{1}{\lambda}\mu(\widetilde{N_T}(\mathbf{1})) = \frac{1}{\lambda}\mu(\Theta_T(\mathbf{1}) + B_\beta^*(\mathbf{1})) = 0$$

(see (4.1.6) and (5.5.2)).

Observe that, if $h \in A_0(K)$, then

$$h \circ \beta = ah + bh(z)1,$$

and, thus, if $m = 1$ and $h_1 \in A_0(K)$, then

$$\mu(h_1) = \frac{1}{\lambda}(\mu(\Theta_T(h_1)) + \mu(B_\beta^*(h_1))) = \frac{1}{\lambda}(a\mu((h_1)) + bh_1(z)\mu(\mathbf{1})) = \frac{a}{\lambda}\mu(h_1),$$

so that, also in this case, $\mu(h_1) = 0$.

Assume that $m \geq 2$ and fix $h_1, \dots h_m \in A_0(K)$; then, by means of (5.5.2) and of (5.6.2),

$$B^*_\beta(h_1, \dots, h_m) = \sum_{i=1}^{m}(ah_i + bh_i(z)1)\prod_{\substack{j=1\\ j\neq i}}^{m} h_j = ma\prod_{i=1}^{m} h_i + b\sum_{i=1}^{m} h_i(z)\prod_{\substack{j=1\\ j\neq i}}^{m} h_j. \quad (2)$$

Hence, if $m = 2$, then, taking (4.1.6), (5.6.3) and (2) into account, we have that

$$\begin{aligned}\mu(h_1h_2) &= \frac{1}{\lambda}(\mu(T(h_1h_2)) - \mu(h_1h_2) + \mu(B^*_\beta(h_1, h_2)))\\ &= \frac{1}{\lambda}(-\mu(h_1h_2) + 2a\mu(h_1h_2) + bh_1(z)\mu(h_2) + bh_2(z)\mu(h_1))\\ &= \frac{1}{\lambda}(-\mu(h_1h_2) + 2a\mu(h_1h_2)),\end{aligned}$$

and therefore $\mu(h_1h_2) = 0$.

Let us now fix $m > 2$ and suppose that $\mu = 0$ on $P_m(K)$; we shall prove that $\mu = 0$ on $P_{m+1}(K)$. To this end, consider $h_1, \dots, h_{m+1} \in A_0(K)$ and set $f = \prod_{i=1}^{m+1} h_i$; then, by virtue of (4.1.6), (5.6.3) and (2),

$$\begin{aligned}\mu(f) &= \frac{1}{\lambda}(\mu(\Theta_T(h_1, \dots, h_{m+1})) + \mu(B^*_\beta(h_1, \dots, h_{m+1})))\\ &= \frac{1}{\lambda}\mu\left(\sum_{1\leq i<j\leq m+1} T(h_ih_j)\prod_{\substack{k=1\\ k\neq i,j}}^{m+1} h_k - \binom{m+1}{2} f\right)\\ &\quad + \frac{1}{\lambda}\left((m+1)a\mu(f) + b\sum_{i=1}^{m+1} h_i(z)\mu\left(\prod_{\substack{j=1\\ j\neq i}}^{m+1} h_j\right)\right)\\ &= \frac{1}{\lambda}\left(-\frac{m(m+1)}{2} + a(m+1)\right)\mu(f).\end{aligned}$$

We notice that $-\frac{m(m+1)}{2} + a(m+1) < 0$ for every $m > 2$. Accordingly, $\mu(f) = 0$; hence, by induction, $\mu = 0$ on each $P_m(K)$, $m \geq 1$, and thus $\mu = 0$ on $P_\infty(K)$.

By virtue of Theorem 2.2.1 there exists a contractive C_0-semigroup $(S(t))_{t\geq 0}$ on $\mathscr{C}(K)$ such that, for every $t \geq 0$ and $f \in \mathscr{C}(K)$,

$$S(t)(f) = \lim_{n\to\infty} L_n^{k(n)}(f) \quad (3)$$

uniformly on K, for every sequence $(k(n))_{n\geq 1}$ of positive integers such that $\lim_{n\to\infty} \frac{k(n)}{n} = t$, and the generator $(W, D(W))$ of $(S(t))_{t\geq 0}$ is the closure of $(\Omega, D(\Omega))$. From the approximation formula (3) it also follows that each $S(t)$ $(t \geq 0)$ is positive.

Moreover, $W = \Omega$ on $P_\infty(K)$. Consequently, it follows that

$$\overline{(I-W)(P_\infty(K))} = \overline{(I-\Omega)(P_\infty(K))} = \mathscr{C}(K)$$

with respect to $\|\cdot\|_\infty$ and thus $P_\infty(K)$ is a core for $(W, D(W))$.

Consider now the bounded operator Γ on $\mathscr{C}(K)$ defined by $\Gamma(f) := \gamma f$ ($f \in \mathscr{C}(K)$) and set

$$C_T^* := W + \Gamma$$

defined on $D(C_T^*) := D(W)$.

Then, by Theorem 2.1.13, the operator $(C_T^*, D(C_T^*))$ generates a C_0-semigroup $(T(t))_{t\geq 0}$ such that

$$\|T(t)\| \leq \exp(\|\gamma\|_\infty t) \qquad (t \geq 0)$$

(see (2.1.10)).

If $\gamma \leq 0$, then Γ, as well as W, satisfies the positive maximum principle (see Definition 2.3.3). Hence, also C_T^* satisfies the positive maximum principle, so that $\|T(t)\| \leq 1$ for every $t \geq 0$ (see Theorem 2.3.4).

Moreover, as it can be easily checked, $P_\infty(K)$ is a core for $(C_T^*, D(C_T^*))$ and hence $(\lambda I - C_T^*)(P_\infty(K))$ is dense in $\mathscr{C}(K)$ for any $\lambda > \|\gamma\|_\infty$ (see Remark 2.1.5).

Therefore, as

$$Z(f) = N_T(f) = \Omega(f) + \gamma f = W(f) + \gamma f = C_T^*(f)$$

for every $f \in P_\infty(K)$, also $(\lambda I - Z)(P_\infty(K))$ is dense in $\mathscr{C}(K)$ for any $\lambda > \|\gamma\|_\infty$.

Furthermore, since $\|M_n\| \leq 1 + \frac{\|\gamma\|_\infty}{n} \leq \exp\left(\frac{\|\gamma\|_\infty}{n}\right)$ for every $n \geq n_0$,

$$\|M_n^k\| \leq \exp\left(\frac{\|\gamma\|_\infty k}{n}\right) \qquad (k \geq 1).$$

By applying again Theorem 2.2.1 we deduce that the operator $(Z, D(Z))$ is closable and its closure $(B_T^*, D(B_T^*))$ generates a C_0-semigroup $(\widetilde{T}(t))_{t\geq 0}$ on $\mathscr{C}(K)$ satisfying (5.6.4) and (5.6.5).

Finally, the statement will be completely proved once we recognize that $\widetilde{T}(t) = T(t)$ for every $t \geq 0$ or, equivalently, that $(B_T^*, D(B_T^*)) = (C_T^*, D(C_T^*))$; but this last assertion easily follows from the fact that $P_\infty(K)$ is a core both for $(B_T^*, D(B_T^*))$ and $(C_T^*, D(C_T^*))$ (see Proposition 2.1.7, (g)), which implies, in particular, that $(B_T^*, D(B_T^*))$ is the closure of $(N_T; P_\infty(K))$. □

We now prove that the only case when (5.6.3) is satisfied is in the setting of Bauer simplices (see Theorem 1.1.12).

Proposition 5.6.2. *Let K be a metrizable convex compact subset of some locally convex space X and assume that $0 \in K$. Consider a Markov operator $T : \mathscr{C}(K) \longrightarrow \mathscr{C}(K)$ satisfying* (3.1.2).

Moreover, assume that (5.6.3) *holds true, i.e.,*

$$T(h_1h_2) \in A_0(K) \text{ if } h_1, h_2 \in A(K) \text{ and } h_1 \text{ or } h_2 \text{ belongs to } A_0(K).$$

Then $T(h_1h_2) \in A(K)$ for every $h_1, h_2 \in A(K)$. Consequently, K is a Bauer simplex and T is the canonical projection on $\mathscr{C}(K)$.

Conversely, if K is a Bauer simplex and $0 \in \partial_e K$, then (5.6.3) *holds true for the canonical projection on K.*

Proof. Fix $h_1, h_2 \in A(K)$. Then

$$h_1h_2 = (h_1 - h_1(0)\mathbf{1})(h_2 - h_2(0)\mathbf{1}) + h_1(0)h_2 + h_2(0)h_1 - h_1(0)h_2(0)\mathbf{1}.$$

Since $T(h) = h$ for every $h \in A(K)$ and $T(\mathbf{1}) = \mathbf{1}$, from (5.6.3), the result easily follows. Finally, K is a Bauer simplex and T is the canonical projection associated with it by virtue of Theorem 4.3.3.

Conversely, fix $h_1, h_2 \in A(K)$ and assume that $h_1(0) = 0$ or $h_2(0) = 0$. Then, by assumption, $T(h_1h_2) \in A(K)$; moreover, since $0 \in \partial_e K$, by (3.1.3), we have that

$$T(h_1h_2)(0) = h_1(0)h_2(0) = 0$$

and this completes the proof. □

We point out that the assumption $0 \in \partial_e K$ (or even $0 \in K$) in infinite-dimensional settings is needed only if β is not the constant function of constant value 0 (see Remark 5.5.4); actually, this assumption may be relaxed in finite-dimensional settings.

Assume, indeed, that K is a convex compact subset of $\mathbf{R}^d$, $d \geq 1$, with non-empty interior, consider $\beta_1, \dots, \beta_d, \gamma \in \mathscr{C}(K)$ satisfying (5.2.1) and fix a Markov operator $T : \mathscr{C}(K) \longrightarrow \mathscr{C}(K)$ such that $T(h) = h$ for every $h \in A(K)$. After setting (as in (4.1.16)), for every $i, j = 1, \dots, d$,

$$\alpha_{ij}(x) := T((pr_i - x_i\mathbf{1})(pr_j - x_j\mathbf{1}))(x) = T(pr_i pr_j)(x) - x_i x_j,$$

consider the elliptic second-order differential operator defined by

$$V_T(u)(x) = \frac{1}{2}\sum_{i,j=1}^{d} \alpha_{ij}(x)\frac{\partial^2 u}{\partial x_i \partial x_j}(x) + \sum_{i=1}^{d} \beta_i(x)\frac{\partial u}{\partial x_i}(x) + \gamma(x)u(x)$$

($u \in \mathscr{C}^2(K), x \in K$), which is a first-order perturbation of the operator W_T defined by (4.1.15).

In what follows, we shall prove that $(V_T, \mathscr{C}^2(K))$ is closable and its closure generates a C_0-semigroup.

From now on we shall assume that, for every $i = 1, \dots, d$,

$$\beta_i \text{ is the restriction to } K \text{ of a polynomial of degree at most 1.} \tag{5.6.8}$$

Then the following result holds.

Theorem 5.6.3. *Let K be a convex compact subset of $\mathbf{R}^d$, $T : \mathscr{C}(K) \longrightarrow \mathscr{C}(K)$ be a Markov operator satisfying (3.1.2), $\beta_1, \dots, \beta_d, \gamma \in \mathscr{C}(K)$ satisfying (5.2.1) and (5.6.8) and consider the sequence $(M_n)_{n \geq n_0}$ of modified Bernstein-Schnabl operators associated with T as in (5.2.3). Furthermore, assume that, for every $m \geq 2$,*

$$T(P_m(K)) \subset P_m(K). \tag{5.6.9}$$

Then the operator $(V_T, \mathscr{C}^2(K))$ is closable and its closure $(B_T, D(B_T))$ generates a positive C_0-semigroup $(T(t))_{t \geq 0}$ on $\mathscr{C}(K)$ satisfying (5.6.4) and (5.6.5).

If $\gamma \leq 0$, then $(T(t))_{t \geq 0}$ is, in fact, a Feller semigroup.

Moreover, the generator $(B_T, D(B_T))$ coincides with the closure of the linear operator $Z : D(Z) \longrightarrow \mathscr{C}(K)$ defined by (5.6.6) and (5.6.7).

Finally, $P_\infty(K)$ and $\mathscr{C}^2(K)$ are cores for $(B_T, D(B_T))$ and, if γ is constant, $T(t)(P_m(K)) \subset P_m(K)$ for every $t \geq 0$ and $m \geq 1$.

Proof. We preliminarily consider the sequence $(L_n)_{n \geq n_0}$ of positive linear contractions on $\mathscr{C}(K)$ defined by

$$L_n(f) := B_n \left(f \circ \left(id + \frac{\beta}{n} \right) \right)$$

($f \in \mathscr{C}(K)$, $n \geq n_0$.)

We prove that, under assumption (5.6.9), to the sequence $(L_n)_{n \geq 0}$ it is possible to associate a (uniquely determined) Feller semigroup $(S(t))_{t \geq 0}$ such that

$$S(t)(f) = \lim_{n \to \infty} L_n^{k(n)}(f) \tag{1}$$

uniformly on K, for every $f \in \mathscr{C}(K)$ and for every sequence $(k(n))_{n \geq 1}$ of positive integers such that $\lim\limits_{n \to \infty} k(n)/n = t$.

If (5.6.9) holds true, then $B_n(P_m(K)) \subset P_m(K)$ for every $n, m \geq 1$ (see (4.1.5)). Moreover, since each β_i is a polynomial of degree at most 1, $f \circ \left(id + \frac{\beta}{n} \right) \in P_m(K)$ provided that $f \in P_m(K)$ ($n \geq n_0, m \geq 1$). Therefore $L_n(P_m(K)) \subset P_m(K)$ for every $n \geq n_0$ and $m \geq 1$.

From Theorem 5.5.6, setting $\gamma = 0$, it follows that

$$\lim_{n \to \infty} n(L_n(u) - u) = \frac{1}{2} \sum_{i,j=1}^{d} \alpha_{ij} \frac{\partial^2 u}{\partial x_i \partial x_j} + \sum_{i=1}^{d} \beta_i \frac{\partial u}{\partial x_i}$$

uniformly on K, for every $u \in \mathscr{C}^2(K)$ and hence for every $u \in P_\infty(K)$.

Accordingly, by Theorem 2.2.9, there exists a strongly continuous contraction semigroup $(S(t))_{t \geq 0}$ on $\mathscr{C}(K)$ satisfying (1). Futher, each $S(t)$ $(t \geq 0)$ is positive because of (1).

Moreover, the generator $(L, D(L))$ of the semigroup $(S(t))_{t \geq 0}$ is the closure of the linear operator

$$\Omega(f) := \lim_{n \to \infty} n(L_n(f) - f)$$

defined on

$$D(\Omega) := \left\{ g \in \mathscr{C}(K) \mid \lim_{n\to\infty} n(L_n(g) - g) \text{ exists in } \mathscr{C}(K) \right\}.$$

Note that, if $f \in P_m(K)$ for some $m \geq 1$, then $L_n^k(f) \in P_m(K)$ for every $n \geq n_0$ and $k \geq 1$, so that, given $t \geq 0$ and considered a sequence $(k(n))_{n\geq 1}$ of positive integers such that $\lim_{n\to\infty} k(n)/n = t$, $S(t)(f) = \lim_{n\to\infty} L_n^{k(n)}(f) \in P_m(K)$ since $P_m(K)$ is closed. This reasoning shows that $S(t)(P_\infty(K)) \subset P_\infty(K)$ for every $t \geq 0$ and hence $P_\infty(K)$ is a core for $(L, D(L))$ (see Proposition (2.1.7), (f)).

From Theorem 5.5.6 (with $\gamma = 0$) it also follows that $\mathscr{C}^2(K) \subset D(\Omega) \subset D(L)$ and, for every $u \in \mathscr{C}^2(K)$,

$$L(u) = \Omega(u) = \frac{1}{2}\sum_{i,j=1}^{d} \alpha_{ij}\frac{\partial^2 u}{\partial x_i \partial x_j} + \sum_{i=1}^{d} \beta_i \frac{\partial u}{\partial x_i}.$$

Consider now the bounded operator Γ on $\mathscr{C}(K)$ defined by $\Gamma(f) := \gamma f$ ($f \in \mathscr{C}(K)$) and notice that

$$C_T := L + \Gamma \quad \text{and} \quad D(C_T) := D(L).$$

At this point, the proof runs as the proof of Theorem 5.6.1.

Finally notice that, if γ is constant, then $M_n(P_m(K)) \subset P_m(K)$ for every $n \geq n_0$ and $m \geq 1$, and hence $T(t)(P_m(K)) \subset P_m(K)$ because of (5.6.5). □

Remarks 5.6.4.

1. A thorough inspection of the proof of Theorems 5.6.1 and 5.6.3 shows that conditions (5.6.1) and (5.6.2) (resp., (5.6.8) in the finite dimensional setting) were involved only to prove that the operator $(\Omega, D(\Omega))$ defined by

$$D(\Omega) := \left\{ g \in \mathscr{C}(K) \mid \lim_{n\to\infty} n\left(B_n\left(g \circ \left(id + \frac{\beta}{n}\right)\right) - g\right) \text{ exists in } \mathscr{C}(K) \right\}$$

and, for every $f \in D(\Omega)$,

$$\Omega(f) := \lim_{n\to\infty} n\left(B_n\left(f \circ \left(id + \frac{\beta}{n}\right)\right) - f\right)$$

is closable, its closure generates a C_0-semigroup of positive linear contractions and $P_\infty(K)$ is a core for it.

Therefore, if in some particular cases the above-mentioned generation property is already known to be satisfied, then Theorems 5.6.1 and 5.6.3 hold true without the required assumptions on the function β.

2. In the light of Theorems 5.6.1 and 5.6.3, for the sake of simplicity, let us now denote by $(A, D(A))$ the operators $(B_T^*, D(B_T^*))$ or $(B_T, D(B_T))$ and let us consider the abstract Cauchy problem

$$\begin{cases} \dfrac{\partial u}{\partial t}(x,t) = A(u(\cdot,t))(x) & x \in K, \quad t \geq 0; \\[2ex] u(x,0) = u_0(x) & u_0 \in D(A), \quad x \in K. \end{cases}$$

Then, for every $t \geq 0$, we may approximate its solutions $u(\cdot, t)$ in terms of iterates of the modified Bernstein-Schnabl operators, namely

$$u(x,t) = T(t)(u_0)(x) = \lim_{n\to\infty} M_n^{k(n)}(u_0)(x), \tag{5.6.10}$$

where the limit is uniform with respect to $x \in K$ and $(k(n))_{n\geq 1}$ is a sequence of positive integers such that $\lim_{n\to\infty} k(n)/n = t$.

We recall that, under the assumptions of Theorem 5.6.3, A coincides with the elliptic second-order differential operator V_T defined by (5.5.10) on $\mathscr{C}^2(K)$.

Therefore, if γ is constant and $u_0 \in P_m(K)$ $(m \geq 1)$, then $u(x,t)$ is the unique solution to the Cauchy problem

$$\begin{cases} \frac{\partial u}{\partial t}(x,t) = \frac{1}{2} \sum\limits_{i,j=1}^{d} \alpha_{ij}(x) \frac{\partial^2 u(x,t)}{\partial x_i \partial x_j} + \sum\limits_{i=1}^{d} \beta_i(x) \frac{\partial u(x,t)}{\partial x_i} + \gamma u(x,t) & x \in K, t \geq 0; \\ u(x,0) = u_0(x) & x \in K. \end{cases}$$

Moreover, from the corresponding properties of the modified Bernstein-Schnabl operators we may infer some spatial regularity properties of the solutions (5.6.10).

Below we show a simple example.

Proposition 5.6.5. *Under the assumptions of Theorems 5.6.1 or 5.6.3, further suppose that γ is constant.*

(i) Assume that $T(\mathrm{Lip}(1,1)) \subset \mathrm{Lip}(1,1)$ and $u_0 \in \mathrm{Lip}(M,1)$ for some $M \geq 0$. Then, for every $t \geq 0$,

$$u(\cdot,t) \in \mathrm{Lip}(M \exp(|\gamma| + B)t, 1), \tag{5.6.11}$$

for some $B \geq 0$.

(ii) Assume that

(a) T maps continuous convex functions into (continuous) convex functions.

(b) For every convex function $f \in \mathscr{C}(K)$ and for every $x, y \in K$,

$$\Delta(\tilde{f}; x, y) \geq 0$$

(see (3.4.2) and (3.4.3)).

If $f \in \mathscr{C}(K)$ is a convex function, then $T(t)(f)$ is convex for any $t \geq 0$.

Proof. (i). First of all, we notice that, if (5.6.2) or (5.6.8) holds true, then $\beta \in \mathbb{L}\mathrm{ip}_K(B,1)$ (see (5.4.1)), for some $B \geq 0$. In particular, $B = a$ if (5.6.2) holds true.

Now, it is sufficient to consider, for every $t \geq 0$, a sequence $(k(n))_{n\geq 1}$ of positive integers such that $\lim_{n\to\infty} k(n)/n = t$; unless to replace u_0 with $\frac{1}{M}u_0$, we may assume that $u_0 \in \mathrm{Lip}(1,1)$. Then, by Proposition 5.4.1, $M_n^{k(n)}(u_0) \in$

$\mathrm{Lip}\left(\left(1+\frac{|\gamma|}{n}\right)^{k(n)}\left(1+\frac{B}{n}\right)^{k(n)},1\right)$ and hence, by passing to the limit, $u(\cdot,t) = \lim_{n\to\infty} M_n^{k(n)}(u_0) \in \mathrm{Lip}(\exp(|\gamma|+B)t,1)$.

(ii). It is a consequence of Proposition 5.4.4, since, under assumptions (5.6.2) or (5.6.8), β is an affine function. □

5.7 The unit interval

In this section we present some results about the generation of C_0-semigroups for complete degenerate second-order differential operators defined on suitable domains of $\mathscr{C}([0,1])$ which include several initial-boundary (or lateral) conditions. Particular attention will be paid in finding some cores for such operators. Moreover, in the same spirit of the preceding sections, we discuss the approximation of such semigroups by means of iterates of positive linear operators. As a matter of fact, in this particular setting other tools are available, such as Feller theory (see Subsection 2.3.3), so that it is possible to relax some assumptions. For instance, Theorem 5.6.3 may be obtained without assuming that condition (5.6.8) or (5.6.9) hold true.

For the convenience of the reader, we split up the section into two subsections.

5.7.1 Complete degenerate second-order differential operators on $[0,1]$

Consider the second-order differential operator defined by setting

$$L(u)(x) := \frac{\alpha(x)}{2}u''(x) + \beta(x)u'(x) + \gamma(x)u(x) \tag{5.7.1}$$

($u \in \mathscr{C}^2([0,1])$, $x \in [0,1]$), where $\beta, \gamma \in \mathscr{C}([0,1])$ and $\alpha \in \mathscr{C}([0,1])$ satisfies

$$0 < \alpha(x) \leq x(1-x) \quad \text{for every } 0 < x < 1. \tag{5.7.2}$$

In particular, $\alpha(0) = \alpha(1) = 0$.

We recall that, as we have seen in Section 4.5, if $\alpha \in \mathscr{C}([0,1])$ satisfies (5.7.2), then there always exists a Markov operator T on $\mathscr{C}([0,1])$ such that $\alpha = T(e_2) - e_2$.

This special expression of α will enter into play in Subsection 5.7.2 where we discuss the approximation of the semigroups generated by L on suitable domains.

We preliminarily consider the differential operator

$$A(u)(x) := \begin{cases} \dfrac{\alpha(x)}{2}u''(x) + \beta(x)u'(x) & \text{if } 0 < x < 1; \\ 0 & \text{if } x = 0, 1, \end{cases} \tag{5.7.3}$$

defined on

$$D_V A) := \left\{u \in \mathscr{C}([0,1] \cap \mathscr{C}^2(]0,1[) \mid \lim_{x \to 0^+} A(u)(x) = \lim_{x \to 1^-} A(u)(x) = 0\right\}. \tag{5.7.4}$$

The boundary (lateral) conditions which are incorporated in the domain $D_V(A)$ are also referred to as Ventcel's boundary conditions.

Under suitable assumptions, we shall prove that $(A, D_V(A))$ is the generator of a Markov semigroup on $\mathscr{C}([0,1])$ and that it possesses a subspace of $\mathscr{C}^2([0,1])$ as a core.

To this end, fix $x_0 \in [0,1]$ and set

$$W(x) := \exp\left(-2\int_{x_0}^{x} \frac{\beta(t)}{\alpha(t)}\,dt\right) \qquad (0 \le x \le 1). \tag{5.7.5}$$

We begin with the following preliminary result.

Lemma 5.7.1. *Assume that the function* $\frac{1}{\alpha W}$ *is not integrable near 0 and 1. Fix* $u \in D_V(A)$ *and set* $f := A(u)$. *Then, for every* $\varepsilon > 0$, *there exist* $a, b \in \mathbf{R}$, $0 < a < b < 1$ *and* $g \in \mathscr{C}([0,1])$ *such that*
(i) $\operatorname{Supp}(g) \subset [a,b]$;
(ii) $\|f - g\|_\infty \le \varepsilon$;
(iii) $\int_a^{x_0} \frac{2g(s)}{\alpha(s)W(s)}\,ds = \int_{x_0}^{b} \frac{2g(s)}{\alpha(s)W(s)}\,ds = -u'(x_0)$.

Proof. Indeed, since $f(0) = f(1) = 0$, for a given $\varepsilon > 0$, there exist $a_0, b_0 \in \mathbf{R}$, $0 < a_0 < x_0 < b_0 < 1$, such that $|f(x)| \le \varepsilon/3$ for every $x \in [0, a_0[\cup]b_0, 1]$.

For every $a_1, b_1, a_2, b_2 \in \mathbf{R}$ such that $a_0 < a_1 < b_1 < x_0 < b_0 < a_2 < b_2$, consider the function $\varphi_{1,2} : [0,1] \longrightarrow \mathbf{R}$ such that

$$\varphi_{1,2}(x) := \begin{cases} 1 & \text{if } b_1 \le x \le a_2; \\ e^{-1/(x-a_2)}(1 + e^{-1/(b_2-x)}) & \text{if } a_2 < x < b_2; \\ e^{-1/(x-a_1)}(1 + e^{-1/(b_1-x)}) & \text{if } a_1 < x < b_1; \\ 0 & \text{if } b_2 \le x \le 1 \text{ or } 0 \le x \le a_1. \end{cases}$$

Then $\varphi_{1,2} \in \mathscr{C}^\infty([0,1])$, $\operatorname{Supp}(\varphi_{1,2}) \subset [a_1, b_2]$ and $\varphi_{1,2}$ continuously depends on the parameters a_1, b_1, a_2, b_2.

Analogously, for every $r_1, r_2 \in \mathbf{R}$ such that $0 < r_1 < a_0 < b_0 < r_2 < 1$, consider some functions $\psi_i \in \mathscr{C}^\infty([0,1])$ which continuously depend on the parameters r_i, $i = 1,2$, and such that $\psi_1(x) = 1$ if $x \in [0, r_1]$, $\psi_2(x) = 1$ if $x \in [r_2, 1]$, $\psi_1(x) = 0$ if $x \in [a_0, 1]$ and $\psi_2(x) = 0$ if $x \in [0, b_0]$.

Finally, for every $\gamma_1, \gamma_2 \in [-1, 1]$, set

$$g_{1,2} := \varphi_{1,2}\left(f + \frac{2}{3}\varepsilon\gamma_1\psi_1 + \frac{2}{3}\varepsilon\gamma_2\psi_2\right).$$

Then, for every choice of the parameters a_i, b_i, r_i, γ_i $(i = 1,2)$, $\operatorname{Supp}(g_{1,2}) \subset [a_1, b_2]$ and $\|f - g_{1,2}\|_\infty \le \varepsilon$.

We prove that there exists a suitable choice of the parameters a_i, b_i, r_i, γ_i $(i = 1,2)$ such that, for the corresponding function $g_{1,2}$, one gets

$$\int_{a_1}^{x_0} \frac{2g_{1,2}(s)}{\alpha(s)W(s)}\,ds = \int_{x_0}^{b_2} \frac{2g_{1,2}(s)}{\alpha(s)W(s)}\,ds = -u'(x_0).$$

Indeed, let r_1, a_1, b_1, γ_1 be arbitrarily chosen and consider the function

$$\chi(r_2, a_2, b_2, \gamma_2) := \int_{x_0}^{1} \frac{2g_{1,2}(x)}{\alpha(x)W(x)}\,dx = \int_{x_0}^{b_2} \frac{2g_{1,2}(x)}{\alpha(x)W(x)}\,dx,$$

with $b_0 < r_2 < x_0 < a_2 < b_2$; then, for $\gamma_2 = 1$,

$$\int_{r_2}^{a_2} \frac{2g_{1,2}(x)}{\alpha(x)W(x)}\,dx = 2\int_{r_2}^{a_2} \frac{f(x)+\frac{2}{3}\varepsilon}{\alpha(x)W(x)}\,dx \geq \frac{2}{3}\varepsilon \int_{r_2}^{a_2} \frac{1}{\alpha(x)W(x)}\,dx.$$

Accordingly,

$$\lim_{b_2\to 1^-} \int_{x_0}^{b_2} \frac{2g_{1,2}(x)}{\alpha(x)W(x)}\,dx = \lim_{a_2\to 1^-} \int_{x_0}^{a_2} \frac{2g_{1,2}(x)}{\alpha(x)W(x)}\,dx = +\infty,$$

since $\int_{x_0}^{b_2} \frac{|g_{1,2}(x)|}{\alpha(x)W(x)}\,dx < +\infty$.

Analogously, for $\gamma_2 = -1$,

$$\lim_{b_2\to 1^-} \int_{x_0}^{b_2} \frac{2g_{1,2}(x)}{\alpha(x)W(x)}\,dx = -\infty.$$

This yields that χ is surjective, so that there exist a_2, b_2, r_2, γ_2 such that

$$\int_{x_0}^{b_2} \frac{2g_{1,2}(x)}{\alpha(x)W(x)}\,dx = -u'(x_0).$$

Analogously, there exist a_1, b_1, r_1, γ_1 such that

$$\int_{a_1}^{x_0} \frac{2g_{1,2}(x)}{\alpha(x)W(x)}\,dx = -u'(x_0).$$

For $a := a_1$, $b := b_2$ and $g := g_{1,2}$, we get statements (i), (ii) and (iii) above. □

After these preliminaries, keeping in mind Feller classification to which we refer for unexplained symbols (see Subsection 2.3.3), we can state the following result.

Theorem 5.7.2. *Under the above notation and assumptions, if 0 and 1 are not entrance end-points, then $(A, D_V(A))$ generates a Markov semigroup on $\mathscr{C}([0,1])$.*

Moreover, if 0 and 1 are both exit end-points, then the subspace

$$\mathscr{C}_*^2([0,1]) := \{u \in \mathscr{C}^2([0,1]) \mid u \textit{ is constant on a neighborhood of } 0 \textit{ and } 1\} \tag{5.7.6}$$

is a core for $(A, D_V(A))$.

Proof. The first part of the statement is a direct consequence of Theorem 2.3.19.

By passing to the second part, assume that 0 and 1 are exit end-points. In order to show that $\mathscr{C}_*^2([0,1])$ is a core for $(A, D_V(A))$, first of all we remark that the function $\frac{1}{\alpha W}$ is not integrable near 0 and 1.

Otherwise, since 0 (resp., 1) is an exit end-point, we have that $R \in L^1(0, x_0)$ (resp., $R \in L^1(x_0, 1)$) so that, by means of Remark 2.3.18, we get that $W \in$

$L^1(0,x_0)$ (resp., $W \in L^1(x_0,1)$). On the other hand, $W \in L^1(0,x_0)$ and $\frac{1}{\alpha W} \in L^1(0,x_0)$ (resp., $W \in L^1(x_0,1)$ and $\frac{1}{\alpha W} \in L^1(x_0,1)$) imply that 0 (resp., 1) is a regular end-point (see, again, Remark 2.3.18), hence a contradiction.

Consider now $u \in D_V(A)$ and set $f := A(u)$; then, since

$$f = A(u) = \frac{\alpha}{2} W \frac{\mathrm{d}}{\mathrm{d}x}\left(\frac{u'}{W}\right), \tag{1}$$

integrating both sides of (1) between x_0 and x, we get

$$u'(x) = W(x)\left(u'(x_0) + \int_{x_0}^{x} \frac{2f(t)}{\alpha(t)W(t)}\,dt\right) \qquad (0 < x < 1)$$

and, integrating again between x_0 and x,

$$u(x) = u(x_0) + u'(x_0)\int_{x_0}^{x} W(t)\,dt + \int_{x_0}^{x} W(t)\int_{x_0}^{t} \frac{2f(s)}{\alpha(s)W(s)}\,ds\,dt. \tag{2}$$

Fix $\varepsilon > 0$; then, by means of Lemma 5.7.1, there exist $a, b \in \mathbf{R}$, $0 < a < b < 1$ and $g \in \mathscr{C}([0,1])$ such that

(i) $\operatorname{Supp}(g) \subset [a,b]$;

(ii) $\|f - g\|_\infty \le \varepsilon$;

(iii) $\int\limits_a^{x_0} \frac{2g(s)}{\alpha(s)W(s)}\,ds = \int\limits_{x_0}^{b} \frac{2g(s)}{\alpha(s)W(s)}\,ds = -u'(x_0)$.

Set now, for every $x \in [0,1]$,

$$v(x) = u(x_0) + u'(x_0)\int_{x_0}^{x} W(t)\,dt + \int_{x_0}^{x} W(t)\int_{x_0}^{t} \frac{2g(s)}{\alpha(s)W(s)}\,ds\,dt.$$

Then $v \in \mathscr{C}([0,1]) \cap \mathscr{C}^2(]0,1[)$ and

$$v'(x) = W(x)\left(u'(x_0) + \int\limits_{x_0}^{x} \frac{2g(t)}{\alpha(t)W(t)}\,dt\right) \qquad (0 < x < 1).$$

Since $g = 0$ on $]0,a[$ and $]b,1[$, then $v'(x) = 0$ for any $x \in]0,a[$ and $x \in]b,1[$, so that $v \in \mathscr{C}^2_*([0,1]) \subset D_V(A)$.

Moreover, obviously,

$$\|A(u) - A(v)\|_\infty = \|f - g\|_\infty \le \varepsilon.$$

On the other hand, taking (2) into account, it is easy to prove that, for every $x \in [0,1]$,

$$|u(x) - v(x)| \le \int_{x_0}^{x} W(t)\left|\int_{x_0}^{t} \frac{2|f(s) - g(s)|}{\alpha(s)W(s)}\,ds\right|\,dt \le C\varepsilon,$$

where $C := \int\limits_{x_0}^{x} W(t)\left|\int\limits_{x_0}^{t} \frac{2}{\alpha(s)W(s)}\,ds\right|\,dt < +\infty$. □

Examples 5.7.3.

1. *Assume that there exists a neighborhood* U *of* 0 *such that* $\max\limits_{x\in U}\beta(x)\le 0$*; then* 0 *is not an entrance end-point.*

Indeed, set $M := \max\limits_{x\in[0,x_0]}\beta(x)\le 0$, for a suitable $x_0\in]0,1]$. By means of Remark 2.3.18, we prove that $(\alpha W)^{-1}\notin L^1(0,x_0)$, where the function W is given by (5.7.5).

First of all, we notice that, for every $t\in]0,x_0[$,

$$-\frac{2\beta(t)}{\alpha(t)}\ge-\frac{2M}{t(1-t)}.$$

Consequently, there exists a constant $K\ge 0$ such that, for every $x\in]0,x_0[$,

$$\frac{1}{\alpha(x)W(x)}\ge\frac{K}{x^{1-2M}(1-x)^{1+2M}}.$$

Since $M\le 0$, it follows that 0 is not an entrance end-point.

2. *Analogously, if there exists* $x_1\in[0,1[$ *such that* $m_1 := \min\limits_{x\in[x_1,1]}\beta(x)\ge 0$*, then* 1 *is not an entrance point.*

Consider now the operator

$$L(u)=A(u)+\gamma u,$$

(see (5.7.1)) defined on $D_V(L) := D_V(A)$.

The following result, which is a straightforward consequence of Theorems 2.1.13, 2.3.8 and 5.7.2, holds.

Theorem 5.7.4. *Under the assumptions of Theorem 5.7.2,* $(L, D_V(L))$ *is the generator of a* C_0*-semigroup* $(T(t))_{t\ge0}$ *on* $\mathscr{C}([0,1])$ *such that* $\|T(t)\|\le\exp(\|\gamma\|_\infty t)$ *for every* $t\ge 0$*. Moreover,* $\mathscr{C}^2_*([0,1])$ *is a core for* $(L, D_V(L))$*. Finally, if* $\gamma\le 0$*, then* $(T(t))_{t\ge0}$ *is a Feller semigroup.*

In the final part of this subsection, we will investigate under what conditions the operator A (and, consequently, the operator L) is the generator of a C_0-semigroup on more general domains.

More precisely, consider the operator

$$A(u)(x) := \frac{\alpha(x)}{2}u''(x)+\beta(x)u'(x)$$

($u\in\mathscr{C}([0,1])\cap\mathscr{C}^2(]0,1[)$, $0<x<1$), defined, indifferently, on

$$D_M(A) := \left\{u\in\mathscr{C}([0,1])\cap\mathscr{C}^2(]0,1[)\mid A(u)\in\mathscr{C}([0,1])\right\},\tag{5.7.7}$$

$$D_{VM}(A) := \left\{u\in D_M(A)\mid \lim_{x\to0^+}A(u)(x)=0\right\}\tag{5.7.8}$$

or

$$D_{MV}(A) := \left\{ u \in D_M(A) \mid \lim_{x \to 1^-} A(u)(x) = 0 \right\}. \tag{5.7.9}$$

The subspace $D_M(A)$ is referred to as the maximal domain for A, whereas $D_{MV}(A)$ and $D_{VM}(A)$ are often referred to as mixed domains for A.

Theorem 2.3.19 immediately yields the following result.

Theorem 5.7.5. *Under the above notation and assumptions, if 0 and 1 are entrance or natural end-points, then $(A, D_M(A))$ generates a Markov semigroup on $\mathscr{C}([0,1])$.*

If 0 is not an entrance end-point and 1 is an entrance or a natural end-point, then $(A, D_{VM}(A))$ generates a Markov semigroup on $\mathscr{C}([0,1])$.

Finally, if 0 is an entrance or a natural end-point and 1 is not an entrance end-point, then $(A, D_{MV}(A))$ generates a Markov semigroup on $\mathscr{C}([0,1])$.

Consider the complete operator

$$L(u) = A(u) + \gamma u, \tag{5.7.10}$$

(see, again, (5.7.1)) defined indifferently on $D_M(L) := D_M(A)$, $D_{VM}(L) := D_{VM}(A)$ and $D_{MV}(L) := D_{MV}(A)$.

Then, from Theorems 2.1.13 and 2.3.8, we get the following result.

Theorem 5.7.6. *Under the assumptions of Theorem 5.7.5, the operators $(L, D_M(L))$, $(L, D_{VM}(L))$ and $(L, D_{MV}(L))$ generate C_0-semigroups $(T(t))_{t\geq 0}$ on $\mathscr{C}([0,1])$ such that $\|T(t)\| \leq \exp(\|\gamma\|_\infty t)$ for every $t \geq 0$.*

Moreover, if $\gamma \leq 0$, then $(T(t))_{t\geq 0}$ are Feller semigroups.

In order to present some particular situations in which Theorem 5.7.6 applies, we have to introduce some further assumptions on the function α.

As a matter of fact, from now on, we assume that α is differentiable at 0 and at 1, with $\alpha'(0) \neq 0$ and $\alpha'(1) \neq 0$. Clearly, if this is the case, then $\alpha'(0) > 0$ and $\alpha'(1) < 0$.

Under these assumptions it is possible to continuously extend the function λ defined by (4.2.8) to the whole interval $[0,1]$. We continue to denote by λ this continuous extension; more precisely, we have that

$$\lambda(x) := \begin{cases} \alpha'(0) & \text{if } x = 0; \\[2ex] \dfrac{\alpha(x)}{x(1-x)} & \text{if } 0 < x < 1; \\[2ex] -\alpha'(1) & \text{if } x = 1. \end{cases} \tag{5.7.11}$$

Moreover, $\alpha(x) = \lambda(x)x(1-x)$ $(0 \leq x \leq 1)$ and $0 \leq \lambda \leq 1$.

Finally, for the sake of brevity, we set

$$D(L) := D(A) := \begin{cases} D_M(A) & \text{if } \beta(0) \geq \frac{1}{2}\alpha'(0) \text{ and } \beta(1) \leq \frac{1}{2}\alpha'(1); \\ D_{VM}(A) & \text{if } \beta(0) < \frac{1}{2}\alpha'(0) \text{ and } \beta(1) \leq \frac{1}{2}\alpha'(1); \\ D_{MV}(A) & \text{if } \beta(0) \geq \frac{1}{2}\alpha'(0) \text{ and } \beta(1) > \frac{1}{2}\alpha'(1). \end{cases} \tag{5.7.12}$$

Consider now the differential operator

$$B(u)(x) := \frac{x(1-x)}{2}u''(x) + \eta(x)u'(x) \quad (0 < x < 1), \tag{5.7.13}$$

where $\eta := \frac{\beta}{\lambda}$, defined on $D(B) := D(A)$.

With the following result, due to Attalienti and Campiti (see [42, pp. 120-121]), we present sufficient conditions in order that, under assumptions (5.7.12), the operator $(A, D(A))$ generates a Feller semigroup on $\mathscr{C}([0,1])$.

To this aim, we recall that, if f, g are real functions defined on a subset X of $\mathbf{R}$, y_0 is a cluster point of X and $g \neq 0$ in a suitable neighborhood of y_0, we say that f and g are *equivalent infinitesimals* in y_0, and we write $f \approx g$ as $x \to y_0$, if

(a) There exist $\lim_{x \to y_0} f(x) = \lim_{x \to y_0} g(x)$;

(b) $\lim_{x \to y_0} \frac{f(x)}{g(x)} = 1$.

Theorem 5.7.7. *Assume that $\frac{\beta}{\lambda}$ is Hölder continuous at 0 and at 1. Then, under assumptions (5.7.12), the operator $(A, D(A))$ is the generator of a Markov semigroup on $\mathscr{C}([0,1])$.*

Therefore, $(L, D(L))$ is the generator of a C_0-semigroup $(T(t))_{t\geq 0}$ on $\mathscr{C}([0,1])$, which is a Feller semigroup if $\gamma \leq 0$.

Proof. Under assumptions (5.7.12), it was proved in [42, pp. 120-121] that the operator $(B, D(B))$ (see (5.7.13)) is the generator of a Feller semigroup on $\mathscr{C}([0,1])$.

Indeed, as $x \to 0^+$, using Feller terminology for $x_0 = \frac{1}{2}$ (see (2.3.17) and (2.3.18)), since η is Hölder continuous at 0, we have that

$$W(x) \approx \frac{K}{x^{2\eta(0)}}, \tag{1}$$

$$Q(x) \approx \begin{cases} \frac{2}{1-2\eta(0)}\left(1 - \frac{M}{x^{1-2\eta(0)}}\right) & \text{if } \eta(0) \neq \frac{1}{2}; \\ 2(\log x + \log 2) & \text{if } \eta(0) = \frac{1}{2}, \end{cases} \tag{2}$$

and

$$R(x) \approx \begin{cases} \dfrac{1}{\eta(0)} - \dfrac{N}{\eta(0)x^{2\eta(0)}} & \text{if } \eta(0) \neq 0; \\ \\ 2(\log x + \log 2) & \text{if } \eta(0) = 0, \end{cases} \tag{3}$$

where K, M, N are strictly positive constants depending on $\eta(0)$, only.

By replacing x with $1-x$, similar asymptotic relations for W, R and Q hold as $x \to 1^-$.

Accordingly, if $\beta(0) \geq \frac{1}{2}\alpha'(0)$, then $Q \in L^1\left(0, \frac{1}{2}\right)$ and $R \notin L^1\left(0, \frac{1}{2}\right)$, so that 0 is an entrance end-point, as well as, if $\beta(1) \leq \frac{1}{2}\alpha'(1)$, then 1 is an entrance end-point.

Moreover, if $0 < \beta(0) < \frac{1}{2}\alpha'(0)$, then $Q, R \in L^1\left(0, \frac{1}{2}\right)$ and, consequently, 0 is a regular end-point and, analogously, if $\frac{1}{2}\alpha'(1) < \beta(1) < 0$, then 1 is a regular end-point.

Finally, if $\beta(0) \leq 0$ (resp., $\beta(1) \geq 0$), then $Q \notin L^1\left(0, \frac{1}{2}\right)$ and $R \in L^1\left(0, \frac{1}{2}\right)$ and 0 is an exit end-point (resp., 1 is an exit end-point).

Hence, under assumptions (5.7.12), Theorem 2.3.19 applies and, accordingly, $(B, D(B))$ is the generator of a Markov semigroup on $\mathscr{C}([0,1])$.

Moreover, $A = \lambda B$ and $0 < \lambda(x) \leq 1$ for every $x \in [0,1]$. Consequently, by applying Theorem 2.3.11, $(A, D(A))$ is the generator of a Markov semigroup on $\mathscr{C}([0,1])$.

The last part of the statement is a consequence of Theorems 2.1.13 and 2.3.8. □

Finally, we prove that, under suitable assumptions, $\mathscr{C}^2([0,1]) \cap D(A)$ is a core for $(L, D(L))$. The next result will be useful in this respect.

Proposition 5.7.8. *Suppose that $\frac{\beta}{\lambda}$ is Hölder continuous at 0 and at 1. Then, for every $u \in \mathscr{C}([0,1] \cap \mathscr{C}^2(]0,1[)$ such that $B(u)$ (see (5.7.13)) is continuous at 0 (resp., at 1),*

$$\lim_{x\to 0^+} xu'(x) = 0 \qquad (\textit{resp.,} \lim_{x\to 1^-} (1-x)u'(x) = 0). \tag{5.7.14}$$

Proof. For the sake of brevity, we prove only the first part of (5.7.14). The proof for the limit as $x \to 1^-$ is similar.

Indeed, near $x = 0$, we have $B(u)(x) \approx \frac{x}{2}u''(x) + \eta(0)u'(x)$; therefore, integrating by parts,

$$\int_x^{\frac{1}{2}} B(u)(t)\,dt \approx \frac{1}{4}u'\left(\frac{1}{2}\right) - \frac{x}{2}u'(x) - \left(\eta(0) - \frac{1}{2}\right)\left(u\left(\frac{1}{2}\right) - u(x)\right),$$

where the term on the left hand side is convergent as $x \to 0^+$, since $B(u)$ is continuous.

Then there exists $l := \lim_{x\to 0^+} xu'(x) \in \mathbf{R}$. Moreover, $l = 0$; otherwise, if $l \neq 0$, then $u'(x) \approx \frac{l}{x}$ as $x \to 0^+$, so that $u(x) \approx M \log x$ as $x \to 0^+$ and this contradicts the fact that $u \in \mathscr{C}([0,1])$. □

Proposition 5.7.9. *Suppose that $\frac{\beta}{\lambda}$ is Hölder continuous and that $\beta(0) \geq \frac{\alpha'(0)}{2}$ (resp., $\beta(1) \leq \frac{\alpha'(1)}{2}$)). If $u \in \mathscr{C}([0,1]) \cap \mathscr{C}^2(]0,1[)$ and $B(u)$ is continuous at 0 (resp., $B(u)$ continuous at 1), then u is differentiable at 0 (resp., at 1) and*

$$\lim_{x\to 0^+} xu''(x) = 0 \quad (\textit{resp.}, \lim_{x\to 1^-} (1-x)u''(x) = 0). \tag{5.7.15}$$

Proof. For the sake of brevity, we limit ourselves to show the first part of the statement; arguing in a similar way, one can prove also the remaining part.

So, assume that $\beta(0) \geq \frac{\alpha'(0)}{2}$; fix $u \in \mathscr{C}([0,1]) \cap \mathscr{C}^2(]0,1[)$ with $B(u)$ continuous at 0 and set $f := B(u)$; then it is easy to prove that, for every $0 < x < 1$,

$$f(x) = \frac{x(1-x)}{2} W(x) \left(\frac{u'(x)}{W(x)}\right)'$$

so that, integrating from a to x, where $0 < a < x$, we get

$$\int_a^x \frac{2f(t)}{t(1-t)W(t)}\, dt = \frac{u'(x)}{W(x)} - \frac{u'(a)}{W(a)}. \tag{1}$$

Since $\beta(0) \geq \frac{\alpha'(0)}{2}$, then 0 is an entrance end-point (see the proof of Theorem 5.7.7). Thus the function $\frac{1}{t(1-t)W(t)}$ $(0 < t < 1)$ is integrable in a neighborhood of 0 (see Remark 2.3.18).

Accordingly, there exists $\lim_{a\to 0^+} \frac{u'(a)}{W(a)} \in \mathbf{R}$ and, in particular, $l = 0$.

In fact, otherwise, as $a \to 0^+$,

$$u'(a) \approx l\,W(a) \approx l \frac{K}{a^{2\eta(0)}}$$

(see formula (1) in the proof of Theorem 5.7.7).

Therefore, we would have that, for a suitable constant $M \in \mathbf{R}$, $M \neq 0$, as $a \to 0^+$,

$$u(a) \approx \begin{cases} M a^{1-2\eta(0)} & \text{if } \eta(0) > \frac{1}{2}; \\ \\ M \log a & \text{if } \eta(0) = \frac{1}{2} \end{cases}$$

and this leads to a contradiction, since $u \in \mathscr{C}([0,1])$.

Then, passing to the limit as $a \to 0^+$ in (1), we have

$$u'(x) = W(x) \int_0^x \frac{2f(t)}{t(1-t)W(t)}\, dt \approx \frac{K}{x^{2\eta(0)}} \int_0^x \frac{2f(t)}{t(1-t)W(t)}\, dt$$

for every $0 < x < 1$. By de L'Hôspital rule, $\lim_{x\to 0^+} u'(x) = \frac{f(0)}{\eta(0)}$, so that u is differentiable at 0.

From this it follows that there exists $m := \lim_{x\to 0^+} xu''(x) \in \mathbf{R}$ since $B(u)$ is continuous at 0. Moreover, $m = 0$; otherwise, $u''(x) \approx \frac{m}{x}$ as $x \to 0^+$ and, hence, $u'(x) \approx N \log x$ as $x \to 0^+$ (for a suitable constant $N \in \mathbf{R}$), and this contradicts the fact that u' is differentiable at 0. □

Remark 5.7.10. Under the assumptions above, if $\beta(0) \geq \frac{\alpha'(0)}{2}$ (resp., $\beta(1) \leq \frac{\alpha'(1)}{2}$), since $u \in \mathscr{C}([0,1]) \cap \mathscr{C}^2(]0,1[)$ and u is differentiable at 0 (resp., at 1), then there exists $\delta > 0$ such that $u \in \mathscr{C}^1([0,\delta])$ (resp., $u \in \mathscr{C}^1([\delta,1])$).

In particular, if $\beta(0) \geq \frac{\alpha'(0)}{2}$, $\beta(1) \leq \frac{\alpha'(1)}{2}$ and $u \in D_M(B)$, then $u \in \mathscr{C}^1([0,1])$.

We pass now to consider other classes of boundary conditions. To this end, we recall that, given two real functions f, g defined on a subset X of $\mathbf{R}$ and a cluster point y_0 of X, the symbol $f(x) = \mathcal{O}(g(x))$ as $x \to y_0$ denotes the usual *Landau symbol.*

Proposition 5.7.11. *Suppose that $\frac{\beta}{\lambda}$ is Hölder continuous at 0 and at 1 and fix $u \in \mathscr{C}([0,1]) \cap \mathscr{C}^2(]0,1[)$. Then*

(i) If $\beta(0) < 0$ and $\lim_{x\to 0^+} B(u)(x) = 0$, then u is differentiable at 0, $u'(0) = 0$ and $\lim_{x\to 0^+} xu''(x) = 0$. Analogously, if $\beta(1) > 0$ and $\lim_{x\to 1^-} B(u)(x) = 0$, then u is differentiable at 1, $u'(1) = 0$ and $\lim_{x\to 1^-} (1-x)u''(x) = 0$.

(ii) If $\beta(0) = 0$, $\beta(x) = \mathcal{O}(x)$ as $x \to 0^+$ and $\lim_{x\to 0^+} B(u)(x) = 0$, then $\lim_{x\to 0^+} xu''(x) = 0$. Analogously, if $\beta(1) = 0$, $\beta(x) = \mathcal{O}(1-x)$ as $x \to 1^-$ and $\lim_{x\to 1^-} B(u)(x) = 0$, then $\lim_{x\to 1^-} (1-x)u''(x) = 0$.

Proof. We shall prove the first part of statement (i).

To this end, assume that $\beta(0) < 0$; hence, there exists $x_0 \in]0,1[$ such that $\beta(x) < 0$ for every $x \in]0, x_0[$.

Fix now $u \in \mathscr{C}([0,1]) \cap \mathscr{C}^2(]0,1[)$ such that $\lim_{x\to 0^+} B(u)(x) = 0$; then, for every $x \in]0, x_0[$,

$$|u'(x)| \leq W(x)\left(|u'(x_0)| + \int_x^{x_0} \frac{2|B(u)(t)|}{\alpha(t)W(t)}\,dt\right).$$

From formulae (2) and (3) in the proof of Theorem 5.7.7, it follows that 0 is an exit end-point, so that, in particular, $(\alpha W)^{-1} \notin L^1(0, x_0)$.

Accordingly, $\lim_{x\to 0^+} \int_x^{x_0} \frac{|B(u)(t)|}{\alpha(t)W(t)}\,dt = +\infty$; moreover, formula (1) in the proof of Theorem 5.7.7 implies that $\lim_{x\to 0^+} W(x) = 0$; by applying de L'Hôspital rule,

$$\lim_{x\to 0^+} W(x)\left(u'(x_0) + \int_x^{x_0} \frac{2|B(u)(t)|}{\alpha(t)W(t)}\,dt\right) = \lim_{x\to 0^+} \frac{2|B(u)(x)|}{\beta(x)} = 0.$$

Hence u is differentiable in 0 and $u'(0) = 0$.

From this and the fact that $\lim_{x\to 0^+} B(u)(x) = 0$, it obviously follows that $\lim_{x\to 0^+} xu''(x) = 0$.

The proof of (ii) is straightforward, taking Proposition 5.7.8 into account. □

Before stating the next result, the following lemma will be needed.

Lemma 5.7.12. *For every $v \in \mathscr{C}^2([a,b])$ and $c > 0$, $c \le b - a$,*

$$\|v'\|_\infty \le c\|v''\|_\infty + \frac{6}{c}\|v\|_\infty. \tag{5.7.16}$$

Proof. Fix $v \in \mathscr{C}^2([a,b])$ and $c > 0$, $c \le b-a$. If $x_0 \in [a,b]$, consider $s,t \in [a,b] \cap \left]x_0 - \frac{c}{2}, x_0 + \frac{c}{2}\right[$ such that $|s-t| > \frac{c}{3}$.

Then, by Lagrange theorem, there exists $x \in]a,b[\cap \left]x_0 - \frac{c}{2}, x_0 + \frac{c}{2}\right[$ such that

$$v'(x) = \frac{v(t)-v(s)}{t-s}.$$

Accordingly,

$$v'(x_0) = v'(x) + \int_x^{x_0} v''(r)\,dr = \frac{v(t)-v(s)}{t-s} + \int_x^{x_0} v''(r)\,dr$$

and

$$|v'(x_0)| \le \frac{2\|v\|_\infty}{|t-s|} + \|v''\|_\infty |x - x_0| \le \frac{6}{c}\|v\|_\infty + c\|v''\|_\infty.$$

Since x_0 was arbitrarily chosen, this completes the proof. □

Theorem 5.7.13. *Suppose that $\frac{\beta}{\lambda}$ is Hölder continuous and that one of the following assumptions holds true:*
(i) $\beta(0) \ge \frac{\alpha'(0)}{2}$ and $\beta(1) \le \frac{\alpha'(1)}{2}$.
(ii) $\beta(1) \le \frac{\alpha'(1)}{2}$ and $\beta(0) < 0$.
(iii) $\beta(1) \le \frac{\alpha'(1)}{2}$, $\beta(0) = 0$ and $\beta(x) = \mathscr{O}(x)$ as $x \to 0^+$.
(iv) $\beta(0) \ge \frac{\alpha'(0)}{2}$ and $\beta(1) > 0$.
(v) $\beta(0) \ge \frac{\alpha'(0)}{2}$, $\beta(1) = 0$ and $\beta(x) = \mathscr{O}(1-x)$ as $x \to 1^-$.
Then $\mathscr{C}^2([0,1]) \cap D(A)$ is a core for $(A, D(A))$ and, hence, for $(L, D(L))$.

Proof. Let us consider the operator $(B, D(B))$ introduced in (5.7.13) and defined on $D(B) = D(A)$; then, under the above assumptions, $\mathscr{C}^2([0,1]) \cap D(B)$ is a core for $(B, D(B))$.

Assume that (i) holds true and fix $u \in D_M(B)$. Then $u \in \mathscr{C}^1([0,1])$ according to Proposition 5.7.9 and Remark 5.7.10. Moreover, set, for every $n \ge 1$ and $0 \le$

$x \le 1$,

$$u_n(x) := \begin{cases} u\left(\frac{1}{n}\right) + u'\left(\frac{1}{n}\right)\left(x - \frac{1}{n}\right) + \frac{1}{2}u''\left(\frac{1}{n}\right)\left(x - \frac{1}{n}\right)^2 & 0 \le x \le \frac{1}{n}; \\ u(x) & \frac{1}{n} \le x \le 1 - \frac{1}{n}; \\ u\left(1 - \frac{1}{n}\right) + u'\left(1 - \frac{1}{n}\right)\left(x - 1 + \frac{1}{n}\right) & \\ +\frac{1}{2}u''\left(1 - \frac{1}{n}\right)\left(x - 1 + \frac{1}{n}\right)^2 & 1 - \frac{1}{n} \le x \le 1. \end{cases} \tag{1}$$

Then $u_n \in \mathscr{C}^2([0,1]) \cap D(B)$ for every $n \ge 1$.

Given $\varepsilon > 0$, since both u and $B(u)$ are uniformly continuous, there exists $\delta > 0$ such that for every $x, y \in [0,1]$, $|x - y| \le \delta$,

$$|u(x) - u(y)| \le \frac{\varepsilon}{3} \text{ and } |B(u)(x) - B(u)(y)| \le \frac{\varepsilon}{3}.$$

By Propositions 5.7.8 and 5.7.9, there exists a positive integer $\nu \ge 1$, $\nu \ge 1/\delta$, such that, for every $n \ge \nu$,

$$\frac{1}{n}\left|u'\left(\frac{1}{n}\right)\right| \le \frac{\varepsilon}{3}, \quad \frac{1}{n}\left|u'\left(1 - \frac{1}{n}\right)\right| \le \frac{\varepsilon}{3}.$$

$$\frac{1}{2n^2}\left|u''\left(\frac{1}{n}\right)\right| \le \frac{\varepsilon}{3}, \quad \frac{1}{2n^2}\left|u''\left(1 - \frac{1}{n}\right)\right| \le \frac{\varepsilon}{3}.$$

Accordingly, $\lim_{n\to\infty} \|u_n - u\|_\infty = 0$.

We pass now to prove that $\lim_{n\to\infty} \|B(u_n) - B(u)\|_\infty = 0$.

Fix $\varepsilon > 0$ and set $K := \|\beta\|_\infty$; since $u \in \mathscr{C}^1([0,1])$ and (5.7.15) holds true, there exists $\delta > 0$ such that $|u'(x) - u'(y)| \le \frac{\varepsilon}{4(K+1)}$ for every $x, y \in [0,1]$, $|x - y| \le \delta$, $x|u''(x)| \le \frac{\varepsilon}{4}$ for every $0 \le x \le \delta$ and $(1 - x)|u''(x)| \le \frac{\varepsilon}{4}$ for every $\delta \le x \le 1$.

In particular, there exists $\nu \ge 1$, $\nu \ge 1/\delta$, such that, for every $n \ge \nu$,

$$\frac{1}{n}\left|u''\left(\frac{1}{n}\right)\right| \le \frac{\varepsilon}{4(K+1)}, \quad \frac{1}{n}\left|u''\left(1 - \frac{1}{n}\right)\right| \le \frac{\varepsilon}{4(K+1)}.$$

Now fix $n \ge \nu$ and $x \in \left[0, \frac{1}{n}\right]$. Then

$$\begin{aligned} |B(u_n)(x) - B(u)(x)| &\le x\left|u''\left(\frac{1}{n}\right) - u''(x)\right| \\ &+ |\beta(x)|\left|u'\left(\frac{1}{n}\right) - u'(x)\right| + |\beta(x)|\left|u''\left(\frac{1}{n}\right)\left(x - \frac{1}{n}\right)\right| \\ &\le \frac{1}{n}\left|u''\left(\frac{1}{n}\right)\right| + x|u''(x)| + K\left|u'\left(\frac{1}{n}\right) - u'(x)\right| + \frac{K}{n}\left|u''\left(\frac{1}{n}\right)\right| \le \varepsilon. \end{aligned}$$

Analogously, if $x \in \left]\frac{1}{n}, 1\right]$, we have

$$|B(u_n)(x) - B(u)(x)| \le \varepsilon$$

and this completes the proof.

Assume that (ii) holds true and fix $u \in D_{VM}(B)$. Moreover, for every $n \geq 1$, set $\overline{u}_n(x) := g_n(x) + u_n(x)$, where u_n is defined as above and

$$g_n(x) := \begin{cases} \frac{1}{3}n\left(u''\left(\frac{1}{n}\right) - nu'\left(\frac{1}{n}\right)\right)\left(x - \frac{1}{n}\right)^3 & 0 \leq x \leq \frac{1}{n}; \\ 0 & \frac{1}{n} \leq x \leq 1. \end{cases}$$

It is easy to check that $\overline{u}_n \in \mathscr{C}^2([0,1]) \cap D_{VM}(B)$ and $\overline{u}_n'(0) = 0$.

Moreover, for every $x \in [0,1]$,

$$|\overline{u}_n(x) - u(x)| \leq |g_n(x)| + |u_n(x) - u(x)|$$

and

$$|B(\overline{u}_n)(x) - B(u)(x)| \leq |B(g_n)(x)| + |B(u_n)(x) - B(u)(x)|.$$

Fix now $\varepsilon > 0$; since $u \in D_{VM}(B)$, by virtue of Propositions 5.7.8 and 5.7.11, there exists $\nu \geq 1$ such that, for every $n \geq \nu$,

$$u'\left(\frac{1}{n}\right) \leq \varepsilon, \quad \frac{1}{n}u''\left(\frac{1}{n}\right) \leq \varepsilon \quad \text{and } \frac{1}{n}u'\left(\frac{1}{n}\right) \leq \varepsilon.$$

Hence, for every $n \geq \nu$ and $0 \leq x \leq \frac{1}{n}$,

$$|g_n(x)| \leq \frac{1}{3n^2}\left|u''\left(\frac{1}{n}\right)\right| + \frac{1}{3n}\left|u'\left(\frac{1}{n}\right)\right| \leq \varepsilon$$

and

$$|B(g_n)(x)| \leq \frac{2}{n}\left|u''\left(\frac{1}{n}\right)\right| + \left|u'\left(\frac{1}{n}\right)\right| + \frac{K}{n}\left|u''\left(\frac{1}{n}\right)\right| + K\left|u'\left(\frac{1}{n}\right)\right| \leq (2K+3)\varepsilon,$$

where $K := \|\beta\|_\infty$.

This shows that $\lim_{n\to\infty} \overline{u}_n = u$ and $\lim_{n\to\infty} B(\overline{u}_n) = B(u)$, since $u \in D_{VM}(B) \subset D_M(B)$ and $(u_n)_{n\geq 1}$ converges to u in the graph norm.

Assume now that $\beta(1) \leq \frac{\alpha'(1)}{2}$, $\beta(0) = 0$ and $\beta(x) = \mathcal{O}(x)$ as $x \to 0^+$.

Fix $u \in D_{VM}(B)$ and consider the sequence $(u_n)_{n\geq 1}$ defined by (1). Then, for every $n \geq 1$, $u_n \in \mathscr{C}^2([0,1]) \cap D_{VM}(B)$ and, arguing as in the first part of the proof and using statement (ii) of Proposition 5.7.11 instead of Proposition 5.7.9, one can show that $\lim_{n\to\infty} \|u_n - u\|_\infty = 0$.

In order to prove that $\lim_{n\to\infty} \|B(u_n) - B(u)\|_\infty = 0$, fix $\varepsilon > 0$; then, by means of statement (ii) of Proposition 5.7.11, there exists $\delta > 0$ such that

$$x|u''(x)| \leq \varepsilon \text{ for every } x \in [0,\delta].$$

Moreover, according to the hypothesis, we may assume that there exists $L > 0$ such that

$$|\beta(x)| \leq Lx \text{ for every } x \in [0,\delta]. \tag{2}$$

As in the first part of the proof, we can show that there exists $\nu \geq 1$, $\nu \geq 1/\delta$ such that, for every $n \geq \nu$, $\|u_n - u\|_\infty \leq \varepsilon$ and, moreover, for every $x \in \left]\frac{1}{n}, 1\right]$,

$$|B(u_n)(x) - B(u)(x)| \leq \varepsilon.$$

Consider now $x \in \left]0, \frac{1}{n}\right]$. Then, taking (2) into account,

$$\begin{aligned}
&|B(u_n)(x) - B(u)(x)| \leq x\left|u''\left(\frac{1}{n}\right) - u''(x)\right| + Lx\left|u_n'(x) - u'(x)\right| \\
&\leq \frac{1}{n}\left|u''\left(\frac{1}{n}\right)\right| + x|u''(x)| + Lx\left|u_n'(x) - u'(x)\right| \leq 2\varepsilon + Lx \sup_{x \leq t \leq \frac{1}{n}} |u_n'(t) - u'(t)| \\
&\leq 2\varepsilon + 6Lx\left(\sup_{x \leq t \leq \frac{1}{n}} |u_n''(t) - u''(t)| + \sup_{x \leq t \leq \frac{1}{n}} |u_n(t) - u(t)|\right) \\
&\leq 2\varepsilon + 6L\|u_n - u\|_\infty + 6Lx\left|u''\left(\frac{1}{n}\right)\right| + 6Lx \sup_{x \leq t \leq \frac{1}{n}} |u''(t)| \\
&\leq 2\varepsilon + 6L\|u_n - u\|_\infty + 6L\frac{1}{n}\left|u''\left(\frac{1}{n}\right)\right| + 6L \sup_{x \leq t \leq \frac{1}{n}} t|u''(t)| \leq (2 + 18L)\varepsilon;
\end{aligned}$$

in the third inequality, we have used Lemma 5.7.12 for $c = 1$.

Since $B(u_n)$ and $B(u)$ are continuous at 0, the above estimate holds true also at 0, and this completes the proof.

Finally, if (iv) or (v) holds true, the proof runs as before, by just considering, in the case $\beta(0) \geq \frac{\alpha'(0)}{2}$ and $\beta(1) > 0$ the sequence $(v_n)_{n\geq 1}$ defined as follows: for every $n \geq 1$ and $x \in [0, 1]$, $v_n(x) = u_n(x) + h_n(x)$, where $(u_n)_{n\geq 1}$ is defined as in (1) and

$$h_n(x) := \begin{cases} 0 & 0 \leq x \leq 1 - \frac{1}{n}; \\ \frac{1}{3}n\left(u''\left(1 - \frac{1}{n}\right) - nu'\left(1 - \frac{1}{n}\right)\right)\left(x - 1 + \frac{1}{n}\right)^3 & 1 - \frac{1}{n} \leq x \leq 1. \end{cases}$$

Hence, if one of the assumptions (i)-(v) holds true, $\mathscr{C}^2([0, 1]) \cap D(B)$ is a core for $(B, D(B))$ and, hence, for $(A, D(A))$ and for $(L, D(L))$, since $A = \lambda B$. □

5.7.2 Approximation properties by means of modified Bernstein-Schnabl operators

In this subsection we shall tackle the problem of the approximation of the semigroups studied in Subsection 5.7.1. By means of such an approximation we shall state some preservation properties of the semigroups themselves.

So, consider $\alpha, \beta, \gamma \in \mathscr{C}([0, 1])$ as in the beginning of Subsection 5.7.1 and consider a Markov operator T on $\mathscr{C}([0, 1])$ such that $\alpha = T(e_2) - e_2$ (see Section 4.5). Let $(B_n)_{n\geq 1}$ be the sequence of Bernstein-Schnabl operators associated with T.

Furthermore, we assume that

$$-\infty < \inf_{0<x\leq 1} \frac{\beta(x)}{x} := -\beta_0, \qquad \beta_1 := \sup_{0\leq x<1} \frac{\beta(x)}{1-x} < +\infty; \tag{5.7.17}$$

thus, necessarily $\beta(0) \geq 0$ and $\beta(1) \leq 0$.

Choosing $n_0 \geq \max\{\beta_0, \beta_1, \|\gamma\|_\infty\}$, for $n \geq n_0$ (5.2.1) holds true and so we can consider the positive linear operator $M_n : \mathscr{C}([0,1]) \longrightarrow \mathscr{C}([0,1])$ defined by (5.2.3), i.e., for every $f \in \mathscr{C}([0,1])$,

$$M_n(f) := B_n\left(\left(1+\frac{\gamma}{n}\right) \times \left(f \circ \left(e_1 + \frac{\beta}{n}\right)\right)\right). \tag{5.7.18}$$

We already know that, for any $n \geq n_0$,

$$\|M_n\| \leq 1 + \frac{\|\gamma\|_\infty}{n}. \tag{5.7.19}$$

In Section 5.2 we have presented some examples of this modification of Bernstein-Schnabl operators on $[0,1]$.

Among their properties, we explicitly mention that, for every $f \in \mathscr{C}([0,1])$,

$$\lim_{n\to\infty} M_n(f) = f \quad \text{uniformly on } [0,1]. \tag{5.7.20}$$

Moreover, from Theorem 5.5.6, we get that, for every $u \in \mathscr{C}^2([0,1])$,

$$\lim_{n\to\infty} n(M_n(u)(x) - u(x)) = L(u)(x) \tag{5.7.21}$$

uniformly w.r.t. $x \in [0,1]$, where L is defined by (5.7.10).

From now on, let us denote indifferently by $D(L)$ one of the domains defined by (5.7.4), (5.7.7)-(5.7.9), and by $(T(t))_{t\geq 0}$ the C_0-semigroup generated by $(L, D(L))$ under the assumptions of Theorems 5.7.4 and 5.7.7.

The following theorem shows that such a semigroup can be approximated by means of iterates of modified Bernstein-Schnabl operators.

Theorem 5.7.14. *Let us assume that the functions α, β and γ satisfy* (5.7.2) *and* (5.7.17). *Moreover, suppose that either*

(i) 0 *and* 1 *are both exit end-points*

or

(ii) α is differentiable at 0 and at 1, $\alpha'(0) \neq 0 \neq \alpha'(1)$ and one of the following properties holds true:

(a) $\beta(0) \geq \frac{\alpha'(0)}{2}$ and $\beta(1) \leq \frac{\alpha'(1)}{2}$;

(b) $\beta(0) = 0$, $\beta(x) = \mathcal{O}(x)$ as $x \to 0^+$ and $\beta(1) \leq \frac{\alpha'(1)}{2}$;

(c) $\beta(0) \geq \frac{\alpha'(0)}{2}$, $\beta(1) = 0$ and $\beta(x) = \mathcal{O}(1-x)$ as $x \to 1^-$.

Then, if $t \geq 0$ and if $(k(n))_{n\geq 1}$ is an arbitrary sequence of positive integers such that $\lim_{n\to\infty} \frac{k(n)}{n} = t$, for every $f \in \mathscr{C}([0,1])$,

$$T(t)(f) = \lim_{n\to\infty} M_n^{k(n)}(f) \quad \textit{uniformly on } [0,1]. \tag{5.7.22}$$

Proof. From (5.7.19) it follows that, for any $n \geq n_0$ and $k \geq 1$,

$$\|M_n^k\| \leq \left(1 + \frac{\|\gamma\|_\infty}{n}\right)^k \leq \exp\left(\|\gamma\|_\infty \frac{k}{n}\right).$$

Moreover, if 0 and 1 are exit end-points, then by Theorem 5.7.4, $\mathscr{C}_*^2([0,1])$ is a core for $(L, D(L))$; on the other hand, if α is differentiable at 0 and at 1 and if one among (a)-(c) holds true, then, by Proposition 5.7.13, $\mathscr{C}^2([0,1]) \cap D(A)$ is a core for $(L, D(L))$.

Consequently, since for every $u \in \mathscr{C}^2([0,1])$ the asymptotic formula (5.7.21) holds, from Theorem 2.2.3 we get the required assertion. □

Having (5.7.22) at our disposal, we are in a position to study some preservation properties of the semigroup $(T(t))_{t\geq 0}$.

To this end, we first present a preservation property of modified Bernstein-Schnabl operators on the interval $[0,1]$.

Proposition 5.7.15. *Assume that T maps continuous increasing functions into continuous increasing functions. Moreover, assume that γ is constant and $\left(e_1 + \frac{\beta}{n}\right)$ are increasing functions ($n \geq n_0$). Then, for any $n \geq n_0$, $M_n(f)$ is increasing on $[0,1]$ provided that $f \in \mathscr{C}([0,1])$ is increasing on $[0,1]$.*

Proof. In Theorem 4.5.9 it was proved that, if T maps continuous increasing functions into continuous increasing functions, then for every $n \geq 1$, $B_n(f)$ is increasing on $[0,1]$ provided that f is increasing on $[0,1]$, and this completes the proof, since, under the prescribed assumptions on β and γ, the function $\left(1 + \frac{\gamma}{n}\right) f \circ \left(e_1 + \frac{\beta}{n}\right)$ is increasing whenever f is increasing. □

After these preliminaries and taking the results of Section 5.4 into account, we can show some preservation properties of the semigroup $(T(t))_{t\geq 0}$.

Corollary 5.7.16. *Suppose that γ is constant. Under the same assumptions of Theorem 5.7.14, the following statements hold true:*

(i) Under the same assumptions of Proposition 5.7.15, if $f \in \mathscr{C}([0,1])$ is an increasing function, then $T(t)(f)$ is increasing for every $t \geq 0$.

(ii) If $\beta \in \mathrm{Lip}\,(B,1)$, for some $B \geq 0$, $T(\mathrm{Lip}(1,1)) \subset \mathrm{Lip}(1,1)$ and $f \in \mathrm{Lip}(M,1)$ for some $M \geq 0$, then $T(t)(f) \in \mathrm{Lip}(M\exp(|\gamma| + B)t, 1)$ for every $t \geq 0$.

(iii) Assume that

(a) T maps continuous convex functions into (continuous) convex functions.

(b) For every convex function $f \in \mathscr{C}([0,1])$ and for every $x, y \in [0,1]$,

$$\Delta(\tilde{f}; x, y) \geq 0$$

(see (3.4.2) and (3.4.3)).

If $f \in \mathscr{C}([0,1])$ and if $f \circ \left(e_1 + \frac{\beta}{n}\right)$ is convex for every $n \geq n_0$, then $T(t)(f)$ is convex for any $t \geq 0$.

In particular, the result holds true whenever β *is affine (resp., convex) and* f *is convex (resp., convex and increasing).*

Proof. Keeping in mind (5.7.22), statement (i) follows at once from Proposition 5.7.15, whereas statement (ii) is a direct consequence of Theorem 5.4.1. Finally, statement (iii) is a consequence of Theorem 5.4.4. □

5.8 The d-dimensional simplex and hypercube

In this section we deal with some differential operators which have been object of interest in the preceding Subsection 2.3.4 and which have been defined in the special case where K is the d-dimensional simplex K_d (see (1.1.45)) or the d-dimensional hypercube Q_d.

First of all, we consider a Fleming-Viot operator (see (2.3.21)) of the form

$$L_{K_d}(u)(x) := \sum_{i,j=1}^{d} \frac{x_i(\delta_{ij}-x_j)}{2} \frac{\partial^2 u}{\partial x_i \partial x_j}(x) + \sum_{i=1}^{d} \beta_i(x) \frac{\partial u}{\partial x_i}(x) + \gamma(x)u(x) \quad (5.8.1)$$

($u \in \mathscr{C}^2(K_d), x \in K_d$), where $\gamma \in \mathscr{C}(K_d)$, the functions β_i are Lipschitz continuous and, setting $\beta(x) := (\beta_1(x), \dots, \beta_d(x))$ and $\alpha_{ij}(x) := \frac{x_i(\delta_{ij}-x_j)}{2}$, ($x = (x_1, \dots, x_d) \in K_d, i,j = 1, \dots, d$), for every $x \in \partial K_d$,

$$\langle \beta(x), \nabla \det(\alpha_{ij}(x)) \rangle \geq 0$$

(see (2.3.22)).

As we have already mentioned in Remark 2.3.23, the condition above is equivalent to the following one: for every $x = (x_1, \dots, x_d) \in \partial K_d$

$$x_h = 0 \Rightarrow \beta_h(x_1, \dots, x_d) \geq 0, \qquad h = 0, 1, \dots d,$$

where $\beta_0 := -\sum_{i=1}^{d} \beta_i$ and $x_0 := 1 - \sum_{i=1}^{d} x_i$.

We also point out that the operator $(L_{K_d}, \mathscr{C}^2(K_d))$ is a first-order perturbation of the operator defined by (4.2.10), which is the second-order differential operator associated with the Markov projection T_d (see (1.1.48)).

Moreover, the Bernstein-Schnabl operators associated with T_d are the classical Bernstein operators (3.1.18).

We have the following result.

Theorem 5.8.1. *Assume that* (2.3.22) *and* (5.2.1) *hold true. Then* $(L_{K_d}, \mathscr{C}^2(K_d))$ *is closable, its closure* $(B_{K_d}, D(B_{K_d}))$ *generates a positive* C_0*-semigroup* $(T(t))_{t\geq 0}$ *on* $\mathscr{C}(K_d)$ *and* $\mathscr{C}^2(K_d)$ *is a core for* $(B_{K_d}, D(B_{K_d}))$.

Moreover, if $t \geq 0$ *and if* $(k(n))_{n\geq 1}$ *is a sequence of positive integers such that* $\lim_{n\to\infty} \frac{k(n)}{n} = t$, *then*

$$\lim_{n\to\infty} M_n^{k(n)}(f) = T(t)(f) \qquad \textit{uniformly on } K_d$$

for every $f \in \mathscr{C}(K_d)$, *where the operators* M_n *are the modified Bernstein-Schnabl operators on the d-dimensional simplex (see* (5.2.8) *for* $d = 2$*).*

Proof. By Theorems 2.3.22 and 2.1.13, we immediately infer that, under assumption (2.3.22), the differential operator $(L_{K_d}, \mathscr{C}^2(K_d))$ is closable, its closure

$(B_{K_d}, D(B_{K_d}))$ generates a C_0-semigroup $(T(t))_{t\geq 0}$ on $\mathscr{C}(K_d)$ and $\mathscr{C}^2(K_d)$ is a core for $(B_{K_d}, D(B_{K_d}))$.

Therefore (see Remark 5.6.4, 1), under assumption (5.2.1), Theorem 5.6.3 applies (even if the β_i's are not polynomials of first degree), so that, for every $t \geq 0$ and for every sequence $(k(n))_{n\geq 1}$ of positive integers such that $\lim_{n\to\infty} k(n)/n = t$, then

$$\lim_{n\to\infty} M_n^{k(n)}(f) = T(t)(f)$$

for every $f \in \mathscr{C}(K_d)$. In particular, this implies that each $T(t)$ $(t \geq 0)$ is positive. □

Some qualitative properties of the semigroup $(T(t))_{t\geq 0}$, which are consequences of (3.3.14) and (5.6.11), may be presented as well.

Corollary 5.8.2. *Under the assumptions of Theorem 5.8.1, if γ is constant and if $\beta_i \in \mathrm{Lip}\left(\frac{B}{\sqrt{d}}, 1\right)$ $(i = 1, \dots, d)$ for some $B \geq 0$, then for every $t \geq 0$ and $f \in \mathscr{C}(K_d)$ which is Lipschitz continuous on K_d with respect to $\|\cdot\|_1$ (where $\|x\|_1 := \sum_{i=1}^{d} |x_i|, x \in K_d$) and with constant M, $T(t)(f)$ is Lipschitz continuous (with respect to $\|\cdot\|_1$) with constant $M\exp(|\gamma| + B)t$.*

As we have seen in Subsection 2.3.4, among the more general class of Fleming-Viot operators, a special role is played by second-order differential operators of the following form: for every $u \in \mathscr{C}^2(K_d)$ and $x = (x_1, \dots, x_d) \in K_d$,

$$V_1(u)(x) := \sum_{i,j=1}^{d} \frac{x_i(\delta_{ij} - x_j)}{2} \frac{\partial^2 u}{\partial x_i \partial x_j}(x) + \sum_{i=1}^{d} \frac{\mu_i - \nu x_i}{2} \frac{\partial u}{\partial x_i}(x)$$

(see (2.3.24)), where $\mu_i \geq 0$ for every $i = 1, \dots, d$ and $\nu \geq \sum_{i=1}^{d} \mu_i$.

We recall that the operator $(V_1, \mathscr{C}^2(K_d))$ is related to the diffusion approximation of a discrete stochastic model arising in population dynamics, the so-called Wright-Fisher model; for more details we refer the interested reader to Subsection 2.3.4 and to the references quoted therein.

For the operator V_1, condition (5.2.1) is satisfied for $n \geq \frac{\nu}{2}$ (because $\beta(x) = -\rho x + \rho c$ $(x \in K_d)$, where $\rho = \frac{\nu}{2}$ and $c = \left(\frac{\mu_1}{\nu}, \dots, \frac{\mu_d}{\nu}\right) \in K_d$).

Since (5.6.8) and (5.6.9) are satisfied, by Theorem 5.6.3 we obtain not only the well-known result that the differential operator $(V_1, \mathscr{C}^2(K_d))$ is closable and its closure is the generator of a Feller semigroup $(T(t))_{t\geq 0}$ on $\mathscr{C}(K_d)$, but we also get the approximation formula

$$T(t)(f) = \lim_{n\to\infty} M_n^{k(n)}(f) \qquad \text{uniformly on } K_d, \tag{5.8.2}$$

for every $f \in \mathscr{C}(K_d)$, $t \geq 0$ and for every sequence $(k(n))_{n\geq 1}$ of positive integers such that $\lim_{n\to\infty} k(n)/n = t$, where the operators M_n are defined again as in (5.2.8)

with $\gamma = 0$. In this particular case we also point out that $T(t)(\mathbf{1}) = \mathbf{1}$ for every $t \geq 0$, i.e., $(T(t))_{t\geq 0}$ is a Markov semigroup.

Below we state some consequences of the representation (5.8.2).

Corollary 5.8.3. *The following properties hold true:*
(1) $T(t)(P_m(K)) \subset P_m(K)$ for every $t \geq 0$ and $m \geq 1$.
(2) If $f \in \mathscr{C}(K_d)$ is Lipschitz continuous on K_d with respect to $\|\cdot\|_1$ and with constant M, then, for every $t \geq 0$, $T(t)(f)$ is Lipschitz continuous (with respect to $\|\cdot\|_1$) with constant $M\exp\left(\frac{\nu}{2}\right)t$.
(3) For any T-axially convex function $f \in \mathscr{C}(K)$ (and, hence, T-convex in the sense of Definition 3.5.7), $T(t)(f)$ is T-axially convex for every $t \geq 0$.

Proof. Statement (1) is a consequence of Remark 5.6.4 and of the fact that $T_d(P_m(K_d)) \subset P_m(K_d)$ for every $m \geq 1$ (see (1.1.48)), whereas statement (2) is a result of (3.3.14) and (5.6.11).

Finally, by Theorem 3.5.9, under the assumptions of statement (3), also $f \circ \left(id + \frac{\beta}{n}\right)$ is T-axially convex and each $M_n(f)$ is T-axially convex ($n \geq n_0$) (see Example 5.4.6, 1). Accordingly, the result is a direct consequence of (5.8.2). □

For the sake of completeness it may be worthwhile stressing that, as a consequence of (5.8.2), for any $u_0 \in P_\infty(K)$, the solution $u(\cdot, t)$ to the diffusion problem

$$\begin{cases} \frac{\partial u}{\partial t}(x,t) = \sum\limits_{i,j=1}^{d} \frac{x_i(\delta_{ij}-x_j)}{2} \frac{\partial^2 u}{\partial x_i \partial x_j}(x) + \sum\limits_{i=1}^{d} \frac{\mu_i - \nu x_i}{2} \frac{\partial u}{\partial x_i}(x) & x \in K_d, t \geq 0; \\ \\ u(x,0) = u_0(x) & x \in K_d \end{cases} \tag{5.8.3}$$

may be expressed by means of suitable iterates of the operators M_n, i.e., as

$$u(x,t) = \lim_{n\to\infty} M_n^{k(n)}(u_0)(x) \tag{5.8.4}$$

uniformly with respect to $x \in K_d$, for any $t \geq 0$ and for every sequence $(k(n))_{n\geq 1}$ of positive integers such that $\lim\limits_{n\to\infty} k(n)/n \to t$.

Moreover, by virtue of Corollary 5.8.3, if the initial datum u_0 is a polynomial of some degree $m \geq 1$, then the solution $u(\cdot, t)$ is also a polynomial of the same degree; finally, if u_0 is Lipschitz continuous with Lipschitz constant M, then $u(\cdot, t)$ is Lipschitz continuous with Lipschitz constant $M\exp(\frac{\nu}{2})t$.

Finally, we point out that, as shown in [66, Section 3] (see also [18, Theorem 6.2.6, formula (6.2.20)]), for every $f \in \mathscr{C}(K_d)$,

$$\lim_{t\to+\infty} T(t)(f) = \begin{cases} \mu_\infty(f) & \text{if } \nu > 0; \\ \\ T_d(f) & \text{if } \nu = 0 \end{cases} \tag{5.8.5}$$

uniformly on K_d, where μ_∞ is a probability Borel measure on K_d and T_d is defined by (1.1.48).

Fleming-Viot operators have been object of interest also in other settings rather than the d-dimensional simplex.

In particular, in Subsection 2.3.4 we have presented some results about particular Fleming-Viot type operators on the d-dimensional hypercube Q_d (see (2.3.28)).
Below we shall consider the second-order differential operator $(L_{Q_d}, \mathscr{C}^2(Q_d))$ defined as

$$L_{Q_d}(u)(x) := \sum_{i=1}^{d} \frac{x_i(1-x_i)}{2} \frac{\partial^2 u}{\partial x_i^2}(x) + \sum_{i=1}^{d} \beta_i(x) \frac{\partial u}{\partial x_i}(x) + \gamma(x)u(x) \tag{5.8.6}$$

where $\gamma \in \mathscr{C}(Q_d)$ and, for every $i = 1, \dots, d$, the functions β_i satisfy the following assumptions: there exists $\delta \in (0, 1)$ such that, for every $i = 1, \dots, d$, $\beta_i \in \mathscr{C}^\delta(Q_d)$, $\beta_i(0, \dots, 0) > 0$ and $\beta_i(1, \dots, 1) < 0$ (see (2.3.27) for the definition of the space $\mathscr{C}^\delta(Q_d)$).

Then, by means of Theorems 2.3.24 and 2.1.13, and of Remark 5.6.4, 1, if also (5.2.1) holds true, we have the following result.

Theorem 5.8.4. *Under the above assumptions, $(L_{Q_d}, \mathscr{C}^2(Q_d))$ is closable, its closure $(B_{Q_d}, D(B_{Q_d}))$ generates a positive C_0-semigroup $(T(t))_{t\geq 0}$ on $\mathscr{C}(Q_d)$ and $\mathscr{C}^2(Q_d)$ is a core for $(B_{Q_d}, D(B_{Q_d}))$.*

Moreover, if $t \geq 0$ and if $(k(n))_{n\geq 1}$ is a sequence of positive integers such that $\lim_{n\to\infty} \frac{k(n)}{n} = t$, then

$$T(t)(f) = \lim_{n\to\infty} M_n^{k(n)}(f) \quad \textit{uniformly on } Q_d$$

for every $f \in \mathscr{C}(Q_d)$, where the operators M_n are the modified Bernstein-Schnabl operators on the d-dimensional hypercube Q_d (see (5.2.9) for $d = 2$).

5.9 Notes and comments

As pointed out in Section 3.6, Lototsky-Schnabl operators were first introduced by Schempp ([183]) and, in a more general form, by Grossmann ([112]). In the context of Bauer simplices they were investigated by Nishishiraho ([154], [157]). In all the above quoted papers, the authors were interested in studying the approximation properties of these operators. Altomare ([9]) first recognized their significant role in connection with the study of multiplicative perturbations of special classes of (abstract) differential operators by associating them with Markov projections. Further investigations have been carried out in [2], [35], [167], [168], [170], [176] (see also [18, Chapter 6 and the references contained in the relevant notes]).

The results of Section 5.1 are new and they represent a full extension to an arbitrary Markov operator on $\mathscr{C}(K)$.

Modified Bernstein-Schnabl operators were introduced, on the unit interval $[0,1]$, by Altomare and Amiar in [14] (see also [35]).

A first attempt to extend this definition for arbitrary convex compact subsets of $\mathbf{R}^d$ $(d \geq 1)$ and, more generally, for convex compact subset of a (possibly, infinite-dimensional) locally convex space may be found in [19] and [69]. Nevertheless, both papers deal with Bernstein-Schnabl operators associated with a positive projection, rather than an arbitrary Markov operator.

In particular, in [19], the authors sketched the outlines for the study of a wide class of second-order differential operators of the form

$$V(u)(x) := \frac{1}{2}\sum_{i,j=1}^{d} \alpha_{ij}(x)\frac{\partial^2 u(x)}{\partial x_i \partial x_j} + \sum_{i=1}^{d} \beta_i(x)\frac{\partial u(x)}{\partial x_i} + \gamma(x)u(x),$$

through approximation theory techniques (see, for other results, [63], [66], [95]).

Sections 5.2-5.6 contain the extension of the above results to an arbitrary Markov operator. All the results are new.

The one-dimensional differential operators in Section 5.7, not only for the unit interval but also for more general intervals, have been investigated by many authors also from the point of view of approximation of the semigroup (see, e.g., [15], [17], [25], [27], [42], [65], [93, Chapter VI, Section 4], [141], [144] and the references therein).

Lemma 5.7.1 and Theorem 5.7.2 have been established in [65]. Theorems 5.7.7 and 5.7.13 as well as most of the preliminary propositions and lemmas are due to Attalienti and Campiti ([42]). Statement (i) of Proposition 5.7.11 is a particular case of [144, Proposition 3.9] (see, also, [141, Proposition 3.2]).

The results of Subsection 5.7.2 are essentially due to [15, Subsection 3.1].

Theorem 5.8.1 is essentially due to [19] while Theorem 5.8.4 seems to be new. However, for additional comments and references concerning Fleming-Viot operators we refer to Subsection 2.3.4 and to Section 2.4.

Appendices

A.1 A classification of Markov operators on two dimensional convex compact subsets

In this appendix, given a convex compact subset K of $\mathbf{R}^2$, we shall throughly discuss a special class of Markov projections T on $\mathscr{C}(K)$ (if any) which satisfy (3.1.2), (4.3.5) and the following additional property

$$h_{z,\alpha} \in H \quad \text{for every } h \in H, z \in K \text{ and } \alpha \in [0,1], \tag{A.1.1}$$

where $H := T(\mathscr{C}(K))$ and $h_{z,\alpha}$ is defined as in (3.1.32).

The importance of the projections essentially relies on Theorems 4.3.4 and 4.4.7 (see also Remark 4.4.8 and [18, Chapter 6]).

All the positive projections described in Subsections 3.1.1, 1), 3.1.2, 3.1.4 and 3.1.5 satisfy (A.1.1). Moreover, by using these particular assumptions, it can be proved that

$$\partial_T K = \bigcup_{x \in \partial_T K} F_x \tag{A.1.2}$$

where, for every $x \in \partial_T K$,

$$F_x := \{y \in K \mid \text{ there exist } z \in K \text{ and } \alpha \in]0,1] \text{ such that } x = \alpha y + (1-\alpha)z\} \tag{A.1.3}$$

([18, Proposition 6.1.1]).

Non-trivial Markov projections satisfying (3.1.2), (4.3.5) and (A.1.1) will be briefly referred to as *A-projections*.

The main aim of the section is to present a complete characterization of those convex compact subsets K of $\mathbf{R}^2$ such that there exist A-projections on $\mathscr{C}(K)$. Such subsets will be referred to as *P-admissible*. Moreover, if K is P-admissible, we shall describe all the A-projections on $\mathscr{C}(K)$. In particular, we shall show that condition (4.3.5) can be relaxed by only requiring that $T(P_2(K)) \subset P_2(K)$. It would be desirable to investigate whether such an implication is still true for arbitrary convex compact subsets of $\mathbf{R}^d$, $d \geq 3$, as Theorem 4.3.8 seems to suggest, and we leave this question as an open problem.

All the results of this section are taken from [34, Section 4]. We begin by studying similar problems for real compact intervals.

From Theorem 4.3.3 it follows that the canonical projection T_1 (see (1.1.49)) is the unique Markov operator on $\mathscr{C}([0,1])$ satisfying (3.1.2) and (4.3.4). On the other hand, Lemma 4.1.1 shows that there is plenty of Markov operators on $\mathscr{C}([0,1])$ satisfying (3.1.2) and (4.3.5).

The next result shows that T_1 is the unique Markov projection that satisfies (3.1.2) and (4.3.5).

Proposition A.1.1. *Let T be a non-trivial Markov projection on $\mathscr{C}([0,1])$ satisfying (3.1.2) and such that $T(P_2([0,1])) \subset P_2([0,1])$. Then $T = T_1$.*

Proof. Consider the function $e_2(x) = x^2$ $(0 \le x \le 1)$. Since $T \ne I$, by the Korovkin theorem (see Theorem 1.2.10), it follows that $T(e_2) \ne e_2$. On the other hand, $T(e_2) \in P_2([0,1])$ and hence, by Theorem 1.3.1, $\partial_T[0,1]$ cannot contain more than two points. From (3.1.4) it also follows that $\{0,1\} \subset \partial_T[0,1]$ so that $\{0,1\} = \partial_T[0,1] = \partial_H[0,1]$, where $H := T(\mathscr{C}([0,1]))$. Consider now $f \in \mathscr{C}([0,1])$ and set $g := T_1(f) \in A([0,1])$. Since $f = g$ on $\{0,1\}$, by (1.3.10) and (3.1.2) we infer that $T(f) = T(g) = g = T_1(f)$. □

We proceed further by presenting some examples which will be useful in the sequel.

Examples A.1.2.

1. *Let K be a trapezium of $\mathbf{R}^2$ with vertices A, B, C and D. We assume that the segments $\overline{BC}$ and $\overline{AD}$ are parallel (see Fig. 1).*

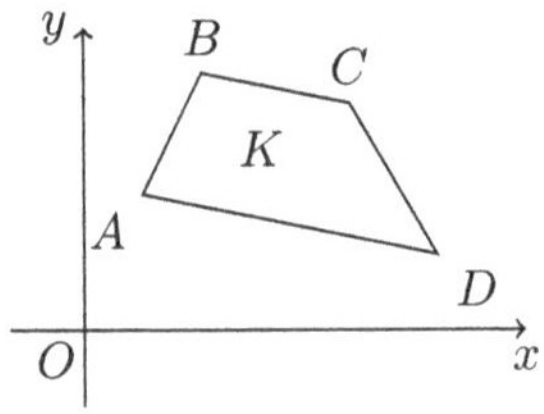

Fig. 1

For every $f \in \mathscr{C}(K)$ denote by $T(f)$ the continuous function on K which is equal to f on the segments $\overline{AB}$ and $\overline{CD}$ and is affine on every segment of K parallel to $\overline{AD}$ (or to $\overline{BC}$).

The operator $T : \mathscr{C}(K) \longrightarrow \mathscr{C}(K)$ is an A-projection (see also the next Example A1.2, 3). It will be called the canonical projection associated with the trapezium *$\overline{ABCD}$.*

The range H of T coincides with the subspace of all functions $h \in \mathscr{C}(K)$ which are affine on every segment of K parallel to $\overline{AD}$. Moreover,

$$\partial_H K = \overline{AB} \cup \overline{CD}. \tag{A.1.4}$$

If we choose a coordinate system Oxy such that $A = (0,0)$, $B = (b_1, b_2)$, $C = (c_1, b_2)$ and $D = (d_1, 0)$, with b_1, b_2, c_1, d_1 strictly positive (see Fig. 2),

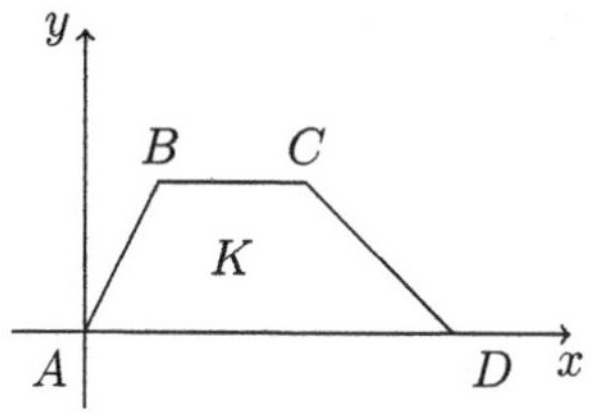

Fig. 2

then for every $f \in \mathscr{C}(K)$ *and* $(x, y) \in K$ *we have*

$$T(f)(x,y) = f(p(y),y) + \frac{f(q(y),y) - f(p(y),y)}{q(y) - p(y)}(x - p(y)), \tag{A.1.5}$$

where

$$p(y) := \frac{b_1}{b_2}y, \quad q(y) := \frac{c_1 - d_1}{b_2}y + d_1 \qquad (0 \le y \le b_2). \tag{A.1.6}$$

Moreover, if $u \in \mathscr{C}^2(K)$, *then*

$$W_T(u)(x,y) = \frac{1}{2}\left(x(p(y) + q(y)) - p(y)q(y) - x^2\right)\frac{\partial^2 u}{\partial x^2}(x,y) \tag{A.1.7}$$

(see (4.1.15)*).*

Note that the above construction can be equally carried out when $B = C$, *i.e.,* K *is a triangle.*

In this case we shall refer to K *as a* degenerate trapezium *and we shall denote it by* $\overline{ABBD}$.

Thus, on a triangle with vertices A, B *and* C *there are defined at least four positive A-projections (in fact, they are the only ones): the canonical projection associated to it as a simplex (see Subsection 1.1.3) and the three canonical projections associated to the trapezia* $\overline{ABBC}$, $\overline{BCCA}$, $\overline{CAAB}$.

2. *Let* K *be a parallelogram of* $\mathbf{R}^2$ *with vertices* $A = (a_1, a_2)$, $B = (b_1, b_2)$, $C = (c_1, c_2)$, $D = (d_1, d_2)$ *(see Fig. 3).*

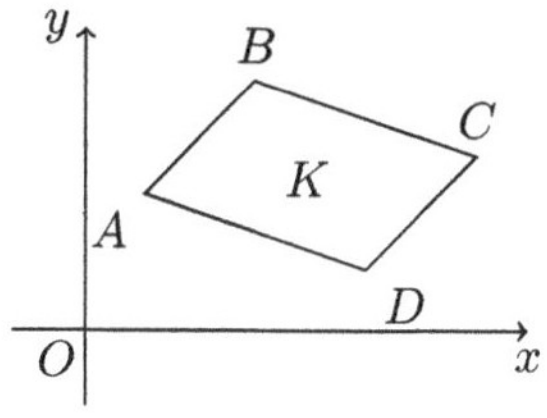

Fig. 3

For every $f \in \mathscr{C}(K)$ *let us consider the function* $T(f)$ *which is equal to* f *on* $\partial_e K$, *i.e., on the vertices of* K, *and is affine on each segment of* K *parallel to a side of* K.

Thus, for every $(x, y) \in K$ *we have*

$$det \begin{pmatrix} T(f)(x,y) & q(x,y) & x & y & 1 \\ f(a_1,a_2) & & & & \\ f(d_1,d_2) & & & & \\ f(c_1,c_2) & & \Gamma & & \\ f(b_1,b_2) & & & & \end{pmatrix} = 0, \tag{A.1.8}$$

where $q(x,y) = ((b_1 - a_1)y - (b_2 - a_2)x)((d_1 - a_1)y - (d_2 - a_2)x)$ *and*

$$\Gamma := \begin{pmatrix} q(a_1,a_2) & a_1 & a_2 & 1 \\ q(d_1,d_2) & d_1 & d_2 & 1 \\ q(c_1,c_2) & c_1 & c_2 & 1 \\ q(b_1,b_2) & b_1 & b_2 & 1 \end{pmatrix}. \tag{A.1.9}$$

The linear operator $T : \mathscr{C}(K) \longrightarrow \mathscr{C}(K)$ *is a positive projection which satisfies* (3.1.2) *and* (A.1.1). *Moreover* $T(P_m(K)) \subset P_2(K)$ *for each* $m \geq 1$ *so that* (4.3.5) *is satisfied as well.*

In this case the range H *of* T *is the subspace of all functions* $h \in \mathscr{C}(K)$ *which are affine on each segment parallel to a side of* K *and*

$$\partial_H K = \partial_e K. \tag{A.1.10}$$

We shall call the operator T *the* canonical projection associated with the parallelogram K.

Finally we point out that the differential operator W_T *is given by*

$$W_T(u)(x,y) = -\left(\frac{\gamma(x,y)}{det(\Gamma)} + xy\right)\frac{\partial^2 u}{\partial x \partial y}(x,y) - \frac{1}{2}\left(\frac{\alpha(x,y)}{det(\Gamma)} + x^2\right)\frac{\partial^2 u}{\partial x^2}(x,y)$$
$$-\frac{1}{2}\left(\frac{\beta(x,y)}{det(\Gamma)} + y^2\right)\frac{\partial^2 u}{\partial y^2}(x,y) \quad (u \in \mathscr{C}^2(K), (x,y) \in K),$$

where

$$\gamma(x,y) := det \begin{pmatrix} 0 & q(x,y) & x & y & 1 \\ a_1 a_2 & & & & \\ d_1 d_2 & & & & \\ c_1 c_2 & & \Gamma & & \\ b_1 b_2 & & & & \end{pmatrix}, \tag{A.1.11}$$

$$\alpha(x,y) := det \begin{pmatrix} 0 & q(x,y) & x & y & 1 \\ a_1^2 & & & & \\ d_1^2 & & & & \\ c_1^2 & & \Gamma & & \\ b_1^2 & & & & \end{pmatrix} \tag{A.1.12}$$

and

$$\beta(x,y) := det \begin{pmatrix} 0 & q(x,y) & x & y & 1 \\ a_2^2 & & & & \\ d_2^2 & & & & \\ c_2^2 & & \Gamma & & \\ b_2^2 & & & & \end{pmatrix}. \tag{A.1.13}$$

Finally, note that on K we can construct two further projections. They are defined exactly as in Example A.1.2, 1), by considering the couples of parallel segments $\overline{BC}$ and $\overline{AD}$, and $\overline{AB}$ and $\overline{DC}$, respectively.

The directions of the segments $\overline{BC}$ and $\overline{AB}$ will be called the main directions *of K and the above defined projections will be called the* canonical projections associated with the main directions.

The corresponding differential operators can be easily described as in (A.1.7).

3. In the next example we shall consider particular convex compact subsets of $\mathbf{R}^2$ which, for the sake of simplicity, we shall call conical.

They are, by definition, those convex compact subsets of $\mathbf{R}^2$ whose boundary is the union of one or two parallel segments with a portion of a conic (see Fig. 4–Fig. 9).

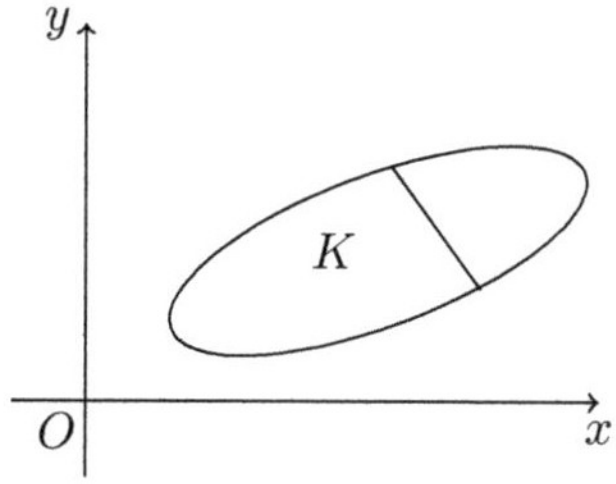

Fig. 4

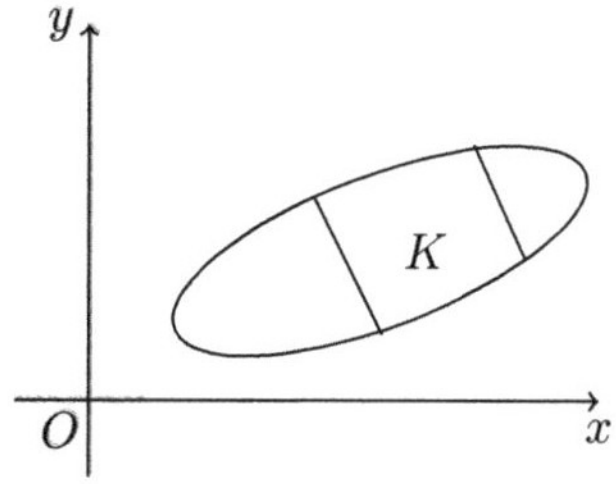

Fig. 5

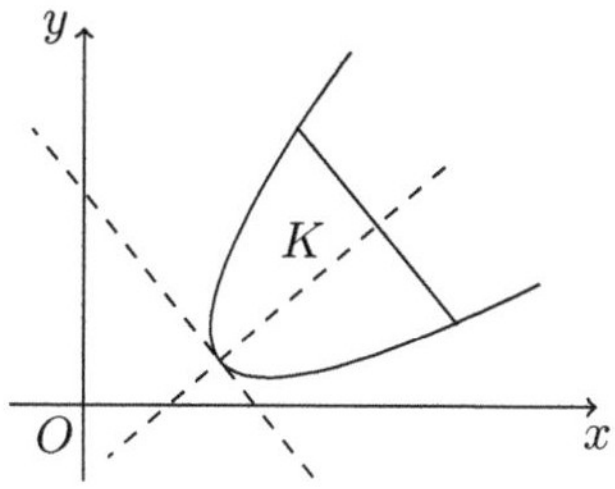

Fig. 6

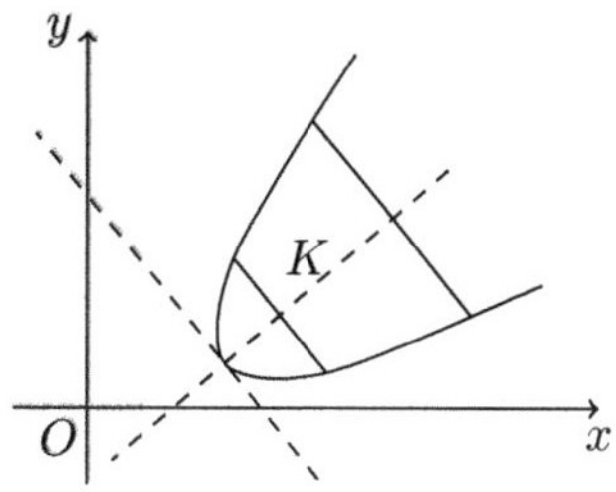

Fig. 7

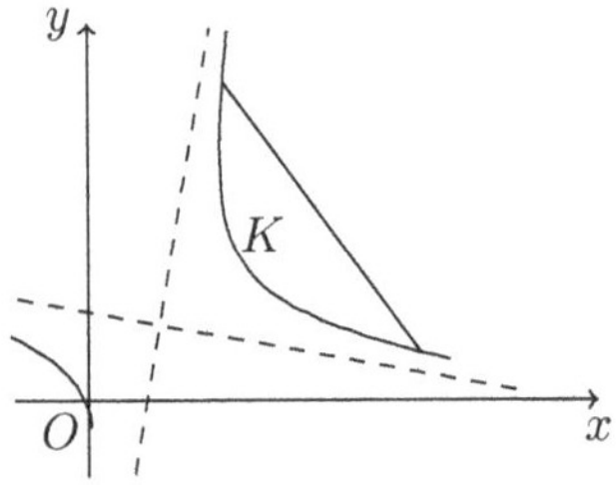

Fig. 8

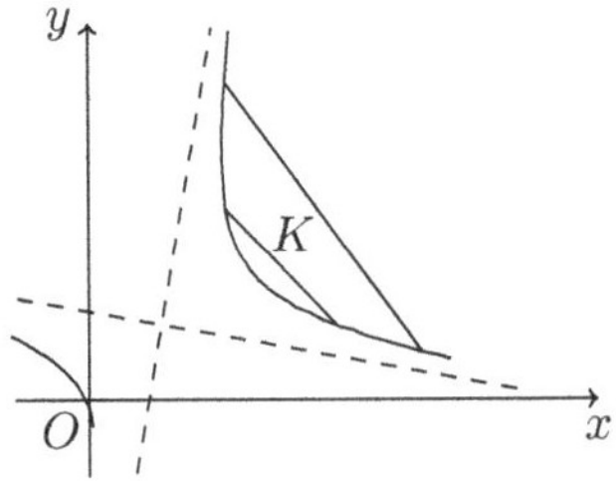

Fig. 9

The direction of the segment(s) of the boundary of K will be called the main direction *of the conical subset.*

Given a conical subset K of $\mathbf{R}^2$, for every $f \in \mathscr{C}(K)$ denote by $T(f)$ the continuous function on K which is equal to f on the portion of the conic of ∂K and is affine on every segment of K parallel to the main direction.

If we choose a coordinate system Oxy such that one of the segments of ∂K lies on the x-axis and if $x = p(y)$ and $x = q(y)$ $(y \in [0, s])$ are the equations of the portions of the conic, where

$$p(y) = a + by - \sqrt{my^2 + 2ny + r} \tag{A.1.14}$$

and

$$q(y) = a + by + \sqrt{my^2 + 2ny + r} \tag{A.1.15}$$

$(a, b, m, n, r \in \mathbf{R},\ n^2 - mr > 0)$ *(see Fig. 10 and Fig. 11),*

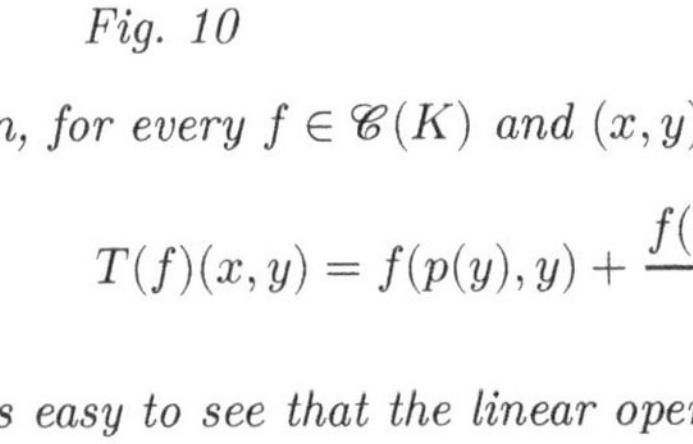

Fig. 10

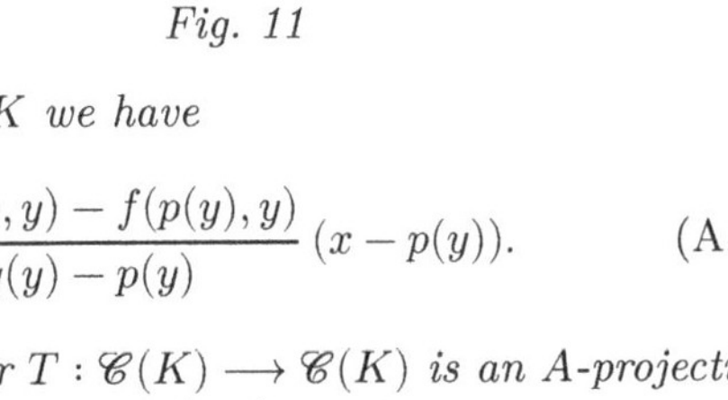

Fig. 11

then, for every $f \in \mathscr{C}(K)$ and $(x, y) \in K$ we have

$$T(f)(x, y) = f(p(y), y) + \frac{f(q(y), y) - f(p(y), y)}{q(y) - p(y)}(x - p(y)). \tag{A.1.16}$$

It is easy to see that the linear operator $T : \mathscr{C}(K) \longrightarrow \mathscr{C}(K)$ is an A-projection.

Indeed, for each $h, k \in \mathbf{N}$ and for $f_{h,k} := pr_1^h pr_2^k$, we have

$$T(f_{h,k})(x, y) = p(y)^h y^k + \frac{q(y)^h y^k - p(y)^h y^k}{q(y) - p(y)}(x - p(y))$$

$((x, y) \in K)$.

Clearly, for every $(x,y) \in K$, $T(f_{0,k})(x,y) = y^k$ *and* $T(f_{1,k})(x,y) = xy^k$. *If* $h \geq 2$, *then*

$$
\begin{aligned}
&T(f_{h,k})(x,y) \\
&= y^k(p(y)^h + (q(y)^{h-1} + q(y)^{h-2}p(y) + \ldots + p(y)^{h-1})(x - p(y))) \\
&= y^k(xS_{h-1}(y) - p(y)q(y)S_{h-2}(y))
\end{aligned}
$$

$((x,y) \in K)$, *where*

$$
S_n(y) := \begin{cases} p(y)^n + p(y)^{n-1}q(y) + \ldots + p(y)q(y)^{n-1} + q(y)^n & n \geq 1; \\ \\ 1 & n = 0. \end{cases}
$$

Note that $S_1 = p + q$ *is a polynomial of degree* 1 *and* $S_2 = (p+q)^2 - pq$ *is a polynomial of degree* 2.

Since $S_n = (p+q)S_{n-1} - pqS_{n-2}$ $(n \geq 3)$, *by induction we see that* S_n *is a polynomial of degree* $\leq n$ *so that* $T(f_{h,k})$ *is a polynomial of degree* $\leq h + k$.

Finally, note that the range H *of* T *is the subspace of all functions* $h \in \mathscr{C}(K)$ *which are affine on every segment of* K *parallel to the main direction.*

Moreover,

$$
\partial_{H_T} K = \partial_e K = \{x \in \partial K \mid x \text{ lies on the conic}\}. \tag{A.1.17}
$$

If T *is defined by (A.1.16), then, for every* $u \in \mathscr{C}^2(K)$ *and* $(x,y) \in K$,

$$
W_T(u)(x,y) = \frac{1}{2}\left(x(p(y) + q(y)) - p(y)q(y) - x^2\right)\frac{\partial^2 u}{\partial x^2}(x,y). \tag{A.1.18}
$$

The projection T *will be called the* canonical projection associated with the conical subset K.

4. *Let* ∂K *be an ellipse and consider a fixed direction* s. *Then we may regard* K *as a conical subset with main direction* s *and the two segments reduced to points. The corresponding projection* T *will be called the* canonical projection associated with the ellipse and the given direction.

In the next results we shall prove that the unique P-admissible subsets of $\mathbf{R}^2$ are those described in the above Examples A.1.2 and in Section 4.3.

To this end, we need some preliminaries.

Let K be a P-admissible convex compact subset of $\mathbf{R}^2$ having non-empty interior and let $T : \mathscr{C}(K) \longrightarrow \mathscr{C}(K)$ be an A-projection.

From (A.1.3) we know that

$$
\partial_H K = \bigcup_{x \in \partial_H K} F_x,
$$

where for every $x \in K$, F_x is defined by (A.1.3).

Thus, F_x is the smallest face of K containing x. We recall that the *face* of a general convex compact subset K of a locally convex space is a convex subset F of K such that, if $x, y \in K$ and $\lambda x + (1-\lambda)y \in F$ for some $\lambda \in \mathbf{R}$, $0 < \lambda < 1$, then $x, y \in F$.

In our case, the faces of K are the singletons $\{x\}$, with $x \in \partial_e K$, the maximal segments contained in ∂K and the whole set K.

Note also that, if F is a (closed) face of K, then

$$F \subset \partial_H K \text{ if and only if } H|_F = \mathscr{C}(F), \tag{A.1.19}$$

where $H|_F := \{h|_F \mid h \in H\}$ (see, e.g., [18, Remark to Proposition 6.1.1]).

By using (A.1.2), it is easy to show that, if F is a maximal (closed) segment contained in ∂K, then the following dichotomy holds true:

$$F \subset \partial_H K \quad \text{or} \quad F \cap \partial_H K = \partial_e F. \tag{A.1.20}$$

We now show that, under the above assumptions on F,

$$F \cap \partial_H K = \partial_e F \quad \text{if and only if} \quad H|_F = A(F). \tag{A.1.21}$$

Indeed, assume that $F \cap \partial_H K = \partial_e F$. Clearly $A(F) \subset H|_F$ because of (3.1.2). Conversely, let $\varphi \in H|_F$ and assume that $\varphi \notin A(F)$. Denote by S the linear subspace of $\mathscr{C}(F)$ generated by $A(F)$ and φ and let $\partial_S F$ be its corresponding Choquet boundary. Since $A(F) \subset S \subset H|_F$, we get

$$\partial_e F \subset \partial_S F \subset \partial_{H|_F} F = \partial_H K \cap F = \partial_e F$$

and hence $\partial_e F = \partial_S F$. Now, choose $g \in A(F)$ such that $g = \varphi$ on $\partial_e F$. Then $\varphi - g \in S$ and $\varphi - g = 0$ on $\partial_S F$ so that $\varphi - g = 0$ on F and we reach a contradiction.

The reverse implication is obvious, because

$$F \cap \partial_H K = \partial_{H|_F} F = \partial_{A(F)} F = \partial_e F.$$

Lemma A.1.3. *Let F be a maximal segment contained in ∂K such that $F \cap \partial_H K = \partial_e F$. Then every $h \in H$ is affine on every segment of K parallel to F.*

Proof. Let G be a segment of K parallel to F and preliminarily assume that $G \setminus \partial_e G$ is contained in the interior $\operatorname{int}(K)$ of K.

Given $x \in G \setminus \partial_e G$ choose $z \in K \setminus G$ and $\alpha \in [0,1]$ so that

$$x \in F_{z,\alpha} \setminus \partial_e F_{z,\alpha} \subset G,$$

where $F_{z,\alpha} := \alpha F + (1-\alpha)z$.

If $h \in H$, then $h_{z,\alpha} \in H$ by (A.1.1) and hence, by virtue of (A.1.21), $h_{z,\alpha}|_F$ is affine, i.e., h is affine on $F_{z,\alpha}$ and hence on G, because $F_{z,\alpha}$ is not a singleton.

Finally, if G is a segment contained in ∂K and parallel to F, then a given $h \in H$ is affine on each segment G' parallel to F (and to G) such that $G' \setminus \partial_e G' \subset int(K)$; by continuity it must be affine on G as well. □

Theorem A.1.4. *Let K be a convex compact subset of $\mathbf{R}^2$ having non-empty interior and let $T : \mathscr{C}(K) \longrightarrow \mathscr{C}(K)$ be a non-trivial positive projection satisfying* (3.1.2) *and* (A.1.1) *such that $T(P_2(K)) \subset P_2(K)$.*

Then $\partial_H K = \partial_e K$ or K is a trapezium (possible degenerate). In this last case T is the canonical projection associated with K.

Proof. Assume that $\partial_H K \neq \partial_e K$ and fix $y \in \partial_H K$, $y \notin \partial_e K$. Let F_y be the smallest face of K containing y (see (A.1.3)). Since $y \notin \partial_e K$, $F_y \neq \{y\}$. From (A.1.2) it follows that $F_y \subset \partial_H K$ and hence $F_y \neq K$ because T is non-trivial. Therefore, F_y is a maximal segment contained in ∂K. Clearly $F_y \neq \partial_H K$, otherwise $\partial_e K$ would consist of two points and hence $\text{int}(K) = \emptyset$.

Choose a coordinate system Oxy such that O is an extreme point of F_y and K lies in the half-plane $y \geq 0$. Denote by $A = (0, a_1)$ the other extreme point of F_y and assume, for example, that $a_1 > 0$. If we set $f := pr_1^2 \in P_2(K)$, then $T(f) \in P_2(K)$ so that

$$T(f)(x,y) = \alpha x^2 + \beta xy + \gamma y^2 + \delta x + \eta y + \theta \qquad ((x,y) \in K)$$

for some $\alpha, \beta, \gamma, \delta, \eta, \theta \in \mathbf{R}$.

Since $T(f) = f$ on $\partial_H K$ and, hence, on $\overline{OA}$, we get

$$\alpha x^2 + \delta x + \theta = x^2 \quad \text{for each } x \in [0, a_1],$$

so that $\alpha = 1$ and $\delta = \theta = 0$.

Therefore on $F := \partial_H K \setminus F_y$ we also have

$$x^2 = T(f)(x,y) = x^2 + \beta xy + \gamma y^2 + \eta y,$$

i.e., $\beta xy + \gamma y^2 + \eta y = 0$. Since F is contained in the half-plane $y > 0$, we infer that, in fact, F is contained in the straight line $s : \beta x + \gamma y + \eta = 0$.

Since $\partial_e K \subset \partial_H K$, $K \cap s$ is a maximal segment of ∂K, say $\overline{BC}$, or it reduces to a single (extreme) point.

We shall discuss separately the five possible cases (because of (A.1.2)):

I) $\{B, C\}$ is disjoint from $\{O, A\}$ and $F = \{B, C\}$.

II) $\{B, C\}$ is disjoint from $\{O, A\}$, $F = \overline{BC}$ and the segment $\overline{OB}$ is not parallel to $\overline{AC}$.

III) $\{B, C\}$ is disjoint from $\{O, A\}$, $F = \overline{BC}$ and the segment $\overline{OB}$ is parallel to $\overline{AC}$.

IV) B is not in $\{O, A\}$, and $C = O$ or $C = A$.

V) $K \cap s$ reduces to a single point.

We first show that cases I) and II) cannot occur. Indeed, in both cases, the segments $\overline{OB}$ and $\overline{AC}$ are maximal segments contained in ∂K which intersect $\partial_H K$ only in their extreme points. Moreover, there are at least two distinct directions

with respect to which every function $h \in H$ must be affine, by virtue of Lemma A.1.3. Thus necessarily every $h \in H$ is a polynomial of degree ≤ 2 and hence $H|_{F_y} \neq \mathscr{C}(F_y)$. By (A.1.19) we get that $F_y \not\subset \partial_H K$ and this is not possible.

In case III), K is a trapezium and, by Lemma A.1.3, every function $h \in H$ is affine on every segment of K parallel to $\overline{OB}$. Thus, T is the canonical projection associated with K.

If case IV) occurs and if, for instance, $C = O$ (the case $C = A$ is analogous) then, necessarily $\partial_H K = \overline{OA} \cup \{B\}$ or $\partial_H K = \overline{OA} \cup \overline{OB}$.

In the first case, by Lemma A.1.3, every $h \in H$ would be affine on each segment parallel to $\overline{OB}$ or to $\overline{AB}$ and, hence, would be a polynomial of degree ≤ 2. Hence, $H|_{F_y} \neq \mathscr{C}(F_y)$ contradicting (A.1.19).

Thus, necessarily $\partial_H K = \overline{OA} \cup \overline{OB}$ and, hence, every $h \in H$ is affine on each segment of K parallel to $\overline{AB}$. Therefore, K is the degenerate trapezium $\overline{AOOB}$ and T is the canonical projection associated with it.

The above reasoning also shows that the case V) cannot occur. □

Corollary A.1.5. *Let K be a convex polygon of $\mathbf{R}^2$ having n extreme points, $n \geq 3$, and let $T\colon \mathscr{C}(K) \longrightarrow \mathscr{C}(K)$ be a non-trivial positive projection satisfying* (3.1.2) *and* (A.1.1) *such that $T(P_2(K)) \subset P_2(K)$.*

Then $n \leq 4$. Moreover, we have only the following cases:

1) K is a triangle;

2) K is a parallelogram;

3) K is a trapezium (possible degenerate).

In each case T is the canonical projection associated with K.

Proof. If K is not a trapezium, by Theorem A.1.4 we have $\partial_H K = \partial_e K$. In this case, if $n = 3$, then K is a triangle (i.e., a simplex) and, by Lemma A.1.3, every $h \in H$ is affine on each segment of K parallel to some side of it. Accordingly, $H \subset A(K)$ and hence, by Theorem 1.1.12, T is the canonical projection associated with the simplex K.

Assume, now, that $n = 4$. Then the sides of K are two by two parallel, because, otherwise, they would determine at least three distinct directions and every $h \in H$ would be affine on each segment parallel to one of them. So, $H \subset A(K)$ and, hence, K would be a triangle.

Therefore K is a parallelogram. Moreover, given $f \in \mathscr{C}(K)$, $T(f)$ is affine on each segment parallel to a side of K and $T(f) = f$ on $\partial_H K = \partial_e K$, so that T is the canonical projection associated with K.

The above reasoning also shows that $n \leq 4$, because, otherwise, the sides would always determine at least three different directions and we would obtain a contradiction as before. □

Before stating our final result we proceed to prove the next

Theorem A.1.6. *Let K be a convex compact subset of $\mathbf{R}^2$ having non-empty interior and let $T\colon \mathscr{C}(K) \longrightarrow \mathscr{C}(K)$ be a non-trivial positive projection satisfying* (3.1.2) *and* (A.1.1) *such that $T(P_2(K)) \subset P_2(K)$.*

Assume that ∂K contains only one segment or that ∂K contains only two parallel segments.

Then K is conical and T is the canonical projection associated with K.

Proof. Choose a coordinate system Oxy such that the only segment of ∂K or one of the two segments of ∂K lies on the x-axis. Moreover, assume that O is an extreme point of the segment, which we shall denote by F, and K lies on the half-plane $y \geq 0$.

If $y = s$ is the straight line supporting K ($s > 0$), then there exist functions $p, q\colon [0, s] \longrightarrow \mathbf{R}$, $p \leq q$, p strictly convex and q strictly concave, such that $K = \{(x, y) \in \mathbf{R}^2 \mid p(y) \leq x \leq q(y)\}$ (see Fig. 12 and Fig. 13).

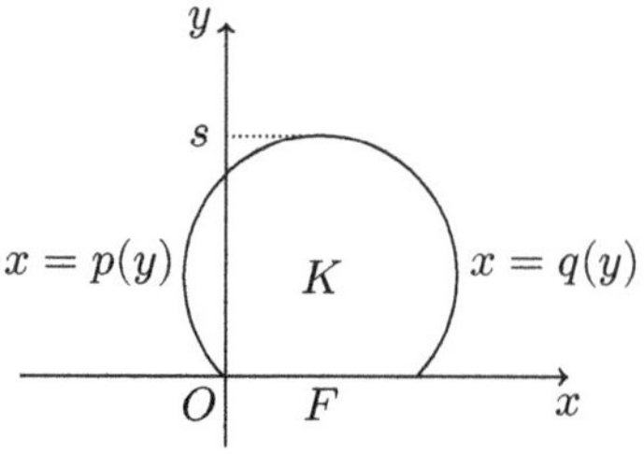

Fig. 12

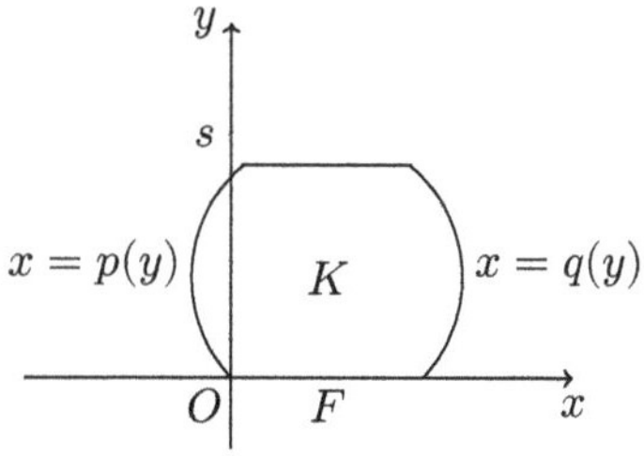

Fig. 13

Since K is not a trapezium, by Theorem A.1.4, we obtain that $\partial_H K = \partial_e K$. Accordingly, the curves $x = p(y)$ and $x = q(y)$ ($0 \leq y \leq s$) are contained in $\partial_H K$.

Since $F \cap \partial_H K = \partial_e F$, from Lemma A.1.3 it follows that for every $f \in \mathscr{C}(K)$, $T(f)$ is affine on each segment parallel to F, and hence, since $T(f) = f$ on $\partial_H K$, we get

$$T(f)(x, y) = f(p(y), y) + \frac{f(q(y), y) - f(p(y), y)}{q(y) - p(y)}\,(x - p(y)) \tag{i}$$

($(x, y) \in K$). In particular, for the function $f = pr_1^2$, we obtain

$$T(f)(x, y) = x(p(y) + q(y)) - p(y)q(y) \tag{ii}$$

($(x, y) \in K$).

On the other hand, $T(f) \in P_2(K)$, so that we can write

$$T(f)(x, y) = \alpha x^2 + 2\beta xy + \gamma y^2 + 2\delta x + 2\eta y + \theta \tag{iii}$$

($(x, y) \in K$).

From (ii) and (iii) it then follows that $\alpha = 0$.

Set $\varphi(y) := p(y) + q(y)$ and $\psi(y) := p(y)q(y)$ ($0 \leq y \leq s$). Then, for every $y \in [0, s]$, choosing $x_1, x_2 \in \mathbf{R}$, $x_1 \neq x_2$, such that (x_1, y), $(x_2, y) \in K$, and

subtracting (iii) from (ii), we obtain

$$(x_1 - x_2)\varphi(y) = 2(x_1 - x_2)(\beta y + \delta).$$

Hence $\varphi(y) = 2(\beta y + \delta)$ $(y \in [0, s])$.

Substituting in (ii) and comparing with the right-side member of (iii), we also get $\psi(y) = -\gamma y^2 - 2\eta y - \theta$ $(y \in [0, s])$.

Therefore, since $p \le q$, we obtain

$$\begin{aligned} p(y) &= \beta y + \delta - \sqrt{(\beta y + \delta)^2 + \gamma y^2 + 2\eta y + \theta} \\ &=: \beta y + \delta - \sqrt{my^2 + 2ny + r} \end{aligned}$$

and

$$\begin{aligned} q(y) &= \beta y + \delta + \sqrt{(\beta y + \delta)^2 + \gamma y^2 + 2\eta y + \theta} \\ &= \beta y + \delta + \sqrt{my^2 + 2ny + r} \end{aligned}$$

$(y \in [0, s])$.

Since q is strictly concave, for every $y \in [0, s]$, we have

$$0 > q''(y) = \frac{mr - n^2}{(my^2 + 2ny + r)^{3/2}}.$$

Hence, $n^2 - mr > 0$ and so, K is conical and T is the canonical projection associated with T. □

The next theorem is our conclusive result.

Theorem A.1.7. *Let K be a convex compact subset of $\mathbf{R}^2$ having non-empty interior. Then the following statements are equivalent:*

(a) K is P-admissible.

(b) There exists a non-trivial positive projection $T\colon \mathscr{C}(K) \longrightarrow \mathscr{C}(K)$ satisfying (3.1.2), (A.1.1) *and* (4.3.4) *or, alternatively,*

$$T(P_2(K)) \subset P_2(K). \tag{A.1.22}$$

(c) For the subset K only the following five cases occur:

(1) K is a triangle.

(2) K is a parallelogram.

(3) K is a trapezium.

(4) K is conical.

(5) the boundary of K is an ellipse.

Moreover, if one of the above statements holds true, then:

- *In case (1), there is a unique positive projection satisfying* (4.3.4), *namely that one associated with K as a simplex, and there are three positive projections satisfying (A.1.22), by regarding K as a degenerate trapezium.*

- *In case (2) there are no positive projections satisfying* (4.3.4) *and there are three positive projections satisfying (A.1.22), namely the canonical one and those associated with the main directions of K.*
- *In cases (3) and (4) there are no positive projections satisfying* (4.3.4) *and only one satisfying (A.1.22), namely the canonical one.*
- *In case (5), there is no projection satisfying* (4.3.4). *Moreover, every positive projection T on $\mathscr{C}(K)$ satisfying (A.1.22) (and* (3.1.2) *and* (A.1.1)*) is either*
  (i) the Poisson operator associated with a suitable strictly elliptic differential operator of the form (4.2.20) *(with* $d = 2$*) whose coefficients are constant and satisfy relation* (4.3.10),

  or

  (ii) the canonical projection associated with the ellipse and a suitable direction (Example A.1.2, 4).

Proof. We only need to show the implication $(b) \Rightarrow (c)$. Assume that statement (b) holds true. If the projection satisfies (4.3.4), by Theorem 4.3.3, K is a simplex, i.e., a triangle. Alternatively, assume that (A.1.22) holds true.

If K is strictly convex, then, by Theorem 4.3.8, ∂K is an ellipse. If K is not strictly convex, then ∂K contains some segments. First, consider the case where ∂K contains at least three segments F_1, F_2, F_3 having distinct directions.

Then K can be a trapezium and hence we have case (3). If K is not a trapezium, then by Theorem A.1.4 we have $\partial_H K = \partial_e K$ so that $F_i \cap \partial_H K = \partial_e F_i$ $(i = 1, 2, 3)$. Therefore, from Lemma A.1.3 it follows that $H = A(K)$, and hence K is a triangle, by Theorem 4.3.3.

Assume, now, that the segments contained in ∂K determine exactly two distinct directions. In this case ∂K contains at most four segments.

Then, necessarily, K is a polygon, because, otherwise, ∂K would contain an arc Γ of some curve and $\Gamma \subset \partial_e K$. Again by Theorem A.1.4, we should have $\partial_H K = \partial_e K$ and hence

$$\Gamma \subset \partial_H K \text{ and } F_i \cap \partial_H K = \partial_e F_i \qquad (i = 1, 2),$$

where F_1 and F_2 are two segments of ∂K parallel to the distinct directions.

By Lemma A.1.3, every function $h \in H$ would be affine on each segment parallel to F_1 or to F_2 and, hence, it would be a polynomial of degree ≤ 2.

Now, if $f \in \mathscr{C}(\Gamma)$ is arbitrarily chosen, denoted by $\tilde{f} \in \mathscr{C}(K)$ an extension of f to K, we would have that $f = \tilde{f} = T(\tilde{f})$ on Γ, i.e., f would be a polynomial of degree ≤ 2 and this is not possible.

Thus K is a polygon, so that, by Corollary A.1.5, K is a parallelogram.

Finally, assume that the segments (at most two) contained in ∂K determine exactly one direction. Then, by Theorem A.1.6, K is conical.

The last part of the statement, concerning the cases (1)–(4), follows from the previous results.

In case (5), suppose that (i) does not hold. Let C be the center of the ellipse. By Theorem 4.3.8 there exists a straight line R passing through C and intersecting ∂K in A and B, and there exist $\alpha, \beta \geq 0$, $\alpha + \beta = 1$, such that for any $f \in \mathscr{C}(K)$ we have that

$$T(f)(C) = \alpha\, f(A) + \beta\, f(B).$$

Since $A(K) \subset H$, from the above identity it follows that $C = \alpha\, A + \beta\, B$ so that $\alpha = \beta = 1/2$.

Accordingly, we obtain $h(A) + h(B) = 2\, h(C)$ for every $h \in H$.

Choose, now, two points $P, Q \in \mathrm{int}(K)$ such that the segment $\overline{PQ}$ is parallel to $\overline{AB}$ and there exists a translation of a homotetic image of K contained in K, with center C_1, whose boundary contains P and Q (see Fig. 14).

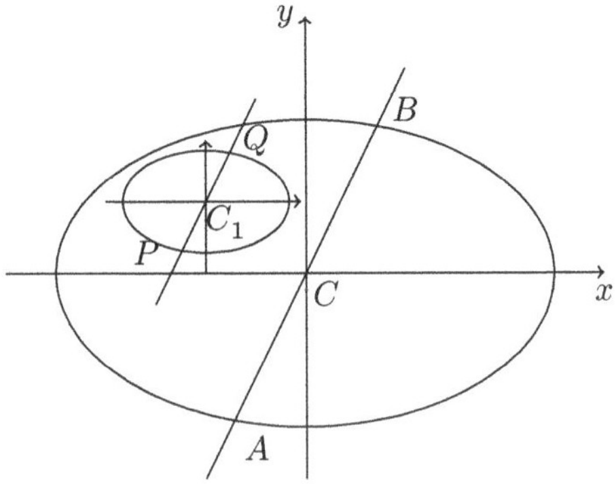

Fig. 14

Then, from (A.1.1) and the above remark, it follows that

$$h(P) + h(Q) = 2\, h(C_1) \qquad \text{for every } h \in H.$$

By using this reasoning, it can be proved that every $h \in H$ is locally affine in the interior of each segment of K parallel to R and, therefore, affine on all these segments. This means that T is associated with the ellipse and the direction R. □

A.2 Rate of convergence for the limit semigroup of Bernstein operators

As we have pointed out at the end of Section 4.4, an interesting and challenging problem which is concerned with Theorems 4.3.1, 4.3.2 and 4.3.4 deals with evaluating the rate of convergence in (4.3.2), i.e., estimating the quantities (4.4.22).

Here we give an answer to this problem in the case of the classical Bernstein operators on $\mathscr{C}([0,1])$.

We recall that the classical Bernstein operators B_n on $\mathscr{C}([0,1])$ are defined by setting, for every $f \in \mathscr{C}([0,1]), x \in [0,1]$ and $n \geq 1$,

$$B_n(f)(x) := \sum_{k=0}^{n} \binom{n}{k} x^k (1-x)^{n-k} f\left(\frac{k}{n}\right)$$

(see (3.1.11)) and that $(B_n)_{n\geq 1}$ is the sequence of the Bernstein-Schnabl operators associated with the Markov projection $T_1 : \mathscr{C}([0,1]) \longrightarrow \mathscr{C}([0,1])$ (see (1.1.49)) defined as

$$T_1(f)(x) := xf(1) + (1-x)f(0)$$

$(f \in \mathscr{C}([0,1]), x \in [0,1])$.

According to (4.2.2) and (4.2.4), the associated differential operator is given by

$$W_{T_1}(u)(x) = \frac{x(1-x)}{2} u''(x), \tag{A.2.1}$$

for every $u \in \mathscr{C}^2([0,1])$ and $x \in [0,1]$.

Moreover, by Corollary 4.5.6, the generator of the semigroup $(T(t))_{t\geq 0}$ associated with the B_n's (as in Theorem 4.3.1) is the closure $(A, D_V(A))$ of $(W_{T_1}, \mathscr{C}^2([0,1]))$, and it can be explicitly described as

$$D_V(A) := \left\{ u \in \mathscr{C}([0,1]) \cap \mathscr{C}^2(]0,1[) \;\middle|\; \lim_{\substack{x\to 0^+ \\ x\to 1^-}} x(1-x)u''(x) = 0 \right\}$$

and, for every $u \in D_V(A)$,

$$A(u)(x) := \begin{cases} \dfrac{x(1-x)}{2} u''(x) & \text{if } x \in]0,1[; \\ \\ 0 & \text{if } x = 0 \text{ or } x = 1. \end{cases}$$

As Theorem 4.5.7 shows, if $t \geq 0$ and if $(k(n))_{n\geq 1}$ is a sequence of positive integers satisfying $\lim\limits_{n\to\infty} \frac{k(n)}{n} = t$, then

$$T(t)(f) = \lim_{n\to\infty} B_n^{k(n)}(f) \quad \text{uniformly on } [0,1],$$

for every $f \in \mathscr{C}([0,1])$.

In order to evaluate the rate of the above convergence we recall that, if I is an interval of $\mathbf{R}$ and $m \geq 1$, we denote by $\mathscr{C}^m(I)$ the space of all m-times continuously differentiable functions on I.

Before stating the next result, we point out that, by direct computations, it is easy to check that, if $x \in [0,1]$ and $n \geq 1$, then

$$B_n(\psi_x)(x) = 0 \qquad B_n(\psi_x^2)(x) = \frac{x(1-x)}{n}, \tag{A.2.2}$$

$$B_n(\psi_x^3)(x) = \frac{x(1-x)(1-2x)}{n^2}, \tag{A.2.3}$$

and

$$B_n(\psi_x^4)(x) = \frac{3x^2(1-x)^2}{n^2} + \frac{x(1-x)(1-6x+6x^2)}{n^3}, \tag{A.2.4}$$

where $\psi_x(y) := y - x$ $(y \in [0,1])$.

Lemma A.2.1. *For all $g \in \mathscr{C}^4([0,1])$ and $n \geq 1$, one has*

$$\left\| B_n(g) - g - \frac{1}{n} W_{T_1}(g) \right\|_\infty \leq \frac{1}{384n^2}\left(16\|g^{(3)}\|_\infty + 7\|g^{(4)}\|_\infty\right). \tag{A.2.5}$$

Proof. Let $x \in [0,1]$ be fixed. By using Taylor's formula, we get

$$\left| g - g(x)e_0 - g'(x)\psi_x - \frac{1}{2}g''(x)\psi_x^2 - \frac{1}{6}g^{(3)}(x)\psi_x^3 \right| \leq \frac{1}{24}\|g^{(4)}\|_\infty \psi_x^4;$$

here we recall that $e_i(x) := x^i$ for every $i \geq 0$ and $x \in [0,1]$.

This entails, taking (A.2.2)-(A.2.4) into account,

$$\left| B_n(g)(x) - g(x) - \frac{1}{2}g''(x)\frac{x(1-x)}{n} - \frac{1}{6}g^{(3)}(x)\frac{x(1-x)(1-2x)}{n^2} \right|$$
$$\leq \frac{1}{24}\|g^{(4)}\|_\infty \left(\frac{3x^2(1-x)^2}{n^2} + \frac{x(1-x)(1-6x+6x^2)}{n^3} \right).$$

It follows that

$$\left| B_n(g)(x) - g(x) - \frac{1}{n}W_{T_1}(g)(x) \right| \leq \frac{x(1-x)|1-2x|}{6n^2}\|g^{(3)}\|_\infty$$
$$+ \frac{1}{24}\|g^{(4)}\|_\infty \left(\frac{3x^2(1-x)^2}{n^2} + \frac{x(1-x)|1-6x+6x^2|}{n^3} \right).$$

In order to get (A.2.5), it suffices to remark that $0 \leq x(1-x) \leq \frac{1}{4}$ and $\frac{|1-6x+6x^2|}{n^3} \leq \frac{1}{n^2}$.

□

With respect to the Markov semigroup $(T(t))_{t\geq 0}$ associated with the B_n's and to the operator W_{T_1} (see (A.2.1)), we can state the following result.

Lemma A.2.2. *For all $g \in \mathscr{C}^4([0,1])$ and $n \geq 1$, one has*

$$\left\|T\left(\frac{1}{n}\right)(g) - g - \frac{1}{n}W_{T_1}(g)\right\|_\infty \leq \frac{1}{128n^2}\left(8\|g''\|_\infty + 8\|g^{(3)}\|_\infty + \|g^{(4)}\|_\infty\right). \tag{A.2.6}$$

Proof. For a fixed $g \in \mathscr{C}^4([0,1])$, $n \geq 1$ and $x \in [0,1]$, we have

$$\begin{aligned} W_{T_1}^2(g)(x) &= W_{T_1}(W_{T_1}(g))(x) \\ &= \frac{x(1-x)}{4}\left(x(1-x)g^{(4)}(x) + 2(1-2x)g^{(3)}(x) - 2g''(x)\right). \end{aligned}$$

This leads to

$$\|W_{T_1}^2(g)\|_\infty \leq \frac{1}{16}\left(\frac{1}{4}\|g^{(4)}\|_\infty + 2\|g^{(3)}\|_\infty + 2\|g''\|_\infty\right). \tag{1}$$

On the other hand, for every $t \geq 0$,

$$\|T(t)(g) - g - tW_{T_1}(g)\|_\infty \leq \frac{t^2}{2}\|W_{T_1}^2(g)\|_\infty \tag{2}$$

(see the proof of Lemma 2.8 at p. 7 of [159]).

Now (A.2.6) is a consequence of (1) and (2). □

Before proving the next result, we need some preliminaries.

Let $m \geq 1$ and fix $a_0 < a_1 < \dots < a_m \in \mathbf{R}$ and $u(x) := (x - a_0)\cdots(x - a_m)$ $(x \in \mathbf{R})$.

If $f : [a_0, a_m] \longrightarrow \mathbf{R}$, the *divided difference* of f with respect to $a_0, \dots, a_m$ is defined by

$$[a_0, \dots, a_m; f] := \sum_{k=0}^{m} \frac{f(a_k)}{u'(a_k)}. \tag{A.2.7}$$

In particular, if $f \in \mathscr{C}^m([a_0, a_m])$, then there exists $\xi \in]a_0, a_m[$ such that

$$[a_0, \dots, a_m; f] = \frac{f^{m)}(\xi)}{m!}. \tag{A.2.8}$$

Definition A.2.3. *A function $f : I \longrightarrow \mathbf{R}$ is said to be* m-convex *on the interval I if for all $a_0 < a_1 < \dots < a_m$ in I one has $[a_0, \dots, a_m; f] \geq 0$.*

Remark A.2.4. Due to (A.2.8), if $f \in \mathscr{C}^m(I)$, f is m-convex if and only if $f^{(m)} \geq 0$ on I.

On the other hand, if $a \in I$ and $h > 0$ are such that $a, a+h, a+2h, \dots, a+mh \in I$, then

$$[a, a+h, a+2h, \dots, a+mh; f] = \frac{1}{m!\,h^m}\Delta_h^m f(a), \tag{A.2.9}$$

where $\Delta_h^m f(a)$ is the m-th difference of f with step h at point a and it is defined by (1.6.1). Hence, if f is m-convex, then $\Delta_h^m f(a) \geq 0$.

More details on m-convex functions may be found in [173].

In the next result, we prove that Bernstein operators preserve the m-convexity.

Proposition A.2.5. *If a function $f \in \mathscr{C}([0,1])$ is m-convex, then $B_n(f)$ is m-convex for every $n \geq 1$.*

Proof. Fix an m-convex function $f \in \mathscr{C}([0,1])$. According to [18, p. 460, Formula (2)], for every $n, m \geq 1$ and $x \in [0,1]$,

$$(B_n(f))^{(m)}(x)$$
$$= n(n-1)\dots(n-m+1)\sum_{h=0}^{n-m} \Delta_{1/n}^m f\left(\frac{h}{n}\right)\binom{n-m}{h} x^h(1-x)^{n-m-h}. \quad (1)$$

Then, taking Remark A.2.4 into account, we get that $(B_n(f))^{(m)} \geq 0$ for every $m \geq 1$ and this completes the proof.

□

After these preliminaries, we can state the following lemma.

Lemma A.2.6. *Let $p \in P_\infty([0,1])$ be a polynomial function, $t \geq 0$ and $m \in \mathbf{N}$. Then*

$$\|(T(t)(p))^{(m)}\|_\infty \leq \|p^{(m)}\|_\infty \exp\left(-\frac{(m-1)m}{2}t\right). \quad (A.2.10)$$

Proof. It is well known (see, e.g., [123]) that, for every $m \in \mathbf{N}$,

$$B_n(e_m) = \frac{n(n-1)\cdots(n-m+1)}{n^m} e_m + \text{terms of lower degree.}$$

Fix $t \geq 0$ and let $(k(n))_{n\geq 1}$ be a sequence of positive integers such that $\lim_{n\to\infty} \frac{k(n)}{n} = t$. Then

$$B_n^{k(n)}(e_m) = \left(\frac{n(n-1)\cdots(n-m+1)}{n^m}\right)^{k(n)} e_m + \text{terms of lower degree.}$$

This entails

$$T(t)(e_m) = e_m \exp\left(-\frac{(m-1)m}{2}t\right) + \text{terms of lower degree}$$

and, consequently,

$$(T(t)(e_m))^{(m)} = m! \exp\left(-\frac{(m-1)m}{2}t\right). \quad (1)$$

On the other hand,

$$\left(\frac{1}{m!}\|p^{(m)}\|_\infty e_m \pm p\right)^{(m)} \geq 0,$$

which means that the functions $\frac{1}{m!}\|p^{(m)}\|_\infty e_m \pm p$ are m-convex. Since each B_n preserves the m-convexity, the functions $B_n^{k(n)}\left(\frac{1}{m!}\|p^{(m)}\|_\infty e_m \pm p\right)$ are m-convex and, consequently, the functions $T(t)\left(\frac{1}{m!}\|p^{(m)}\|_\infty e_m \pm p\right)$ are m-convex. This yields

$$\left(T(t)\left(\frac{1}{m!}\|p^{(m)}\|_\infty e_m \pm p\right)\right)^{(m)} \geq 0. \tag{2}$$

Now (A.2.10) is a consequence of formulae (1) and (2).

□

Next, we state an estimate involving iterates of the B_n's.

Lemma A.2.7. *For any $p \in P_\infty([0,1])$ and $n, m \geq 1$,*

$$\left\|B_n^m(p) - T\left(\frac{m}{n}\right)(p)\right\|_\infty \leq \frac{m}{192n^2}\left(12\|p''\|_\infty + 20\|p^{(3)}\|_\infty + 5\|p^{(4)}\|_\infty\right). \tag{A.2.11}$$

Proof. First, we remark that

$$\left\|B_n^m(p) - T\left(\frac{m}{n}\right)(p)\right\|_\infty = \left\|\sum_{j=0}^{m-1}\left(B_n^{m-j}\, T\left(\frac{j}{n}\right) - B_n^{m-j-1}\, T\left(\frac{j+1}{n}\right)\right)(p)\right\|_\infty$$
$$= \left\|\sum_{j=0}^{m-1}\left(B_n^{m-j-1}\left(B_n - T\left(\frac{1}{n}\right)\right)T\left(\frac{j}{n}\right)\right)(p)\right\|_\infty$$
$$\leq \sum_{j=0}^{m-1}\left\|\left(B_n - T\left(\frac{1}{n}\right)\right)T\left(\frac{j}{n}\right)(p)\right\|_\infty.$$

For a fixed $j = 0, \ldots m-1$, consider the function $g := T\left(\frac{j}{n}\right)(p) \in P_\infty([0,1])$. By using Lemma A.2.1 and Lemma A.2.2, we get

$$\left\|\left(B_n - T\left(\frac{1}{n}\right)\right)T\left(\frac{j}{n}\right)(p)\right\|_\infty \leq \left\|B_n(g) - g - \frac{1}{n}W_{T_1}(g)\right\|_\infty$$
$$+\left\|T\left(\frac{1}{n}\right)(g) - g - \frac{1}{n}W_{T_1}(g)\right\|_\infty \leq \frac{1}{192n^2}\left(12\|g''\|_\infty + 20\|g^{(3)}\|_\infty + 5\|g^{(4)}\|_\infty\right).$$

On the other hand, Lemma A.2.6 shows that

$$\|g^{(i)}\|_\infty = \left\|\left(T\left(\frac{j}{n}\right)(p)\right)^{(i)}\right\|_\infty \leq \|p^{(i)}\|_\infty \qquad (i \geq 0).$$

Summing up, for all $j = 0, \ldots, m-1$,

$$\left\|\left(B_n - T\left(\frac{1}{n}\right)\right)T\left(\frac{j}{n}\right)(p)\right\|_\infty \leq \frac{1}{192n^2}\left(12\|p''\|_\infty + 20\|p^{(3)}\|_\infty + 5\|p^{(4)}\|_\infty\right),$$

which leads immediately to (A.2.11).

□

We are now in a position to prove the following fundamental result which is adapted from [109]. In that paper, the quantities $\|B_n^m(f) - T(t)(f)\|_\infty$ are majorized, for $f \in \mathscr{C}([0,1])$, in terms of moduli of smoothness.

Other results concerning the estimates (4.4.22) may be found, for example, in [138], [107], [109], [110], [120], [121], [108], [67], [68], [44], [45], [43].

Theorem A.2.8. *Let $f \in \mathscr{C}^4([0,1])$, $t \geq 0$ and $n, m \geq 1$. Then*

$$\|B_n^m(f) - T(t)(f)\|_\infty \leq \left(\frac{13m}{48n^2} + \frac{1}{8}\left|\frac{m}{n} - t\right|\right)\|f''\|_\infty + \frac{5m}{64n^2}\|f^{(4)}\|_\infty. \quad \text{(A.2.12)}$$

Proof. For each $k \geq 1$ we have

$$\begin{aligned}
&\|B_n^m(f) - T(t)(f)\|_\infty \\
&\leq \left\|B_n^m(f) - T\left(\frac{m}{n}\right)(f)\right\|_\infty + \left\|T\left(\frac{m}{n}\right)(f) - T(t)(f)\right\|_\infty \\
&\leq \|B_n^m(f) - B_n^m(B_k(f))\|_\infty + \left\|B_n^m(B_k(f)) - T\left(\frac{m}{n}\right)(B_k(f))\right\|_\infty \\
&+ \left\|T\left(\frac{m}{n}\right)(B_k(f)) - T\left(\frac{m}{n}\right)(f)\right\|_\infty + \left\|\int_t^{\frac{m}{n}} T(u)W_{T_1}(f)\,du\right\|_\infty.
\end{aligned}$$

Using Lemma A.2.7 and the fact that each B_n is a contraction and $(T(t))_{t\geq 0}$ is a Markov semigroup, we get

$$\begin{aligned}
&\|B_n^m(f) - T(t)(f)\|_\infty \\
&\leq 2\|B_k(f) - f\|_\infty + \left|\frac{m}{n} - t\right| \|W_{T_1}(f)\|_\infty \\
&+ \frac{m}{192n^2}\left(12\|(B_k(f))''\|_\infty + 20\|(B_k(f))^{(3)}\|_\infty + 5\|(B_k(f))^{(4)}\|_\infty\right).
\end{aligned}$$

From (A.2.8), (A.2.9) and formula (1) Proposition A.2.5 we infer that, for every $i \geq 0$,

$$\|(B_k(f))^{(i)}\|_\infty \leq \frac{k(k-1)\cdots(k-i+1)}{k^i}\|f^{(i)}\|_\infty \leq \|f^{(i)}\|_\infty.$$

Therefore,

$$\begin{aligned}
&\|B_n^m(f) - T(t)(f)\|_\infty \\
&\leq 2\|B_k(f) - f\|_\infty + \frac{1}{8}\left|\frac{m}{n} - t\right| \|f''\|_\infty \\
&+ \frac{m}{192n^2}\left(12\|f''\|_\infty + 20\|f^{(3)}\|_\infty + 5\|f^{(4)}\|_\infty\right).
\end{aligned}$$

Letting $k \to \infty$,

$$\begin{aligned}
&\|B_n^m(f) - T(t)(f)\|_\infty \\
&\leq \frac{1}{16}\left(2\left|\frac{m}{n} - t\right| + \frac{m}{n^2}\right)\|f''\|_\infty + \frac{5m}{192n^2}\left(4\|f^{(3)}\|_\infty + \|f^{(4)}\|_\infty\right).
\end{aligned}$$

Combining the previous inequality with the Landau's inequality (see, e.g., [145, formula (3.2.71)])

$$\|f^{(3)}\|_\infty \le 2\|f''\|_\infty + \frac{1}{2}\|f^{(4)}\|_\infty,$$

we get (A.2.12). □

Let again $(k(n))_{n\ge1}$ be a sequence of positive integers such that $\lim_{n\to\infty} \frac{k(n)}{n} = t$; for $m := k(n)$, (A.2.12) becomes

$$\|B_n^{k(n)}(f) - T(t)(f)\|_\infty \le \left(\frac{13}{48n}\frac{k(n)}{n} + \frac{1}{8}\left|\frac{k(n)}{n} - t\right|\right)\|f''\|_\infty + \frac{5}{64n}\frac{k(n)}{n}\|f^{(4)}\|_\infty. \tag{A.2.13}$$

In particular, for every $n \ge 1$, let $k(n) = [nt]$; then $\frac{k(n)}{n} \le t$ and $\left|\frac{k(n)}{n} - t\right| \le \frac{1}{n}$. Accordingly, from (A.2.13), we get that

$$\|B_n^{[nt]}(f) - T(t)(f)\|_\infty \le \frac{13t+6}{48n}\|f''\|_\infty + \frac{5t}{64n}\|f^{(4)}\|_\infty, \tag{A.2.14}$$

for every $f \in \mathscr{C}^4([0,1])$, $n \ge 1$ and $t \ge 0$.

Bibliography

[1] R. A. Adams, *Sobolev Spaces*, Pure and Applied Mathematics, vol. 65, Academic Press [A subsidiary of Harcourt Brace Jovanovich, Publishers], New York-London, 1975. MR 0450957 (56 #9247).

[2] J. A. Adell, J. de la Cal, and I. Raşa, *Lototsky-Schnabl operators on the unit interval*, Rend. Circ. Mat. Palermo (2) **48** (1999), no. 3, 517–536. MR 1731452 (2000j:41027).

[3] A. A. Albanese, M. Campiti, and E. M. Mangino, *Regularity properties of semigroups generated by some Fleming-Viot type operators*, J. Math. Anal. Appl. **335** (2007), no. 2, 1259–1273. MR 2346904 (2009c:35177).

[4] A. A. Albanese and E. M. Mangino, *Analyticity of a class of degenerate evolution equations on the canonical simplex of* $\mathbb{R}^d$ *arising from Fleming-Viot processes*, J. Math. Anal. Appl. **379** (2011), no. 1, 401–424. MR 2776482 (2012d:35106).

[5] E. M. Alfsen, *Compact Convex Sets and Boundary Integrals*, Ergebnisse der Mathematik und ihrer Grenzgebiete, Band 57, Springer-Verlag, New York-Heidelberg, 1971. MR 0445271 (56 #3615).

[6] C. D. Aliprantis and O. Burkinshaw, *Positive Operators*, Pure and Applied Mathematics, vol. 119, Academic Press, Inc., Orlando, FL, 1985. MR 809372 (87h:47086).

[7] F. Altomare, *Proiettori positivi, famiglie risolventi e problema di Dirichlet*, Ricerche Mat. **26** (1977), no. 1, 63–78 (Italian). MR 0626371 (58 #30099).

[8] ———, *Limit semigroups of Bernstein-Schnabl operators associated with positive projections*, Ann. Scuola Norm. Sup. Pisa Cl. Sci. (4) **16** (1989), no. 2, 259–279. MR 1041898 (91a:47036).

[9] ———, *Lototsky-Schnabl operators on compact convex sets and their associated limit semigroups*, Monatsh. Math. **114** (1992), no. 1, 1–13. MR 1184800 (94c:47044).

[10] ———, *Asymptotic formulae for Bernstein-Schnabl operators and smoothness*, Boll. Unione Mat. Ital. (9) **2** (2009), no. 1, 135–150, Correngendum, Boll. Unione Mat. Ital. (9) **4** (2011), no. 2, 259–262. MR 2493648 (2010a:41025).

[11] ———, *Korovkin-Type Theorems and Approximation by Positive Linear Operators*, Surv. Approx. Theory **5** (2010), 92–164, free available online at http://www.math.techmion.ac.il/sat/papers/13/. MR 2721174 (2011g:41023).

[12] ———, *On some convergence criteria for nets of positive operators on continuous function spaces*, J. Math. Anal. Appl. **398** (2013), no. 2, 542–552. MR 2990079.

[13] F. Altomare and R. Amiar, *Asymptotic formulae for positive linear operators*, Math. Balkanica (N.S.) **16** (2002), no. 1-4, 283–304, Dedicated to the 70th anniversary of Academician Blagovest Sendov. MR 1933998 (2003k:41033).

[14] ———, *Approximation by positive operators of the C_0-semigroups associated with one-dimensional diffusion equations. I*, Numer. Funct. Anal. Optim. **26** (2005), no. 1, 1–15. MR 2128741 (2006f:41029).

[15] ———, *Approximation by positive operators of the C_0-semigroups associated with one-dimensional diffusion equations. II*, Numer. Funct. Anal. Optim. **26** (2005), no. 1, 17–33. MR 2128742 (2006f:41030).

[16] F. Altomare and A. Attalienti, *Degenerate evolution equations in weighted continuous function spaces, Markov processes and the Black-Scholes equation. I*, Results Math. **42** (2002), no. 3-4, 193–211. MR 1946740 (2003k:34103).

[17] ———, *Degenerate evolution equations in weighted continuous function spaces, Markov processes and the Black-Scholes equation. II*, Results Math. **42** (2002), no. 3-4, 212–228. MR 1946741 (2003k:34104).

[18] F. Altomare and M. Campiti, *Korovkin-Type Approximation Theory and its Applications*, de Gruyter Studies in Mathematics, vol. 17, Walter de Gruyter & Co., Berlin, 1994, Appendix A by Michael Pannenberg and Appendix B by Ferdinand Beckhoff. MR 1292247 (95g:41001).

[19] F. Altomare, M. Cappelletti Montano, and S. Diomede, *Degenerate elliptic operators, Feller semigroups and modified Bernstein-Schnabl operators*, Math. Nachr. **284** (2011), no. 5-6, 587–607. MR 2663755 (2012d:47116).

[20] F. Altomare, M. Cappelletti Montano, and V. Leonessa, *Iterates of multidimensional Kantorovich-type operators and their associated positive C_0-semigroups*, Stud. Univ. Babeş-Bolyai Math. **56** (2011), no. 2, 219–235. MR 2843684.

[21] F. Altomare, M. Cappelletti Montano, V. Leonessa, and I. Raşa, *On differential operators associated with Markov operators*, J. Funct. Anal. **266** (2014), no. 6, 3612–3631. MR 3165236.

[22] ———, *On Markov operators preserving polynomials*, J. Math. Anal. Appl. **415** (2014), no. 1, 477–495. MR 3173179.

[23] F. Altomare and S. Diomede, *Asymptotic formulae for positive linear operators on convex subsets of Banach spaces*, J. Concr. Appl. Math. **3** (2005), no. 4, 413–435. MR 2224027 (2007a:41041).

[24] ———, *Asymptotic formulae for positive linear operators: direct and converse results*, Jaen J. Approx. **2** (2010), no. 2, 255–287.

[25] F. Altomare and V. Leonessa, *An invitation to the study of evolution equations by means of positive linear operators*, Lecture Notes of Seminario Interdisciplinare di Matematica. Volume VIII, Lect. Notes Semin. Interdiscip. Mat., vol. 8, Semin. Interdiscip. Mat. (S.I.M.), Potenza, 2009, pp. 1–41.

[26] F. Altomare, V. Leonessa, and S. Milella, *Continuous selections of Borel measures and Bernstein-Schnabl operators*, Numerical analysis and approximation theory, Casa Cărţii de ştiinţă, Cluj-Napoca, 2006, pp. 1–26. MR 2281966 (2007k:41056).

[27] ———, *Cores for second-order differential operators on real intervals*, Commun. Appl. Anal. **13** (2009), no. 4, 477–496. MR 2583584 (2010k:47095).

[28] F. Altomare, V. Leonessa, and I. Raşa, *On Bernstein-Schnabl operators on the unit interval*, Z. Anal. Anwend. **27** (2008), no. 3, 353–379. MR 2411481 (2009c:41042).

[29] F. Altomare and S. Milella, *On the C_0-semigroups generated by second order differential operators on the real line*, Taiwanese J. Math. **13** (2009), no. 1, 25–46. MR 2489306 (2010b:35183).

[30] F. Altomare, S. Milella, and G. Musceo, *On a class of positive C_0-semigroups of operators on weighted continuous function spaces*, Note Mat. **31** (2011), no. 1, 15–27. MR 2943046.

[31] F. Altomare and G. Musceo, *Markov processes and positive semigroups on some classes of weighted continuous function spaces*, Rend. Circ. Mat. Palermo (2) **57** (2008), no. 1, 61–82. MR 2420523 (2009m:47108).

[32] ———, *Positive semigroups generated by degenerate second-order differential operators*, Funkcial. Ekvac. **51** (2008), no. 3, 431–458. MR 2493877 (2010b:35255).

[33] ———, *Degenerate second order differential operators with generalized reflecting barriers boundary conditions*, Math. Rep. (Bucur.) **12(62)** (2010), no. 2, 101–117. MR 2723369 (2011j:47125).

[34] F. Altomare and I. Raşa, *Towards a characterization of a class of differential operators associated with positive projections*, Atti Sem. Mat. Fis. Univ. Modena, Suppl. al Vol. **XLVI** (1998), 3–38, Dedicated to Prof. C. Vinti (Perugia, 1996). MR 1645708 (99h:35072).

[35] ———, *Feller semigroups, Bernstein type operators and generalized convexity associated with positive projections*, New developments in approximation theory (Dortmund, 1998), Internat. Ser. Numer. Math., vol. 132, Birkhäuser, Basel, 1999, pp. 9–32. MR 1724908 (2000m:41036).

[36] ———, *Lipschitz contractions, unique ergodicity and asymptotics of Markov semigroups*, Boll. Unione Mat. Ital. (9) **5** (2012), no. 1, 1–17. MR 2919646.

[37] G. A. Anastassiou, C. Cottin, and H. Gonska, *Global smoothness of approximating functions*, Analysis **11** (1991), no. 1, 43–57. MR 1113067 (92h:41047).

[38] G. A. Anastassiou and S. G. Gal, *Approximation Theory. Moduli of Continuity and Global Smoothness Preservation*, Birkhäuser Boston, Inc., Boston, MA, 2000. MR 1727161 (2000k:41001).

[39] O. Aramă, *Propriétés concernant la monotonie de la suite des polynômes d'interpolation de S. N. Bernšteĭn et leur application à l'étude de l'approximation des fonctions*, Acad. R. P. Rom. Fil. Cluj. Stud. Cerc. Mat. **8** (1957),

195–210. MR 0124674 (23 #A1986a).

[40] W. Arendt, A. Grabosch, G. Greiner, U. Groh, H. P. Lotz, U. Moustakas, R. Nagel, F. Neubrander, and U. Schlotterbeck, *One-Parameter Semigroups of Positive Operators*, Lecture Notes in Mathematics, vol. 1184, Springer-Verlag, Berlin, 1986. MR 839450 (88i:47022).

[41] D. H. Armitage, *A linear function from a space of polynomials onto a space of harmonic polynomials*, J. London Math. Soc. (2) **5** (1972), 529–538. MR 0315152 (47 #3701).

[42] A. Attalienti and M. Campiti, *Degenerate evolution problems and beta-type operators*, Studia Math. **140** (2000), no. 2, 117–139. MR 1784628 (2001i:41028).

[43] A. Attalienti, R. M. Mininni, and I. Raşa, *Gamma-type operators and the Black-Scholes semigroup*, J. Approx. Theory **163** (2011), no. 2, 163–182. MR 2754988 (2011m:41038).

[44] A. Attalienti and I. Raşa, *Overiterated linear operators and asymptotic behaviour of semigroups*, Mediterr. J. Math. **5** (2008), no. 3, 315–324. MR 2465578 (2009k:47112).

[45] ———, *Shape-preserving properties and asymptotic behaviour of the semigroup generated by the Black-Scholes operator*, Czechoslovak Math. J. **58** (2008), no. 2, 457–467. MR 2411101 (2010d:47059).

[46] A. Avez, *Differential Calculus*, A Wiley-Interscience Publication, John Wiley & Sons, Ltd., Chichester, 1986, Translated from the French by D. Edmunds. MR 839642 (87d:58001).

[47] H. Bauer, *Un problème de Dirichlet pour la frontière de Shilov d'un espace compact*, C. R. Acad. Sci. Paris **247** (1958), 843–846.

[48] ———, *Probability Theory and Elements of Measure Theory*, Academic Press, Inc. [Harcourt Brace Jovanovich, Publishers], London-New York, 1981, Second edition of the translation by R. B. Burckel from the third German edition, Probability and Mathematical Statistics. MR 636091 (82k:60001).

[49] ———, *Probability Theory*, de Gruyter Studies in Mathematics, vol. 23, Walter de Gruyter & Co., Berlin, 1996, Translated from the fourth (1991) German edition by Robert B. Burckel and revised by the author. MR 1385460 (97f:60001).

[50] ———, *Measure and Integration Theory*, de Gruyter Studies in Mathematics, vol. 26, Walter de Gruyter & Co., Berlin, 2001, Translated from the German by Robert B. Burckel. MR 1897176 (2003a:28001).

[51] M. Becker and R. J. Nessel, *Iteration von Operatoren und Saturation in lokalkonvexen Räumen*, Forschungsberichte des Landes Nordrhein-Westfalen **2470** (1975), 27–49.

[52] H. Berens and R. DeVore, *Quantitative Korovkin theorems for positive linear operators on L_p-spaces*, Trans. Amer. Math. Soc. **245** (1978), 349–361.

[53] H. Berens and G. G. Lorentz, *Inverse theorems for Bernstein polynomials*, Indiana Univ. Math. J. **21** (1971/72), 693–708. MR 0296579 (45 #5638).

[54] M. Beśka and K. Dziedziul, *Saturation theorems for interpolation and the Bernstein-Schnabl operator*, Math. Comp. **70** (2001), no. 234, 705–717. MR 1677470 (2001g:41034).

[55] E. Bishop and K. de Leeuw, *The representations of linear functionals by measures on sets of extreme points*, Ann. Inst. Fourier. Grenoble **9** (1959), 305–331. MR 0114118 (22 #4945).

[56] R. M. Blumenthal and R. K. Getoor, *Markov Processes and Potential Theory*, Pure and Applied Mathematics, vol. 29, Academic Press, New York-London, 1968. MR 0264757 (41 #9348).

[57] M. Blümlinger, *Approximation durch Bernsteinoperatoren auf kompakten konvexen Mengen*, Österreich. Akad. Wiss. Math.-Natur. Kl. Sitzungsber. II **196** (1987), no. 4-7, 181–215. MR 963217 (89j:41035).

[58] J.-M. Bony, P. Courrège, and P. Priouret, *Semi-groupes de Feller sur une variété à bord compacte et problèmes aux limites intégro-différentiels du second ordre donnant lieu au principe du maximum*, Ann. Inst. Fourier (Grenoble) **18** (1968), no. 2, 369–521 (1969). MR 0245085 (39 #6397).

[59] N. Bourbaki, *Éléments de Mathématique. Fasc. XXXV. Livre VI: Intégration, Ch. 1-9*, Actualités Scientifiques et Industrielles, No. 1343, Hermann, Paris, 1969. MR 0276436 (43 #2183).

[60] B. M. Brown, D. Elliott, and D. F. Paget, *Lipschitz constants for the Bernstein polynomials of a Lipschitz continuous function*, J. Approx. Theory **49** (1987), no. 2, 196–199. MR 874953 (88d:41023).

[61] P. L. Butzer and H. Berens, *Semi-Groups of Operators and Approximation*, Die Grundlehren der Mathematischen Wissenschaften, Band 145, Springer-Verlag New York Inc., New York, 1967. MR 0230022 (37 #5588).

[62] P. L. Butzer and R. J. Nessel, *Fourier Analysis and Approximation. Volume 1: One-Dimensional Theory*, Pure and Applied Mathematics, vol. 40, Academic Press, New York-London, 1971. MR 0510857 (58 #23312).

[63] M. Campiti, *Recursive Bernstein operators and degenerate diffusion processes*, Acta Sci. Math. (Szeged) **68** (2002), no. 1-2, 179–201. MR 1916575 (2003g:41035).

[64] M. Campiti and G. Metafune, *Approximation of solutions of some degenerate parabolic problems*, Numer. Funct. Anal. Optim. **17** (1996), no. 1-2, 23–35. MR 1391871 (97b:41008).

[65] M. Campiti, G. Metafune, and D. Pallara, *General Voronovskaja formula and solutions of second-order degenerate differential equations*, Rev. Roumaine Math. Pures Appl. **44** (1999), no. 5-6, 755–766 (2000), In honour of Academician Romulus Cristescu on his 70th birthday. MR 1839675 (2002e:35011).

[66] M. Campiti and I. Raşa, *Qualitative properties of a class of Fleming-Viot operators*, Acta Math. Hungar. **103** (2004), no. 1-2, 55–69. MR 2047873 (2005b:47084).

[67] M. Campiti and C. Tacelli, *Rate of convergence in Trotter's approximation theorem*, Constr. Approx. **28** (2008), no. 3, 333–341. MR 2453370 (2009i:41018).

[68] ______, *Trotter's approximation of semigroups and order of convergence in $C^{2,\alpha}$-spaces*, J. Approx. Theory **162** (2010), no. 12, 2303–2316. MR 2736885 (2011m:47076).

[69] M. Cappelletti Montano and S. Diomede, *Modified Bernstein-Schnabl operators on convex compact subsets of locally convex spaces and their limit semigroups*, Commun. Appl. Anal. **13** (2009), no. 4, 609–632. MR 2583593 (2010m:41030).

[70] A. S. Cavaretta, Jr. and A. Sharma, *Variation diminishing properties and convexity for the tensor product Bernstein operator*, Functional analysis and operator theory (New Delhi, 1990), Lecture Notes in Math., vol. 1511, Springer, Berlin, 1992, pp. 18–32. MR 1180745 (93k:41010).

[71] S. Cerrai and P. Clément, *On a class of degenerate elliptic operators arising from Fleming-Viot processes*, J. Evol. Equ. **1** (2001), no. 3, 243–276, Dedicated to Ralph S. Phillips. MR 1861222 (2002j:60136).

[72] ______, *Schauder estimates for a degenerate second order elliptic operator on a cube*, J. Differential Equations **242** (2007), no. 2, 287–321. MR 2363317 (2009h:35174).

[73] G. Z. Chang and P. J. Davis, *The convexity of Bernstein polynomials over triangles*, J. Approx. Theory **40** (1984), no. 1, 11–28. MR 728296 (85c:41001).

[74] G. Z. Chang and Y. Y. Feng, *An improved condition for convexity of Bernstein-Bézier surfaces over triangles*, Computer Aided Geometric Design **1** (1984), 279–293.

[75] G. Z. Chang and J. Zhang, *Converse theorems of convexity for Bernstein polynomials over triangles*, J. Approx. Theory **61** (1990), no. 3, 265–278. MR 1054540 (91j:41016).

[76] G. Choquet, *Existence et unicité des représentations intégrales*, Séminaire Bourbaki (1956), 15 pp.

[77] ______, *Lectures on Analysis. Vol. I and Vol. II*, Edited by J. Marsden, T. Lance and S. Gelbart, W. A. Benjamin, Inc., New York-Amsterdam, 1969. MR 0250011 (40 #3252).

[78] P. Clément and C. A. Timmermans, *On C_0-semigroups generated by differential operators satisfying Ventcel's boundary conditions*, Nederl. Akad. Wetensch. Indag. Math. **48** (1986), no. 4, 379–387. MR 869754 (88c:47075).

[79] W. Dahmen, *Convexity and Bernstein-Bézier polynomials*, Curves and surfaces (Chamonix-Mont-Blanc, 1990), Academic Press, Boston, MA, 1991, pp. 107–134. MR 1123725 (92h:65018).

[80] W. Dahmen and C. A. Micchelli, *Convexity and Bernstein polynomials on k-simploids*, Acta Math. Appl. Sinica (English Ser.) **6** (1990), no. 1, 50–66. MR 1049635 (91f:41006).

[81] F. Dai and Y. Xu, *Moduli of smoothness and approximation on the unit sphere and the unit ball*, Adv. Math. **224** (2010), no. 4, 1233–1310. MR 2646298 (2011f:41028).

[82] T. Dankel, Jr., *On moduli of continuity of analytic and harmonic functions*, Rocky Mountain J. Math. **9** (1979), no. 3, 519–526. MR 528749 (81b:31003).

[83] E. B. Davies, *One-Parameter Semigroups*, London Mathematical Society Monographs, vol. 15, Academic Press, Inc. [Harcourt Brace Jovanovich, Publishers], London-New York, 1980. MR 591851 (82i:47060).

[84] K. de Leeuw, *On the degree of approximation by Bernstein polynomials*, J. Analyse Math. **7** (1959), 89–104. MR 0113075 (22 #3916).

[85] R. A. DeVore, *The Approximation of Continuous Functions by Positive Linear Operators*, Lecture Notes in Mathematics, Vol. 293, Springer-Verlag, Berlin-New York, 1972. MR 0420083 (54 #8100).

[86] R. A. DeVore and G. G. Lorentz, *Constructive Approximation*, Grundlehren der Mathematischen Wissenschaften [Fundamental Principles of Mathematical Sciences], vol. 303, Springer-Verlag, Berlin, 1993. MR 1261635 (95f:41001).

[87] A. Dinghas, *Über einige Identitäten vom Bernsteinschen Typus*, Norske Vid. Selsk. Forh., Trondheim **24** (1951), 96–97 (German). MR 0049394 (14,167b).

[88] Z. Ditzian, *Polynomial approximation and $\omega_{\phi}^{r}(f,t)$ twenty years later*, Surv. Approx. Theory **3** (2007), 106–151. MR 2342231 (2008f:41010).

[89] Z. Ditzian and V. Totik, *Moduli of Smoothness*, Springer Series in Computational Mathematics, vol. 9, Springer-Verlag, New York, 1987. MR 914149 (89h:41002).

[90] J. R. Dorroh, *Contraction semi-groups in a function space*, Pacific J. Math. **19** (1966), 35–38. MR 0201983 (34 #1860).

[91] E. B. Dynkin, *Theory of Markov Processes*, Translated from the Russian by D. E. Brown; edited by T. Köváry, Prentice-Hall, Inc., Englewood Cliffs, N.J.; Pergamon Press, Oxford-London-Paris, 1961. MR 0131900 (24 #A1747).

[92] T. Eisner, *Stability of Operators and Operator Semigroups*, Operator Theory: Advances and Applications, vol. 209, Birkhäuser Verlag, Basel, 2010. MR 2681062 (2011f:47073).

[93] K.-J. Engel and R. Nagel, *One-Parameter Semigroups for Linear Evolution Equations*, Graduate Texts in Mathematics, vol. 194, Springer-Verlag, New York, 2000, with contributions by S. Brendle, M. Campiti, T. Hahn, G. Metafune, G. Nickel, D. Pallara, C. Perazzoli, A. Rhandi, S. Romanelli and R. Schnaubelt. MR 1721989 (2000i:47075).

[94] R. Engelking, *General Topology*, Second ed., Sigma Series in Pure Mathematics, vol. 6, Heldermann Verlag, Berlin, 1989, Translated from the Polish by the author. MR 1039321 (91c:54001).

[95] S. N. Ethier, *A class of degenerate diffusion processes occurring in population genetics*, Comm. Pure Appl. Math. **29** (1976), no. 5, 483–493. MR 0428488 (55 #1509).

[96] S. N. Ethier and T. G. Kurtz, *Markov Processes*, Wiley Series in Probability and Mathematical Statistics: Probability and Mathematical Statistics, John Wiley & Sons, Inc., New York, 1986, Characterization and convergence. MR 838085 (88a:60130).

[97] ______, *Fleming-Viot processes in population genetics*, SIAM J. Control Optim. **31** (1993), no. 2, 345–386. MR 1205982 (94d:60131).

[98] W. Feller, *The parabolic differential equations and the associated semi-groups of transformations*, Ann. of Math. (2) **55** (1952), 468–519. MR 0047886 (13,948a).

[99] ______, *On the generation of unbounded semi-groups of bounded linear operators*, Ann. of Math. (2) **58** (1953), 166–174. MR 0055572 (14,1093g).

[100] ______, *Diffusion processes in one dimension*, Trans. Amer. Math. Soc. **77** (1954), 1–31. MR 0063607 (16,150d).

[101] W. H. Fleming and M. Viot, *Some measure-valued Markov processes in population genetics theory*, Indiana Univ. Math. J. **28** (1979), no. 5, 817–843. MR 542340 (81a:60059).

[102] I. Gavrea, H. Gonska, R. Păltănea, and G. Tachev, *General estimates for the Ditzian-Totik modulus*, East J. Approx. **9** (2003), no. 2, 175–194. MR 1991849 (2004e:41006).

[103] I. Gavrea and M. Ivan, *On the iterates of positive linear operators preserving the affine functions*, J. Math. Anal. Appl. **372** (2010), no. 2, 366–368. MR 2678868 (2011g:41027).

[104] ______, *Asymptotic behaviour of the iterates of positive linear operators*, Abstr. Appl. Anal. (2011), Art. ID 670509, 11. MR 2802830 (2012e:41029).

[105] ______, *On the iterates of positive linear operators*, J. Approx. Theory **163** (2011), no. 9, 1076–1079. MR 2832743 (2012h:41031).

[106] J. A. Goldstein, *Semigroups of Linear Operators and Applications*, Oxford Mathematical Monographs, The Clarendon Press, Oxford University Press, New York, 1985. MR 790497 (87c:47056).

[107] H. Gonska, M. Heilmann, and I. Raşa, *Convergence of iterates of genuine and ultraspherical Durrmeyer operators to the limiting semigroup: C^2-estimates*, J. Approx. Theory **160** (2009), no. 1-2, 243–255. MR 2558024 (2010k:41015).

[108] H. Gonska, D. Kacsó, and P. Piţul, *The degree of convergence of over-iterated positive linear operators*, J. Appl. Funct. Anal. **1** (2006), no. 4, 403–423. MR 2220800 (2006m:41036).

[109] H. Gonska and I. Raşa, *The limiting semigroup of the Bernstein iterates: degree of convergence*, Acta Math. Hungar. **111** (2006), no. 1-2, 119–130. MR 2188976 (2007b:41005).

[110] ———, *On infinite products of positive linear operators reproducing linear functions*, Positivity **17** (2013), no. 1, 67–79. MR 3027646.

[111] H. Gonska, I. Raşa, and M.-D. Rusu, *Applications of an Ostrowski-type inequality*, J. Comput. Anal. Appl. **14** (2012), no. 1, 19–31. MR 2920807.

[112] M. W. Grossman, *Note on a generalized Bohman-Korovkin theorem*, J. Math. Anal. Appl. **45** (1974), 43–46. MR 0336171 (49 #947).

[113] ———, *Lototsky-Schnabl functions on compact convex sets*, J. Math. Anal. Appl. **55** (1976), no. 2, 525–530. MR 0447927 (56 #6237).

[114] O. Hájek, *Uniform polynomial approximation*, The American Mathematical Monthly **72** (1965), no. 6, 681.

[115] E. Hille, *Functional Analysis and Semi-Groups*, American Mathematical Society Colloquium Publications, vol. 31, American Mathematical Society, New York, 1948.

[116] E. Hille and R. S. Phillips, *Functional Analysis and Semi-Groups, Rev. Ed.*, American Mathematical Society Colloquium Publications, vol. 31, American Mathematical Society, Providence, R. I., 1957. MR 0089373 (19,664d).

[117] A. Hinkkanen, *Modulus of continuity of harmonic functions*, J. Analyse Math. **51** (1988), 1–29. MR 963148 (89m:31002).

[118] K. Itô, *Stochastic Processes*, Iwanami-Shoten, Tokyo, 1957 (Japanese).

[119] M. A. Jiménez Pozo, *Déformation de la convexité et théorèmes du type Korovkin*, C. R. Acad. Sci. Paris Sér. A-B **290** (1980), no. 5, A213–A215. MR 564313 (81a:41054).

[120] D. Kacsó, *Certain Bernstein-Durrmeyer type operators preserving linear functions*, Habilitationsschrift, Univ. Duisburg-Essen, SM-DU-675, 2006.

[121] ———, *The limiting semigroup of the genuine Bernstein-Durrmeyer iterates*, Results Math. **53** (2009), no. 3-4, 303–310. MR 2524732 (2010g:41008).

[122] S. Karlin and W. J. Studden, *Tchebycheff Systems: with Applications in Analysis and Statistics*, Pure and Applied Mathematics, Vol. XV, Interscience Publishers John Wiley & Sons, New York-London-Sydney, 1966. MR 0204922 (34 #4757).

[123] R. P. Kelisky and T. J. Rivlin, *Iterates of Bernstein polynomials*, Pacific J. Math. **21** (1967), 511–520. MR 0212457 (35 #3328).

[124] R. A. Khan, *Some probabilistic methods in the theory of approximation operators*, Acta Math. Acad. Sci. Hungar. **35** (1980), no. 1-2, 193–203. MR 588894 (81m:41024).

[125] P. P. Korovkin, *On convergence of linear positive operators in the space of continuous functions*, Doklady Akad. Nauk SSSR (N.S.) **90** (1953), 961–964. MR 0057472 (15,236a).

[126] ______, *Linear Operators and Approximation Theory*, Translated from the Russian ed. (1959). Russian Monographs and Texts on Advanced Mathematics and Physics, Vol. III, Gordon and Breach Publishers, Inc., New York; Hindustan Publishing Corp. (India), Delhi, 1960. MR 0150565 (27 #561).

[127] U. Krengel, *Ergodic Theorems*, de Gruyter Studies in Mathematics, vol. 6, Walter de Gruyter & Co., Berlin, 1985. MR 797411 (87i:28001).

[128] T. G. Kurtz, *A general theorem on the convergence of operator semigroups*, Trans. Amer. Math. Soc. **148** (1970), 23–32. MR 0256210 (41 #867).

[129] J. Lamperti, *Stochastic Processes. A Survey of the Mathematical Theory*, Applied Mathematical Sciences, vol. 23, Springer-Verlag, New York-Heidelberg, 1977. MR 0461600 (57 #1585).

[130] T. Lindvall, *Bernstein polynomials and the law of large numbers*, Math. Sci. **7** (1982), no. 2, 127–139. MR 674421 (84d:41009).

[131] G. G. Lorentz, *Approximation of Functions*, Second ed., Chelsea Publishing Co., New York, 1986. MR 917270 (88j:41001).

[132] ______, *Bernstein Polynomials*, Second ed., Chelsea Publishing Co., New York, 1986. MR 864976 (88a:41006).

[133] G. G. Lorentz, M. v. Golitschek, and Y. Makovoz, *Constructive Approximation. Advanced Problems*, Grundlehren der Mathematischen Wissenschaften [Fundamental Principles of Mathematical Sciences], vol. 304, Springer-Verlag, Berlin, 1996. MR 1393437 (97k:41002).

[134] G. G. Lorentz and L. L. Schumaker, *Saturation of positive operators*, J. Approximation Theory **5** (1972), 413–424, Collection of articles dedicated to J. L. Walsh on his 75th birthday, IV (Proc. Internat. Conf. Approximation Theory, Related Topics and their Applications, Univ. Maryland, College Park, Md., 1970). MR 0344756 (49 #9495).

[135] J. Lukeš, J. Malý, I. Netuka, and J. Spurný, *Integral Representation Theory. Applications to Convexity, Banach Spaces and Potential Theory*, de Gruyter Studies in Mathematics, vol. 35, Walter de Gruyter & Co., Berlin, 2010. MR 2589994 (2011e:46002).

[136] G. Lumer and R. S. Phillips, *Dissipative operators in a Banach space*, Pacific J. Math. **11** (1961), 679–698. MR 0132403 (24 #A2248).

[137] R. G. Mamedov, *The asymptotic value of the approximation of differentiable functions by linear positive operators*, Dokl. Akad. Nauk SSSR **128** (1959), 471–474. MR 0110016 (22 #899).

[138] E. M. Mangino and I. Raşa, *A quantitative version of Trotter's approximation theorem*, J. Approx. Theory **146** (2007), no. 2, 149–156. MR 2328177 (2009b:41015).

[139] V. Maz'ya, *Sobolev Spaces with Applications to Elliptic Partial Differential Equations*, Augmented ed., Grundlehren der Mathematischen Wissenschaften [Fundamental Principles of Mathematical Sciences], vol. 342, Springer, Heidelberg, 2011. MR 2777530 (2012a:46056).

[140] R. E. Megginson, *An Introduction to Banach Space Theory*, Graduate Texts in Mathematics, vol. 183, Springer-Verlag, New York, 1998. MR 1650235 (99k:46002).

[141] G. Metafune, *Analyticity for some degenerate one-dimensional evolution equations*, Studia Math. **127** (1998), no. 3, 251–276. MR 1489456 (98k:47082).

[142] C. A. Micchelli, *Convergence of positive linear operators on $C(X)$*, J. Approximation Theory **13** (1975), 305–315, Collection of articles dedicated to G. G. Lorentz on the occasion of his sixty-fifth birthday, III. MR 0382937 (52 #3819).

[143] S. Milella, *On a class of second order differential operators on the real line*, Rev. Roumaine Math. Pures Appl. **53** (2008), no. 2-3, 189–207. MR 2450369 (2009j:47083).

[144] ———, *Cores for generators of C_0-semigroups satisfying the Feller property*, to appear on Semigroup Forum (2014), DOI: 10.1007/S00233-014-9615-Y.

[145] D. S. Mitrinović, *Analytic Inequalities*, Die Grundlehren der Mathematischen Wissenschaften, Band 165, Springer-Verlag, New York-Berlin, 1970, In cooperation with P. M. Vasić. MR 0274686 (43 #448).

[146] I. Miyadera, *Generation of a strongly continuous semi-group operators*, Tôhoku Math. J. (2) **4** (1952), 109–114. MR 0052037 (14,564e).

[147] D. Mugnolo and A. Rhandi, *On the domain of a Fleming-Viot-type operator on an L^p-space with invariant measure*, Note Mat. **31** (2011), no. 1, 139–148. MR 2963964.

[148] G. Mühlbach, *Operatoren vom Bernsteinschen Typ*, J. Approximation Theory **3** (1970), 274–292. MR 0352813 (50 #5299).

[149] J. Nagel, *Asymptotic properties of powers of Kantorovič operators*, J. Approx. Theory **36** (1982), no. 3, 268–275. MR 676457 (84g:41018).

[150] T. Nishishiraho, *A generalization of the Bernstein polynomials and limit of its iterations*, Sci. Rep. Kanazawa Univ. **19** (1974), 1–7. MR 0358189 (50 #10654).

[151] ———, *Saturation of positive linear operators*, Tôhoku Math. J. (2) **28** (1976), no. 2, 239–243. MR 0415153 (54 #3244).

[152] ———, *The degree of convergence of positive linear operators*, Tôhoku Math. J. (2) **29** (1977), no. 1, 81–89. MR 0438003 (55 #10924).

[153] ———, *Saturation of bounded linear operators*, Tôhoku Math. J. (2) **30** (1978), no. 1, 69–81. MR 0493510 (58 #12510).

[154] ———, *Quantitative theorems on approximation processes of positive linear operators*, Multivariate approximation theory, II (Oberwolfach, 1982), Internat. Ser. Numer. Math., vol. 61, Birkhäuser, Basel-Boston, Mass., 1982, pp. 297–311. MR 719916 (84k:41024).

[155] ———, *Convergence of positive linear approximation processes*, Tôhoku Math. J. (2) **35** (1983), no. 3, 441–458. MR 711359 (85i:41024).

[156] ———, *The convergence and saturation of iterations of positive linear operators*, Math. Z. **194** (1987), no. 3, 397–404. MR 879940 (88m:47065).

[157] ———, *The order of approximation by positive linear operators*, Tôhoku Math. J. (2) **40** (1988), no. 4, 617–632. MR 972250 (90c:41033).

[158] R. Păltănea, *Approximation Theory Using Positive Linear Operators*, Birkhäuser Boston, Inc., Boston, MA, 2004. MR 2085239 (2005f:41061).

[159] A. Pazy, *Semigroups of Linear Operators and Applications to Partial Differential Equations*, Applied Mathematical Sciences, vol. 44, Springer-Verlag, New York, 1983. MR 710486 (85g:47061).

[160] R. S. Phillips, *Perturbation theory for semi-groups of linear operators*, Trans. Amer. Math. Soc. **74** (1953), 199–221. MR 0054167 (14,882b).

[161] ———, *Dissipative operators and hyperbolic systems of partial differential equations*, Trans. Amer. Math. Soc. **90** (1959), 193–254. MR 0104919 (21 #3669).

[162] G. Pólya and I. J. Schoenberg, *Remarks on de la Vallée Poussin means and convex conformal maps of the circle*, Pacific J. Math. **8** (1958), 295–334. MR 0100753 (20 #7181).

[163] T. Popoviciu, *Sur l'approximation des fonctions convexes d'ordre supérieur*, Bull. Sect. Sci. Acad. Roum. **20** (1938), 192–195 (French).

[164] I. Raşa, *Sets on which concave functions are affine and Korovkin closures*, Anal. Numér. Théor. Approx. **15** (1986), no. 2, 163–165. MR 889526 (89a:41045).

[165] ———, *Generalized Bernstein operators and convex functions*, Studia Univ. Babeş-Bolyai Math. **33** (1988), no. 2, 36–39. MR 1004816 (90f:41005).

[166] ———, *On the monotonicity of sequences of Bernstein-Schnabl operators*, Anal. Numér. Théor. Approx. **17** (1988), no. 2, 185–187. MR 1027226.

[167] ———, *Altomare projections and Lototsky-Schnabl operators*, Proceedings of the Second International Conference in Functional Analysis and Approximation Theory (Acquafredda di Maratea, 1992), no. 33, 1993, pp. 439–451. MR 1295282 (95h:41046).

[168] ———, *On some properties of Altomare projections*, Confer. Sem. Mat. Univ. Bari (1993), no. 253, ii+19 pp. (1994). MR 1305426 (96a:41027).

[169] ———, *Feller semigroups, elliptic operators and Altomare projections*, Proceedings of the Fourth International Conference on Functional Analysis and Approximation Theory, Vol. I (Potenza, 2000), no. 68, part I, 2002, pp. 133–155. MR 1975501 (2004e:47067).

[170] ———, *Positive operators, Feller semigroups and diffusion equations associated with Altomare projections*, Conf. Semin. Mat. Univ. Bari (2002), no. 284, 1–26. MR 1967992 (2004b:47073).

[171] ———, *Asymptotic behaviour of iterates of positive linear operators*, Jaen J. Approx. **1** (2009), no. 2, 195–204. MR 2597952 (2010m:41032).

[172] ———, *Power series of Bernstein operators and approximation of resolvents*, Mediterr. J. Math. **9** (2012), no. 4, 635–644. MR 2991155.

[173] I. Raşa and T. Vladislav, *Analiză Numerică. Elemente Introductive*, Editura Tehnică, Bucureşti, 1997.

[174] ———, *Altomare projections and Lipschitz constants*, Proceedings of the Third International Conference on Functional Analysis and Approximation Theory, Vol. II (Acquafredda di Maratea, 1996), no. 52, Vol. II, 1998, pp. 749–756. MR 1644908 (99g:41046).

[175] M. Rogalski, *Opérateurs de Lion, projecteurs boréliens et simplexes analytiques*, J. Functional Analysis **2** (1968), 458–488 (French). MR 0236763 (38 #5057).

[176] M. Romito, *Lototsky-Schnabl operators associated with a strictly elliptic differential operator and their corresponding Feller semigroup*, Monatsh. Math. **126** (1998), no. 4, 329–352. MR 1657814 (2000a:35149).

[177] L. A. Rubel, A. L. Shields, and B. A. Taylor, *Mergelyan sets and the modulus of continuity of analytic functions*, J. Approximation Theory **15** (1975), no. 1, 23–40. MR 0412395 (54 #521).

[178] M. D. Rusk, *Korovkin Type Theorems for Finitely Defined Operators*, ProQuest LLC, Ann Arbor, MI, 1975, Thesis (Ph.D.)–University of California, Riverside. MR 2626780.

[179] K. Sato and T. Ueno, *Multi-dimensional diffusion and the Markov process on the boundary*, J. Math. Kyoto Univ. **4** (1964/1965), 529–605. MR 0198547 (33 #6702).

[180] T. Sauer, *Multivariate Bernstein polynomials and convexity*, Comput. Aided Geom. Design **8** (1991), no. 6, 465–478.

[181] ———, *Ein Bernstein-Durrmeyer Operator auf dem Simplex*, Ph.D. thesis, Dissertation, Universität Erlangen-Nürnberg, 1992.

[182] ———, *Axially parallel subsimplices and convexity*, Comput. Aided Geom. Design **12** (1995), no. 5, 491–505. MR 1341704 (96d:65227).

[183] W. Schempp, *Zur Lototsky-Transformation über kompakten Räumen von Wahrscheinlichkeitsmassen*, Manuscripta Math. **5** (1971), 199–211 (German). MR 0352785 (50 #5271).

[184] ———, *A note on Korovkin test families*, Arch. Math. (Basel) **23** (1972), 521–524. MR 0318854 (47 #7400).

[185] R. Schnabl, *Eine Verallgemeinerung der Bernsteinpolynome*, Math. Ann. **179** (1968), 74–82 (German). MR 0236586 (38 #4881).

[186] ———, *Zum Saturationsproblem der Verallgemeinerten Bernsteinoperatoren*, Abstract Spaces and Approximation (Proc. Conf., Oberwolfach, 1968), Birkhäuser, Basel, 1969, pp. 281–289 (German). MR 0271610 (42 #6493).

[187] ———, *Über gleichmäßige Approximation durch positive lineare Operatoren*, Constructive theory of functions (Proc. Internat. Conf., Varna, 1970), Izdat.

Bolgar. Akad. Nauk, Sofia, 1972, pp. 287–296 (German). MR 0370012 (51 #6241).

[188] T. I. Seidman, *Approximation of operator semi-groups*, J. Functional Analysis **5** (1970), 160–166. MR 0254659 (40 #7866).

[189] B. Sendov and V. A. Popov, *The Averaged Moduli of Smoothness. Applications in Numerical Methods and Approximation*, Pure and Applied Mathematics (New York), John Wiley & Sons, Ltd., Chichester, 1988, A Wiley-Interscience Publication. MR 995672 (91b:41013).

[190] N. Shimakura, *Équations différentielles provenant de la génétique des populations*, Tôhoku Math. J. **29** (1977), no. 2, 287–318 (French). MR 0504058 (58 #20621).

[191] ———, *Formulas for diffusion approximations of some gene frequency models*, J. Math. Kyoto Univ. **21** (1981), no. 1, 19–45. MR 606311 (82i:60136).

[192] ———, *Existence and uniqueness of solutions for a diffusion model of intergroup selection*, J. Math. Kyoto Univ. **25** (1985), no. 4, 775–788. MR 810980 (87c:92037).

[193] ———, *Partial Differential Operators of Elliptic Type*, Translations of Mathematical Monographs, vol. 99, American Mathematical Society, Providence, RI, 1992, Translated and revised from the 1978 Japanese original by the author. MR 1168472 (93h:35002).

[194] P. C. Sikkema, *On some linear positive operators*, Nederl. Akad. Wetensch. Proc. Ser. A 73 = Indag. Math. **32** (1970), 327–337. MR 0270038 (42 #4931).

[195] K. Taira, *Diffusion Processes and Partial Differential Equations*, Academic Press, Inc., Boston, MA, 1988. MR 954835 (90m:60089).

[196] ———, *On the existence of Feller semigroups with boundary conditions*, Mem. Amer. Math. Soc. **99** (1992), no. 475, viii+65. MR 1120243 (93f:47050).

[197] ———, *Analytic Semigroups and Semilinear Initial Boundary Value Problems*, London Mathematical Society Lecture Note Series, vol. 223, Cambridge University Press, Cambridge, 1995. MR 1369247 (97g:47035).

[198] ———, *On the existence of Feller semigroups with boundary conditions. III*, Hiroshima Math. J. **27** (1997), no. 1, 77–103. MR 1437927 (98e:35081).

[199] ———, *Semigroups, Boundary Value Problems and Markov Processes*, 2nd ed., Springer Monographs in Mathematics, Springer-Verlag, Berlin, Heidelberg, 2014.

[200] W. B. Temple, *Stieltjes integral representation of convex functions*, Duke Math. J. **21** (1954), 527–531. MR 0062815 (16,22a).

[201] C. A. Timmermans, *On C_0-semigroups in a space of bounded continuous functions in the case of entrance or natural boundary points*, Approximation and optimization (Havana, 1987), Lecture Notes in Math., vol. 1354, Springer, Berlin, 1988, pp. 209–216. MR 996675 (90f:47058).

[202] H. F. Trotter, *Approximation of semi-groups of operators*, Pacific J. Math. **8** (1958), 887–919. MR 0103420 (21 #2190).

[203] J. van Casteren, *Generators of Strongly Continuous Semigroups*, Pitman, Boston-London-Melbourne, 1985.

[204] J. van Neerven, *The Asymptotic Behaviour of Semigroups of Linear Operators*, Operator Theory: Advances and Applications, vol. 88, Birkhäuser Verlag, Basel, 1996. MR 1409370 (98d:47001).

[205] A. D. Ventcel, *On boundary conditions for multi-dimensional diffusion processes*, Theor. Probability Appl. **4** (1959), 164–177. MR 0121855 (22 #12585).

[206] V. I. Volkov, *On the convergence of sequences of linear positive operators in the space of continuous functions of two variables*, Dokl. Akad. Nauk SSSR (N.S.) **115** (1957), 17–19. MR 0094693 (20 #1205).

[207] E. V. Voronovskaja, *The asymptotic properties of the approximation of functions with Bernstein polynomials*, Dokl. Akad. Nauk SSSR **A** (1932), 79–85.

[208] S. Waldron, *Affine generalized barycentric coordinates*, Jaen J. Approx. **3** (2011), no. 2, 209–226. MR 2954373.

[209] K. Yoshida, *On the differentiability and the representation of one-parameter semigroups for linear operators*, J. Math. Soc. Japan **1** (1948), 15–21.

[210] D. X. Zhou, *Converse theorems for multidimensional Kantorovich operators*, Anal. Math. **19** (1993), no. 1, 85–100. MR 1232056 (95d:41047).

Symbol index

3. Mappings, functions, measures

4. Function spaces and operators

5. Approximation processes

Index

De Gruyter Studies in Mathematics

Volume 60
Vladimir A. Mikhailets, Alexandr A. Murach, Peter V. Malyshev
Hörmander Spaces, Interpolation, and Elliptic Problems, 2014
ISBN 978-3-11-029685-3, e-ISBN 978-3-11-029689-1, Set-ISBN 978-3-11-029690-7

Volume 59
Jan de Vries
Topological Dynamical Systems, 2014
ISBN 978-3-11-034073-0, e-ISBN 978-3-11-034240-6,Set-ISBN 978-3-11-034241-3

Volume 58
Lubomir Banas, Zdzislaw Brzezniak, Mikhail Neklyudov, Andreas Prohl
Stochastic Ferromagnetism: Analysis and Numerics, 2014
ISBN 978-3-11-030699-6, e-ISBN 978-3-11-030710-8, Set-ISBN 978-3-11-030711-5

Volume 57
Dmitrii S. Silvestrov
American-Type Options: Stochastic Approximation Methods, Volume 2, 2014
ISBN 978-3-11-032968-1, e-ISBN 978-3-11-032984-1, Set-ISBN 978-3-11-032985-8

Volume 56
Dmitrii S. Silvestrov
American-Type Options: Stochastic Approximation Methods, Volume 1, 2013
ISBN 978-3-11-032967-4, e-ISBN 978-3-11-032982-7, Set-ISBN 978-3-11-032983-4

Volume 55
Lucio Boccardo, Gisella Croce
Elliptic Partial Differential Equations, 2013
ISBN 978-3-11-031540-0, e-ISBN 978-3-11-031542-4, Set-ISBN 978-3-11-031543-1

Volume 54
Yasushi Ishikawa
Stochastic Calculus of Variations for Jump Processes, 2013
ISBN 978-3-11-028180-4, e-ISBN 978-3-11-028200-9, Set-ISBN 978-3-11-028201-6

www.degruyter.com

Bei Fragen zur Produktsicherheit wenden Sie sich bitte an:
If you have any questions regarding product safety,
please contact:

Walter de Gruyter GmbH
Genthiner Straße 13
10785 Berlin
productsafety@degruyterbrill.com

AF775234